SUSTAINABLE FOREST MANAGEMENT

Managing Forest Ecosystems

Volume 1

Series Editors:

Klaus von Gadow

Institute of Forest Management,
University of Göttingen,
Göttingen, Germany

Timo Pukkala

University of Joensuu,
Joensuu, Finland

and

Margarida Tomé

Instituto Superior de Agronomía,
Lisbon, Portugal

The titles published in this series are listed at the end of this volume.

Sustainable Forest Management

Edited by

Klaus von Gadow

Institute of Forest Management,
University of Göttingen,
Göttingen, Germany

Timo Pukkala

University of Joensuu,
Joensuu, Finland

and

Margarida Tomé

Instituto Superior de Agronomía,
Lisbon, Portugal

KLUWER ACADEMIC PUBLISHERS
DORDRECHT / BOSTON / LONDON

A C.I.P. Catalogue record for this book is available from the Library of Congress.

Published by Kluwer Academic Publishers,
P.O. Box 17, 3300 AA Dordrecht, The Netherlands.

Sold and distributed in North, Central and South America
by Kluwer Academic Publishers,
101 Philip Drive, Norwell, MA 02061, U.S.A.

In all other countries, sold and distributed
by Kluwer Academic Publishers,
P.O. Box 322, 3300 AH Dordrecht, The Netherlands.

Printed on acid-free paper

ISBN-13: 978-1-4020-0278-6 e-ISBN-13: 978-94-010-9819-9
DOI: 10.1007/978-94-010-9819-9

Softcover reprint of the hardcover 1st edition 2000

Foreword

During its 200-year history the concept of sustainability has been the object of scientific and political discussion, with varying degrees of intensity, - promoted with vehement fervour during periods of social or economic crisis, and less intensely during periods of stability. Starting during the early 1980's, a new wave of writings and international conferences focussing on the sustainability concept has been sweeping the world of science, economics and politics. In June 1992 the United Nations Conference on the Environment and Development (UNCED) in Rio de Janeiro initiated a world-wide process of negotiation with the aim of ensuring sustainable management, conservation and development of forest resources. Although there seems to be unanimous support for sustainable development from all quarters, there is neither a coherent understanding of human values as they are applied across multiple generations and cultures nor a generally accepted set of indicators which allows comparisons to be made between a given situation and a desirable one. Does the concept of sustainable forest management represent not much more than a magic formula for achieving consensus, a vague idea which makes it difficult to match action to rhetoric? In a recent summary paper prepared by the FAO Forestry and Planning Division Ljungman et al. (1999)[1] find that Forest resources continue to diminish, while being called upon to produce a greater range of goods and services; that markets do not exist for the many social and environmental services that forests provide; that sustainable forest management can be profitable, but that unsustainable forestry practices are often more financially attractive in the short-term; that calls for sustainable forest management will simply go unheeded if the legal, policy and administrative environment do not effectively control undesirable practices. It may take many decades for a forest to develop a preferred habitat or to produce timber of a desired size, and reconciling multiple and sometimes conflicting management objectives is usually compounded by a high degree of uncertainty.

The concept of sustainable forest management is likely to remain an imprecise one, but we can contribute to avoiding management practices that are clearly unsustainable. This volume, which forms part of the book series *Managing Forest Ecosystems*, presents state-of-the-art research results, visions and perceptions as well as specific methods for ensuring sustainable forest ecosystem management. If technical knowledge is a constraint to the

---

[1] Ljungman, C.L.S., Martin, R.M. and Whiteman, A., 1999: Beyond sustainable forest management – opportunities and challenges for improving forest management in the next millenium. FAO, Rome: 45 p.

implementation of sustainable forest management, this book contains a wealth of information which may be useful to company management, the legal and policy environment and forestry administrators. The volume is subdivided into three parts and includes ecological, socio-political and timber supply perspectives.

The first part presents three contributions concentrating on ecosystems perspectives. The focus of the first paper by *Schlaepfer and Elliott* is on the sustainable ecosystem-based management. They argue that foresters need to adapt to the increasing importance of landscape level planning. Certification can be a powerful tool for implementing ecosystem-based management. *Hauhs and Lange* derive lessons from European forest history based on the Rammelsberg mining activities and suggest ways towards operational implementation of sustainable ecosystem utilization. *Dye and Bosch* offer useful insights relating to the problem of sustainable water yields from afforested catchment areas in countries where water is a particularly scarce and valuable resource.

The second part concentrates on social issues. Nearly 500 million people around the world rely on forests for their livelihoods, among them a high number of forest and wood workers. *Bowling* argues that sustainable forest management must include safe, stable jobs with adequate wages and working conditions. Because of their close relation to the forest, forest workers can play a key role in implementing sustainable practices. Making Rio work is *Burger's* aim and he shows how forest certification can help make sustainable development a reality. Whether it will succeed in doing so will depend on a number of factors including legal aspects and financial viability, and acknowledgement by business partners. The contribution prepared by *Hütte* concentrates on the different perceptions and patterns of argumentation used by foresters and nature conservation NGO's in Germany, France and Spain. Forest scientists are called upon to participate in public debate to help clarify issues related to sustainable resource management.

In the third part, the emphasis is on sustainable forest management and timber supply. *Roise, Cubbage, Abt and Siry* present details on forest plantations and sustained yield regulation of industrial wood fiber. A sample of private and public forestry analysts in different parts of the world revealed the most important harvest regulation approaches. With sustainability as guiding principle, several components of yield regulation systems are identified by *Seydack* in tropical/subtropical moist forests, including yield optimisation and stand regeneration strategies and the tactical components of harvesting cycle and harvest tree selection criteria. Sustainable yield regulation systems occupy positions of varying intensity and impact on the forest ecosystem, from manipulation to mimicking of forest structure and dynamics. The final contribution by *Gadow and Puumalainen* specifically deals with scenario planning techniques on the forest management unit level. A series of techniques for generating timber harvest scenarios, including age-class simulation, area change models and multi-period harvest scheduling for even-aged forests and continuous cover forests are presented with numerical examples.

We wish to acknowledge the valuable contributions made by our anonymous referees. Finally, we appreciate the diligent proofreading and editing assistance provided by Sonja Rüdiger and Marga von Gadow.

Klaus v. Gadow　　Timo Pukkala　　Margarida Tomé

Contents

Ecosystem perspectives

Social and political perspectives

Timber supply and forest management

Ecological and Landscape Considerations in Forest Management: The End of Forestry?

R. Schlaepfer[1] *and C. Elliott*[2]

Abstract

This paper presents the evolution of ecological considerations in forestry and its consequences for future forest management practices. Forest management has a long history, but there has been a remarkable increase in interest in the subject over the last 10 years. Much of this interest has come from outside the forestry profession as the roles of forests in issues such as carbon cycling, climate change and the conservation of biodiversity becomes increasingly recognised. Recent emphasis has been on sustainable forest management and ecosystem-based management. Forests should be considered as part of a broader landscape and forest managers need to adjust their horizons to this scale. Forest certification is an important tool to implement ecosystem-based management. The introduction of ecosystem-based management is associated with a number of impacts, and many forestry graduates are ill-equipped to deal with the new types of demands placed on forest managers. Foresters, if they can adapt to the increasing importance of ecosystem-based considerations, will be able to use their undoubted skills at a landscape level beyond the confines of the forests. If they cannot, they will continue to find their role, even within forests, being dramatically reduced.

Keywords: *Ecosystem-based management, sustainable forest management, ecosystem management, certification, forest landscape.*

---

[1] Prof. Rodolphe Schlaepfer, Laboratory of Ecosystem Management, Swiss Federal Institute of Technology , Lausanne (EPFL), CH-1015 Lausanne, Switzerland

[2] Dr. Chris Elliott, Forests for Life Campaign, World Wide Fund For Nature-WWF, CH-1196 Gland, Switzerland

K. von Gadow et al. (eds.), Sustainable Forest Management, 1–67.

Introduction

In 1989 the American historian Francis Fukuyama published an article entitled «The End of History» in *The National Interest*. The article was later expanded into a book (Fukuyama 1993). After the Berlin wall fell, Fukuyama speculated that a paradigm shift had occurred and that with the collapse of communism, liberal democracy constituted the «final form of human government.» Hence the phrase «the end of history.» Perhaps forestry has not quite had its equivalent to the fall of the Berlin wall but there are many voices today calling for a «paradigm shift» in forestry. It is argued that old views based on the primacy of timber above all other functions are outdated and a new ecosystem-based paradigm is needed. Such arguments are based on the perception that foresters remain locked into a paradigm where the production of timber is paramount. As will be demonstrated, this view is in part dated, and major steps have been and are still being taken within the forestry profession to address the need for sustainable forest management and the integrated management of forest resources. This chapter explores the evolution of trends in forest management and discusses ecosystem-based management as a new approach. In the conclusion, we return to the question of whether this new paradigm means the end of forestry as we know it today.

Over the last 10 years, forestry has been faced with fundamental changes, which have been described as follows:

> «The practice of forestry is currently undergoing the most profound and rapid change since its establishment a century ago. The evolution from sustained yield management of a relatively small number of commercial tree species to the protection and sustainable management of forest ecosystems is changing some of the fundamental premises of forest management.» (Sample et.al., 1993, p. 4).

Changes in forestry practices are being accompanied by changes in the role of professional foresters. In 1987 Glück described the dominant belief system of European foresters as being based on the following values: timber primacy, sustained yield, absolute standards and long term views. He noted that:

> «One of the outcomes of the conservative attitude of foresters is their critical view of libertarianism... they distrust human nature and the pluralism of interests... The doctrine of timber primacy finds its justification in the so called wake theory. This states that in general the other forest services should follow in the wake of timber production. The theoretical content of the wake theory is small. It does not contribute to a multi-use analysis, but instead tries to justify the primacy of timber production. The forester who defends his position against conflicting interests by referring to the wake theory plays a losing game for

two reasons. First, there are many cases where the wake theory does not hold, and second, it supports arguments against compensation demands for non-priced forest services because it claims that all forest functions are by-products of timber production.» (Glück, 1987, p. 2-3).

We can see from the two quotes above that forest management and forest policy are in a highly dynamic phase at the end of the twentieth century. The global supply and demand context in which these changes are occurring have been reviewed by Solberg et al. (1996). They studied long-term trends and prospects in the world supply and demand for wood and its implications for sustainable forest management. The team came to the conclusion that the world's forests are biologically capable of supplying the quantity and type of wood consistent with the highest demand projections. Therefore, their assessment is that the outlook for demand and supply of wood does not at present require a warning of impending crisis. However, the global long-term outlook is for a steadily rising demand for wood and its products and for forest services, and a declining area of forest available for their production, because of deforestation and competing demands for forests such as urban development and protected areas. These trends will result in increasing pressures on wood supply arising from the growing importance of non-wood goods and services and demands for environmental protection. This example shows why multi-functional and holistic approaches will increasingly have to be the basis for managing forest resources.

A central and critical element of these approaches is the attempt to integrate ecological, economic and social issues in formulating goals and objectives and in decision making. In particular, the increasing importance of the maintenance of biological diversity gives a new and strong ecological dimension to the use of natural resources, including forest resources. The need to protect forest ecosystems and to maintain or restore their ecological quality is now an important driving force, shared by many conservationists and foresters. We claim that this constitutes a new paradigm in forest management. Although many foresters argue that the history of forestry in their countries demonstrates that they are practising sustainable forestry (Sedjo et al., 1998), we feel that such arguments are based on a misunderstanding of the changes in the way in which forests and forest resources will be managed in the future. In addition, many foresters still seem unaware that the «myth of the omniscient forester» (Bunnell, 1976) was shattered several decades ago.

The aim of this paper lies within this framework. We attempt to:

- outline the development of ecological considerations in the history of forest management;

- give an overview of ecological considerations today; and
- present a concept of ecosystem-based management of forest and landscape resources.

The following questions will be discussed:

- Did ecological considerations play a role in the development of forest management?
- What were the main results of the 1992 Earth Summit and what were their consequences for forest management?
- What is meant today by sustainable forest management?
- What is ecosystem-based management?
- How are forest ecosystems and landscapes related?
- How can forest certification contribute to the better management of forests?
- What are the ecological perspectives associated with forest management?
- What conclusions can we draw?

Development of sustainability and ecological considerations in forest management

Forest management has a long history. Its practices and processes are numerous and vary over time and space. It is therefore difficult to give a global overview of forest management. In the following section, our goal is to show, with some selected examples, how the paradigms of forest management have evolved over time.

Forest historians often consider that forest practices and policies tend to go through a number of stages or paradigms in a given country or region, starting with the unregulated exploitation of timber. When government sees the opportunity to tax forestry revenue this can lead to a phase of «regulation for revenue.» If this leads eventually to resource depletion a phase of «sustained yield production» may begin. When other demands are placed on the forest in addition to timber, «sustainable management» is called for. In this paper we argue that we now have to go beyond «sustainable forest management» to looking at forests as part of broader ecosystems, leading us to call for an «ecosystem-based approach» to forest management.

The first clear record of forest conservation measures are to be found in Assyria in 700 BC, when game reserves were set aside by decree for royal hunts (Dixon and Sherman, 1991). Forests in China were valued for more than just hunting: around 300 BC, when the Chinese philosopher Mencius wrote of his concern about deforestation of Bull Mountain as a result of timber harvesting and overgrazing and its impact on stream flows (Waley, 1939). At a similar time Plato wrote of deforestation in the hills around Athens as a result of the Athenian navy's wood consumption and agricultural clearance (Jowett, 1892; Mather, 1990). This situation led to trade measures designed to conserve forest resources, and by the first century AD the Roman writer Columella described forest management techniques for coppicing chestnut and oak (Rackham, 1986).

In the following centuries, a variety of regulated forest management practices were adopted. These ranged from setting aside large tracts of forests as hunting preserves (such as the New Forest, created in England in the 11th century) to the re-afforestation of deforested hills to prevent flooding in lower catchments (as in Japan in the 9th century). Most of these initiatives were designed with a single problem in mind, with anthropogenic values (prevention of flooding, provision of timber, maintenance of game populations) being paramount.

In the 17th century, forest management was defined as harvesting trees in order to obtain timber or other wooden products. In 1837, Salomon defined forest management as the regulation of the exploitation, in order to assure every year a constant and equal succession of the best possible products. This definition clearly contains an element of sustainable yield. Tassy (1858) explicitly used the term «sustainable.» He defined forest management as the regulation of the harvesting of a forest in order to achieve an annual yield as sustainable («soutenu») and as advantageous as possible.

Henri Biolley, in his famous papers about selection cutting («Le Jardinage Cultural», 1901), considered forest management and silvicultural treatment as closely linked. He believed that the most efficient way to produce wood was with a forest structured to give the maximum photosynthetic activity per unit area. This would be the case when the highest possible volume of the three dimensional space of a forest is occupied with leaves. This idea led him to a forest structure, in which over very small areas (some 100 m^2), all sizes and ages of trees were present, but with only one tree with large height and diameter. For Biolley, silviculture and forest management were tools to achieve this structure with the aim to produce the highest possible quantity and quality of wood, with the lowest possible growing stock.

Biolley considered sustainable yield to be the result of the equilibrium between increment and growing stock and this should be the rule only when the highest possible yield and the necessary and sufficient means to reach it are known. His management concept, presented here in a simplified manner, is therefore based on the periodic observation of increment, the interpretation of its development, the adaptation, if necessary, of the target growing stock and calculation of the allowable cutting. He calculated the periodic increment as the sum of the exploitation over the period and the growing stock difference between the end and the beginning of the period. To do this, he developed appropriate forest inventory, and cutting control methods. In a sense, the techniques proposed by Biolley were forerunners of adaptive management, which is one of the fundamental principles of modern forest management. By using growing stock, increment and exploited volume not only to calculate the allowable cutting volume but also to measure the progress towards sustainability, he was also a precursor of the concept of criteria and indicators of sustainable forest management. We should be aware that Biolley's idea of selection cutting was developed for shade-tolerant tree species and has not been widely tested with other tree species.

Later, Schaeffer et al. (1930) defined forest management as the organisation of the forest to obtain a regular yield. Kramer (1976) gave a more general definition, without specifying the goal: forest management is the middle- and long-term planning of the forest enterprise. Many other definitions of forest management can be found in the literature. The following are mentioned by Mantel (1959):

- organisation in space and in time of the economic sector of a forest, in order to make it possible to reach an economic goal (Judeich, 1922);
- planned organisation of the forest enterprise (C. Wagner, 1928);
- sustainable management and planning (C. Baader, 1942);
- organisation of the forest enterprise (H. Knuchel, 1950);
- analysis, planning and monitoring of the conditions for sustainability (A. Richter, 1952) or
- periodic planning of the forest enterprise (Mantel, 1959).

The calculation of the allowable cut was (and still is in some areas) an important element of forest management. It is expressed in term of forest area or wood volume to be harvested. The goal was to obtain a sustainable yield or the best possible yield. Several mathematical expressions, differing for even-aged and uneven-aged forests, were used for this purpose.

We see from these examples, that sustainable yield has been one of the basic tenets of forest management for a considerable length of time. Initially, the only goal or at least the main goal in the mind of foresters was usually wood production. Later, many foresters such as Biolley recognised the multiple functions of forests, but believed that forests that were organised for wood production were also fulfilling the other functions. This is what was meant by the wake theory as described by Glück (1987).

In the 1970s and 1980s, the need to manage forests to reach multiple objectives emerged, particularly in the USA. For Leuschner (1984), the term forest management refers to the study and application of analytical techniques to aid choosing those management alternatives that contribute most to organisational multiple objectives. His understanding of forest management is based on the belief that mathematical modelling, including linear programming and other similar techniques, are important tools to minimise subjectivity in complex decision making. These techniques were intensively used in forest management, especially in North America. In 1987, Davis and Johnson identified three fundamental shifts in the tenets of forest management: (1) the new dominance of economic and social goals as determinants of forest management choices, (2) the necessity of private and especially public managers to consider all timber management decisions in the context of larger, socially defined, multiple-use management problems, and (3) the need to show on the basis of quantitative analysis that management recommendations, decisions, and plans satisfy owner constraints and are the best of alternative choices. At the same time, Schlaepfer (1987) linked forest management and forest policy and put the emphasis on the integration of the different planning elements. He defined forest management as the set of activities that enabled the organisation of a forest enterprise to achieve a number of forest policy goals, including:

- fixing the general objectives of the forest enterprise, which for him are necessarily multiple;
- collecting information concerning the different forest functions;
- forecasting the middle-term harvesting (estimation of the areas to by treated and the wood volume resulting from the different silvicultural interventions);
- integrated planning of middle and long term activities (land use, silviculture, infrastructure, finance, human resources, investment); and
- control.

The inclusion of broader ecosystem functions in management planning has gained increasing importance in forestry over the last decade. However it is not an entirely new idea. Salomon, in his treatise about forest management (Salomon, 1837), was mainly concerned with when, where and how much to harvest. He already recognised the importance of soil and climate for the development of trees and other vegetation, recommending that the soil be studied carefully in order to evaluate its capacity to produce trees. Earlier, in 1821, Cotta was aware that climate and soil influence the shape of trees and, therefore, the standing volume of forests. He defined different fertility classes according to the standing volume of wood in a forest (Cotta, 1821). Biolley (1901) considered soil, atmosphere, life, stand and time as the natural bases of forest management. These examples can be considered as some of the first ecological thoughts in forest management. However, it is clear that at the time, ecological considerations were not treated as objectives, but as instruments for maximising wood production. Such a view continues in some areas today; the close matching of species to site is often more concerned with maximising growth than achieving close-to-natural forests. In the 19th century, the primary purpose of forestry was timber production; and all other goods and services of the forest were considered as by-products (Glück, 1987).

We should not view these approaches to forest management as failures. In Europe at various times they allowed forests to recover from over-harvesting and deforestation. In many temperate countries today forest area, growing stock and increment are at higher levels than ever before (Spiecker et al., 1997). On the other hand, deforestation, mainly in tropical countries, and the loss of forest quality in the Northern Hemisphere, have led to the critical review of forest management paradigms. It has become clear that the ecological and social basis of forest management needs to be improved (see, for example, Brown and Pearce, 1994). In many cases, forest management practices are unable to fulfil the increasing number of expectations for forest resources. The needs of nature conservation have placed forestry under pressure (Poore, 1995). The debate about the goals of forestry needed to be clarified. The debate about sustainability led to a recognition that the objectives of forest management needed to be broadened away from an exclusive reliance on timber, because it was not clear how to weight the other objectives while moving from the «sustained timber yield» paradigm to «sustainable forest management». As Wiersum noted:

> «Sustainability is invariably used to describe a goal which, superficially at least, is indisputably desirable. On closer examination, however, it is found that the concept is defined so broadly, as to be open to widely disparate interpretations, which creates potential for misunderstandings. Thus, one is faced with a situation that, although the concept of sustainability is increasingly being accepted as an ethical principle in relation to the utilisation of natural resources, it has so far been found difficult to operationalize it as a guide to decision-making within a specific context.» (Wiersum, 1995)

Within this framework, sustainable forest management became the subject of several international efforts, including the Earth Summit of Rio de Janeiro and various follow-up processes, which sought to operationalize sustainable forest management. These are discussed in the next section. With the benefit of hindsight we can now see that many of these efforts have been vain, and a new approach is needed. As noted by Tollefson:

> «Increasingly there are calls for a new paradigm which proceeds from an explicit recognition of the forest as a complex and interdependent ecosystem, and which sets, as its overriding priority, the goal of maintaining that system.» (Tollefson, 1998)

We will return to this new paradigm in subsequent sections.

The Rio «Earth Summit» and its consequences

In the last twenty years we have become aware of ongoing global changes such as climate change, changes in the chemistry of the atmosphere and land use change, and their present and potential effects on the biotic and abiotic environment. This awareness is the basis for the increasing political and scientific importance of the notion of sustainable development.

This was the background for the United Nations Conference on Environment and Development (known as the Earth Summit) held in Rio de Janeiro, Brazil in June 1992 and which resulted in several strategically important contributions. Five important documents resulted from the Earth Summit held in Rio in 1992:

- The *Rio Declaration on Environment and Development* includes 27 principles defining the right and the responsibilities of nations as they pursue human development and well being. The following principles illustrate how environmental protection, including ecological considerations, is the driving force of sustainable development:
  - People are at the centre of concern for sustainable development. They are entitled to a healthy and productive life in harmony with nature (Principle 1).
  - The right to development must be fulfilled so as to equitably meet development and environmental needs of present and future generations (Principle 3).
  - In order to achieve sustainable development, environmental protection shall constitute an integral part of the development process and cannot be considered in isolation from it (Principle 4).

  - States shall co-operate in a spirit of global partnership to conserve, protect and restore the health and integrity of the Earth's ecosystem (Principle 7).
  - States should co-operate to strengthen endogenous capacity-building for sustainable development by improving scientific understanding through exchanges of scientific and technological knowledge, and by enhancing the development, adaptation, diffusion and transfer of technologies, including new and innovative technologies (Principle 9).
  - In order to protect the environment, the precautionary approach shall be widely applied by the States according to their capabilities (Principle 15).
  - Environmental impact assessment, as a national instrument, shall be undertaken for proposed activities that are likely to have a significant adverse impact on the environment and are subject to a decision of a competent national authority (Principle 17).

- *Agenda 21* reflects a global consensus and commitment at the highest political level on how to make development socially, economically, and environmentally sustainable. Among other options, it includes protecting the atmosphere, an integrated approach to planning for and managing land resources, combating deforestation, managing fragile ecosystems: combating desertification and droughts, sustainable mountain development, promoting sustainable agriculture and rural development, conservation of biological diversity, managing biotechnology, protecting and managing the oceans, protecting and managing fresh waters, and supporting science for sustainable development.

- A non-legally-binding, authoritative *Statement of Forest Principles* to guide managing, conserving, and sustainably developing all types of forests, which are essential to economic development and maintaining of all forms of life. In the context of this chapter, an important element of this last document is its principle 2(b):

  «Forest resources and forest lands should be sustainably managed to meet the social, economic, ecological, cultural and spiritual needs of present and future generations. These needs are for forest products and services, such as wood and wood products, water, food, fodder, medicine, fuel, shelter, employment, recreation, habitats for wildlife, landscape diversity, carbon sinks and reservoirs, and for other forest products. Appropriate measures should be taken to protect forests against harmful effects of pollution, including airborne pollution, fires, pests and diseases, in order to maintain their full multiple value.»

- The *United Nations Framework Convention on Climate Change* aim is to stabilise greenhouse gases in the atmosphere at concentrations that will not dangerously upset the global climate system. The 1997 Kyoto Protocol maintained the emphasis on sustainable development, stating in Article 2 that:

> «Each Party included in Annex I in achieving its quantified emission limitation and reduction commitments under Article 3, in order to promote sustainable development, shall implement and/or further elaborate policies and measures in accordance with its national circumstances, such as: (i) Enhancement of energy efficiency in relevant sectors of the national economy; (ii) Protection and enhancement of sinks and reservoirs of greenhouse gases not controlled by the Montreal Protocol, taking into account its commitments under relevant international environmental agreements; promotion of sustainable forest management practices, afforestation and reforestation; (iii) Promotion of sustainable forms of agriculture in light of climate change considerations; (iv) Promotion, research, development and increased use of new and renewable forms of energy, of carbon dioxide sequestration technologies and of advanced and innovative environmentally sound technologies;...»

- The *Convention on Biological Diversity* was signed by more than 150 governments at the Rio «Earth Summit» in 1992. It became effective on 29 December 1993 and has become the centre-piece of international efforts to conserve the planet's biological diversity, to ensure the sustainable use of its components, and to promote the fair and equitable sharing of the benefits arising from the use of genetic resources.

After the Earth Summit, considerable efforts were made by governments and international organizations to operationalize the concept of sustainable forest management through the development of Criteria and Indicators (C&I) for sustainable forest management. These C&I are being developed as a means to monitor progress towards sustainable forest management at the national level. Very few, if any, are appropriate at the scale of the forest management unit.

The Ministerial Conference on the Protection of Forests in Europe, held in Helsinki in 1993, is a follow-up of the Rio Summit. The Conference agreed on two general guidelines and two resolutions for specific action. The *General Guidelines for the Sustainable Management of Forests in Europe* provide a definition for the sustainable management of forests. The *General Guidelines for the Conservation of the Biodiversity of European Forests* recognise the conservation and appropriate enhancement of biodiversity as essential parts of sustainable forestry. The *Resolution for Forestry Cooperation with Countries with Economies in Transition* defines principles for collaboration with eastern European Countries. Finally, the *Resolution for Strategies for a Process of Long-term Adaptation of Forests in Europe to Climate Change* reflects the

European consensus on how to manage forests in view of potential climate change. While the achievement of these resolutions should be recognised, follow-up activities have been distinctly limited and there is some uncertainty over the effectiveness of the Resolutions. In some cases, such as with the Resolution for Strategies for a Process of Long-term Adaptation of Forests in Europe to Climate Change, few if any financial resources have been allocated to the development of suitable strategies to fulfil the Resolution. As a result, their value must be seriously questioned.

The Montreal Process and the Helsinki Process (now called the Pan European Process) are two different and highly visible international series of efforts to implement the UNCED Forest Principles, the Agenda 21, the Convention on Biodiversity, and the Helsinki Resolutions. The first was initiated in Montreal and the second in Helsinki. They defined Criteria and Indicators for evaluating and measuring sustainable forest management at the national level. Both processes recognised the importance of biological diversity as a criterion for sustainable forest management. The fact that different political, socio-economic, and geographically diverse countries apparently agree on the importance of issues such as biological diversity is significant. However the discussions on C&I were still held largely within the sustainable forest management paradigm, and involved limited participation by non-foresters. The concept of forests as ecosystems was given only passing attention. Partly as a result of this and partly because of the increasing attention being given to sustainable forest management, ecosystem-based considerations are increasingly included in management and research.

Sustainable forest management, its potentials and limitations

According to Stankey (1995), sustainability is fundamentally a socio-political construct rather than a scientific concept capable of precise, unequivocal measurements. It reflects a state to which we aspire, it embodies a concern with our ability to exist as a species, and it opens the door for scientists and technical specialists to engage society in an issue of mutual concern.

We have already seen that sustainability is an old notion in forestry. It was mainly used to describe sustainable wood production under the term «sustained yield.» However, the concept is now used more generally to cover other forest functions as well under the title «sustainable forest management».

Leopold, in «A Sand County Almanac», first published in 1949 and now an environmental classic, uses the term sustained carrying capacity to describe the state of land and biotas.

In the Rio Declaration on Environment and Development, sustainability is defined as the equitable meeting of development and environmental needs of present and future generations.

In the «Forest Principles», as stated previously, we find the following important elements concerning sustainable forest management:

> «Forest resources and forest lands should be sustainably managed to meet the social, economic, ecological, cultural and spiritual needs of present and future generations. These needs are for forest products and services, such as wood and wood products, water, food, fodder, medicine, fuel, shelter, employment, recreation, habitats for wildlife, landscape diversity, carbon sinks and reservoirs, and for other forest products. Appropriate measures should be taken to protect forests against harmful effects of pollution, including airborne pollution, fires, pests and diseases, in order to maintain their full multiple value.»

This principle clearly recognises the multiple values of forest, including the ecological dimension, which have to be met by forest management. Also, a new element appears: landscape diversity. This important dimension, as we will see below, is one of the most challenging elements facing forest management today. To stress the necessity for a sustainable use of forest resources, the term «forest management» was replaced by «sustainable forest management». This is also illustrated by the definitions given by ITTO and the Pan-European Process. For ITTO, Sustainable Forest Management is the process of managing permanent forest land to achieve one or more clearly specified objectives of management with regard to the production of a continuous flow of desired forest products and services without undue reduction of its inherent values and future productivity and without undue undesirable effects on the physical and social environment (ITTO, 1992, Criteria for the Measurement of Sustainable Tropical Forest Management). The Helsinki Conference defined Sustainable Management of Forests as the stewardship and use of forests and forest lands in a way, and at a rate, that maintains their biodiversity, productivity, regeneration capacity, vitality, and their potential to fulfil, now and in the future, relevant ecological, economic, and social functions, at local, national, and global scales, and that does not cause damage to other ecosystems (Ministerial Conference on the Protection of European Forests, 1993. Resolution H1).

In the United States, a national conference entitled «Defining Sustainable Forestry» was held in January 1992. Its purpose was to begin developing a common framework on which to base the future development of forestry; the conventional approaches that have defined forestry for the past century were to be replaced by an ecosystem approach that would guide it into the next century. At the conference, no single theme was better developed than the idea that sustainable forestry must be ecologically sound, economically viable, and socially desirable (Aplet et al., 1993).

For Bormann et al. (1994b), ecosystem sustainability is the degree of overlap between what people collectively want — reflecting social values and economic concerns — and what is ecologically possible in the long term. The overlap is dynamic because both societal and ecological capacity continually change. They advocate that the desires of future generations be protected by maintaining options for unexpected future ecosystem goods, services, and states.

The different recent definitions of sustainable forest management seek to take account of ecological, economic and social needs in relation to forests. Some mention an ecosystem-based approach. But, as such, these are not useful to managers. Decision-makers need to measure progress towards sustainability, they also need to monitor the effect of management decisions. In order to do so, criteria and indicators of sustainable forest management have been developed. Several international initiatives have produced different proposals for what could be used. Examples include the criteria adopted by the Montreal and Helsinki Processes.

An important result of the Montreal Process is the list of criteria for conserving and sustainably managing temperate and boreal forests (Montreal Process, 1995):

- Conserving biological diversity;
- Maintaining the productive capacity of forest ecosystems;
- Maintaining forest ecosystem health and vitality;
- Conserving and maintaining soil and water resources;
- Maintaining the forest contribution to the global carbon cycle;
- Maintaining and enhancing long-term multiple socio-economic benefits to meet the needs of societies, and
- Creating legal, institutional and economic frameworks for forest conservation and sustainable management.

The European criteria for sustainable forest management adopted by the Ministerial Conference for the Protection of Forests in Europe, in Lisbon (1998), are:

- Maintaining and appropriately enhancing forest resources and their contribution to global carbon cycles;
- Maintaining forest ecosystem health and vitality;
- Maintaining and encouraging productive functions of forests (wood and non-wood);
- Maintaining, conserving and appropriately enhancing biological diversity in forests ecosystems;
- Maintaining and appropriately enhancing protective functions in forest management (notably soil and water) and
- Maintaining other socio-economic functions and conditions.

Despite these international definitions, the practical application of criteria and indicators in management and policy is still distant. As Brooks (1996) pointed out recently, «although it required a considerable effort to define the criteria and select the preliminary list of indicators, no consensus has been reached on how individual indicators should be interpreted». The reasons may be found in the important limitations of the notion of sustainability and the unanswered open questions that follow:

One problem is the difficulty of defining the spatial and temporal scale at which sustainability should be attained. Should we manage our natural resources to be sustainable at the site level, regionally, nationally, internationally or even globally? Should we work towards sustainability at the short-, mid- or long-term horizon? Most publications avoid these delicate questions.

- Another question is «What should be sustained?» The meaning of «sustainable use of a resource» is not always clear. This can be the case when different and simultaneous uses are conflicting in the same ecosystem, for example, timber harvesting and habitat improvement for endangered species, or when two adjacent ecosystems are strongly interacting, when ecosystems are too small, or when they have an unfavourable patch configuration. Because of the multiple, and often contradictory, objectives of sustainability, finding the optimal solution is a complex process. For the resource manager acting alone, finding the optimal solution is often impossible. Participation of all stakeholders in formulating management objectives can facilitate the process.

- A third obstacle is the scarcity of scientific knowledge on many issues of sustainability. An impressive body of scientific results exists about individual elements and processes of ecosystems at the site scale, but not enough is known about many long-term effects or interactions at the landscape scale. Examples are interactions between wildlife and human interventions, ecological and economic effects of nitrogen deposition on the productivity and health of ecosystems or effects of population increase on the carrying capacity of ecosystems and landscapes. This last example leads us to a fundamental question:

- How should we understand sustainability, given the increasing world population? Kauppi (1995) documented the obvious contradiction between the idea of sustainability and the continuing population growth of the Earth.

- Finally, who decides about sustainability? In countries ranging from New Zealand through India to the USA, professional foresters have come under criticism from environmental and social groups. They have been seen by some as part of the problem rather than part of the solution. Their independence as professionals from their employers has been challenged. Sustainable forest management has been criticised as an approach which while appearing to consider all the goods and services provided by forests, leaves the final decisions and trade-offs to foresters. While some of these criticisms are justified, some arise because foresters have been thrust into a new arena where their decisions have to be defended in the face of the wishes of many different special interest groups. Currently, very few practising foresters have the necessary skills to deal adequately with such situations, placing them at a disadvantage when negotiating in public participation processes. This can lead to resentment, loss of trust and the breakdown of communication between foresters and other groups.

These limitations show that the concept of sustainable forest management is difficult to understand, therefore to implement. Should we still fix our effort on the notion of sustainable forest management or should we not move to a new and broader paradigm, which would allow us to improve management in the sense to include ecological, economic and social needs of present and future generation? This new paradigm could be ecosystem-based management.

Ecosystem-based management of forests and landscapes

Introduction

The management policy that emerged from the US Forest Service's New Perspectives Program was *ecosystem management* (Salwasser, 1999). Although not uncontroversial, ecosystem management is an impressive scientific contribution to sustainable forestry that aims to reinforce the application of ecological principles in management.

There are many different definitions for ecosystem management (Baker et al., 1995). For example:

- Gordon (1993) gave an excellent idiosyncratic overview of ecosystem management. It is, he says, at minimum, people trying to accomplish something in a bounded space. He considers ecosystem management as a tool to achieve sustainability that differs from multiple-use management in focusing on inputs, interactions, and processes, as well as on uses and outputs.
- Grumbine (1994) wrote: «Ecosystem management is integrating scientific knowledge of ecological relationships within a complex sociopolitical and values framework toward the general goal of protecting native ecosystem integrity over the long term».
- The Ecological Society of America (Christenson et al., 1996) considered that :

  «Ecosystem management is management driven by explicit goals, executed by policies, protocols, and practices, and made adaptable by monitoring and research based on our best understanding of the ecological interactions and processes necessary to sustain ecosystem composition, structure, and function. Ecosystem management includes sustainability, goals, sound ecological models and understanding, complexity and connectedness, the dynamic character of ecosystems, context and scale, humans as ecosystem components, adaptability, and accountability.»
- Wood (1994) wrote: «Ecosystem management is the integration of ecological, economic, and social principles to manage biological and physical systems in a manner that safeguards the ecological sustainability, natural diversity, and productivity of the landscape.»
- Salwasser (1999) defined ecosystem management as «the synthesis and skilfull use of knowledge and actions in the stewardship of ecosystems to encourage desired conditions of environments, economies, and human well-being. Communities of people must define the desired conditions and set the course for their attainment.»

- The Interagency Ecosystem Management Task Force (1994) established by the Clinton Administration gave this definition:

  «Ecosystem management is a goal-driven approach to restoring and sustaining healthy ecosystems and their functions and values using the best science available. It entails working collaboratively with state, tribal and local governments, community groups, private landowners, and other interested parties to develop a vision of desired future ecosystem conditions. This vision integrates ecological, economic, and social factors affecting the management unit defined by ecological, not political boundaries. The goal is to restore and maintain the health of ecosystems while supporting economies and communities.»

- The definition used by the USDA Forest Service is:

  «Ecosystem management is a concept of natural resources management wherein national forest activities are considered within the context of economic, ecological, and social interactions within a defined area or region over both short and long-term.» (Thomas and Huke 1996)

These last four definitions, as opposed to the others given here, explicitly relate ecosystem management to the integration of ecological, economic, and social factors. This was emphasised by Overbay (1992) in what is still the philosophy behind the US Forest Service's use of the term:

> «Ecosystem management is the means we use to meet the goals specified in our programs and plans. Ecosystem management is the means to an end. It is not an end itself. We do not manage ecosystems just for the sake of managing them or for some notion of intrinsic ecosystem values. We manage them for specific purposes such as producing, restoring, or sustaining certain ecological conditions; desired resource uses and products; vital environmental services; and aesthetic cultural, or spiritual values.»

The emphasis is first on sustaining healthy land and secondly on providing people with the variety of benefits and options that they require or wish (Salwasser, 1999), and not vice-versa. Ecosystem management is clearly a complex and interdisciplinary type of management that needs good monitoring and relevant scientific contributions. Also, ecology and people are clearly fundamental elements of ecosystem management.

The political background of ecosystem management in the Pacific Northwest

During the last decade, as a result of the discussions about the style of management used before 1990 on public lands – clearcut and short rotation – considerable effort was put into ecosystem management and ecosystem management research in the Pacific Northwest. A number of different studies have looked at how ecosystem management evolved on forest lands; examples include

Kennedy and Quigley (1993), Caldwell et al. (1994), Shands (1994) and Clark et al. (1999). The issues were summarized by Baker et al . (1995) as following:

«By 1993, a series of court orders had brought timber harvesting on federal lands to a virtual half within the range of the northern spotted owl (in western Oregon and Washington and northern California). Federal courts ruled that the U.S. Forest Service (USFS) and Bureau of Land Management (BLM) had failed to produce plans satisfying the requirements of several laws, including the National Forest Management Act of 1976, the Endangered Species Act of 1979, and the National Environmental Policy Act of 1969 (FEMAT 1993). Regulations issued for the USFS under the National Forest Management Act require that "fish and wildlife habitat shall be managed to maintain viable populations of existing native and desired non-native vertebrate species in the planning area" and require provision "for diversity of plant and animal communities and tree species". The Endangered Species Act protects all species formally listed as endangered or threatened.

Legal battles focused initially on the northern spotted owl, listed as threatened by the U.S. Fish and Wildlife Service. Northern spotted owls are closely associated with habitat found most often in old-growth forests, that is, forest stands where many old, large trees remain in the overstory. Gradually, the debate shifted from dealing with one species, the northern spotted owl, to considering all species associated with old-growth forests in the Pacific Northwest.

Historically, timber harvests from federal lands accounted for about one-third of total timber sales in western Oregon and Washington and northern California (FEMAT, 1993). Loss of these timber harvests resulted in severe economic disruption, particularly in small communities heavily dependent on timber from federal land for jobs. In response to this problem, President Clinton convened a day-long conference on 2 April 1993 in Portland, Oregon. He posed the following question: «How can we achieve a balanced and comprehensive policy that recognizes the importance of the forests and timber to the economy and jobs in this region, and how can we preserve our precious old-growth forests, which are part of our national heritage and, once destroyed, can never be replaced?»

Following the conference, President Clinton created three interagency working groups: the Forest Ecosystem Management Assessement Team (FEMAT), the Labor and Community Assessment Team, and the Agency Coordination Team. The FEMAT report was published in July 1993 (FEMAT 1993). The interagency working groups established by President Clinton represent the start of ecosystem management in the Pacific Northwest[2].

The controversies about ecosystem management

Although ecosystem management is generally recognised as a good conceptual basis for sustainable development, it is also controversial. Some people support it vigorously, others reject it categorically. According to Bennett (1966),

> «While many concepts and values are common to most views of ecosystem management, major differences also exist. Some ecosystem management advocates focus almost exclusively on ecological considerations. Others see ecosystem management as a way to reconcile human needs for wood products with environmental values.»

In the USA, the main causes of this phenomenon are to be found in the reactions to the process behind the Northwest Forest Plan, as well as to the adopted solution, which corresponds to a drastic diminution of harvested wood volume. The following citations are examples of sceptical voices:

> «The FEMAT was led by experts who had previously advocated and authored plans that favoured the natural reserve approach to managing forests for biodiversity, old-growth-dependent species (e.g., the northern spotted owl), and other values. There were no scientists on the team who represented landscape management.» (Hampton, 1994)

> «The FEMAT report ignores large and tangible economic losses or masks them by claiming dubious economic benefits.» (McKillop, 1994)

> «Efforts to operationalize the ecosystem idea by making ecosystem protection the basis of land use management and natural resource policy are awash in confusion and pseudoscientific buzzwords because they gloss over the concept's inherent characteristics of spatial and temporal vagueness.» (Fitzsimmons, 1996)

Cortner et al. (1966) noted that:

> «There is little consensus about the new terminology, conceptual categories, and classifications used to discuss ecosystem management. Because of this diversity, it is important that we look at all aspects of ecosystem management to determine the full

---

[2] Quoted from «Baker, J.P.; Landers, D.H.; Lee, II P.; Ringold, P.L.; Sumner, R.R.; Wigington, P.J.; Bennett, R.S.; Preston, E.M.; Frick, W.E.; Sigleo, A.C.; Specht, D.T.; Young, D.R. 1995. Ecosystem management research in the Pacific Northwest: Five-year research strategy. Washington D.C.: US Environmental Protection Agency, EPA/600/R-95/069.»

range of its possibilities and limitations.»

Because of these views we should ask the question «What can be improved?.» At the same time, it is worth noting that the Northwest Forest Plan is still considered to have been a success in that environmental protection has been achieved while maintaining a stable, sustainable timber supply for the last two years (Graybeal, 1999).

Ecosystem-based management

When natural resources are managed for sustainable development, managers recognise that they cannot manage all components, processes, and interactions of the ecosystems. Rather, they are manipulating particular aspects of the ecosystems, with a view to sustaining desirable characteristics and by taking into account, as far as possible, inputs, processes, interactions, and outputs of the ecosystems. For example, they can find out what ecosystem composition and structure is needed for producing wanted commodities, wildlife conservation, erosion protection, water quality, or recreation and try to intervene accordingly, keeping the principles for managing natural resources for sustainable development in mind (section 3.2). Therefore Schlaepfer (1997) prefers to think in terms of ecosystem-based management, rather than ecosystem management. An important advantage of this formulation is that it can be used with all types of natural resource management issues: ecosystem based land-use management, ecosystem based management of forest resources, ecosystem based management of water resources, ecosystem based wildlife management, ecosystem based wetland management, and ecosystem based watershed management, and so on.

He sees ecosystem-based management as a systematic process, based on good judgement and sound science. It aims, for a defined area, to achieve the sustainable use of natural resources, by increasing the ecological sensitivity and content of management practices, and by integrating economic, ecological, social, and technological considerations, over both the short and long terms, from the site to the landscape-scale ecosystem. Ecosystem-based management will help people to use natural resources and simultaneously to maintain productivity for the benefit of future generations, to protect particularly valuable ecosystems, to restore degraded habitat, or to rehabilitate altered ecosystems.

Despite the argument in favour of the term ecosystem-based management, we should not forget that the main issue for sustainable development is how natural resources are managed, and not the terms we use. The next section describes briefly the main instruments of ecosystem-based management and some aspects of its implementation.

The scientific contributions to ecosystem management

Science is one of the driving forces in the shift to ecosystem-based management from a mainly commodity production oriented forestry. The scientific contributions are numerous and diverse. In traditional forest sciences, much research has been conducted in silviculture, tree genetics, insect and disease control, harvesting, engineering, and forest management, often with the goal to maximize and sustain wood production. Ecology, by improving our knowledge about the interactions between plants, animals, and their physical environment, has contributed to a better understanding of the functioning of forest ecosystems and of the consequences of natural and human pressures. The methods and results used in landscape ecology have many applications in forestry, especially the knowledge gained about the relationships among the building blocks of the landscape, and the functioning of the landscape as a system. Conservation biology differs from traditional resource conservation in being motivated not by utilitarian, single-species issues, but by the need for conservation of entire systems and all their biological components and processes (Meffe and Carroll, 1994). Its biology has roots in individual disciplines. Its conservation, based on the island biogeography theory, is the basis for preserve systems and national parks. Ecosystem dynamics, with disturbance and landscape-change oriented scientists, focus on causes and effects of changing landscape patterns. Information on the historical variability of ecosystem conditions and the natural disturbance regime that influence such variability is increasingly used in the design of ecosystem management systems (Swanson et al., 1993). Assessment of management effects is approached in part in terms of alternation of disturbance regimes from the «natural» to «managed» periods. Landscape management design elements include areas with differing frequency and severity of treatments based on natural disturbance regimes.

Three examples can be cited which have looked at patterns and relationships at the landscape scale, a field of increasing importance for managing natural resources.

Ripple et al. (1991), under the assumption that wildlife ecology and behaviour may strongly depend on the nature and pattern of landscape elements, studied the use of spatial statistics to quantify the landscape pattern caused by the patchwork of clear-cuts made over a 15-year period in the western Cascades of Oregon. In their paper, forest patches were digitized and analysed to produce both tabular and mapped information describing patch size, shape, abundance and spacing, and matrix characteristics of a given area. The authors also developed a GIS fragmentation index that is a function of the distances between the cell of a grid to a certain patch type. The GIS index was found to be sensitive to patch abundance and to the spatial distribution of patches. A comparison of the spatial statistics calculated for two years indicates an increase in forest fragmentation as characterized by an increase in mean patch abundance and a decrease in interpatch distance, amount of interior natural forest habitat, and the GIS fragmentation index. The authors conclude that such statistics capable of quantifying patch shape and spatial distribution may prove important in evaluating the changing character of interior and edge habitats for wildlife.

McGarigal and McComb (1995) determined how changes in landscape structure (both composition and configuration) affect bird populations in the spatially and temporally dynamic forest landscape of the Oregon Coast Range. They developed FRAGSTATS software to quantify the configuration of the landscape. The authors conclude that, contrary to the idea that habitat fragmentation is detrimental to species that specialize on a particular habitat, most species that exhibited significant relations with habitat configuration in the study were associated with the more fragmented distribution of habitat. The results suggest that landscape structure at the scale of small 300-ha watersheds in the central Oregon Coast Range may have only a moderate effect on bird abundance. However, vertebrate population dynamics in forests being fragmented by timber management activities are likely to differ from those in forests being fragmented by urbanization and agricultural development.

The work on the Camp Pendleton area, California, edited by Steinitz (1996), explored how urban growth and change in the rapidly developing region between San Diego and Los Angeles might influence its biodiversity. The study was conducted by a team of investigators from several universities and agencies. Future change was studied at four scales: several restoration projects, a subdivision, a third-order watershed, and the region as a whole. Regional change is simulated via six alternative projections of development to the year 2010. The first scenario is based on the current local and regional plans. Five alternative scenarios provide a method for exploring and comparing the effects of different land-use and development policies relating to biodiversity. All alternatives accommodate the population forecast for the region. A set of process models – for soil, hydrology, fire, visual preference, and biodiversity – is used to assess each alternative. The results from the evaluation of the six alternatives can be used by stakeholders as a source of information for making decisions. It is worth noting that the alternative that seeks to protect the most significant habitat areas via private conservation produces the best overall result.

The question, «How should the role of disturbances in the ecosystems at all scales be integrated with management?» was considered by Garman et al. (1995) as an important research emphasis, for several reasons. First, designing landscapes based on the historical variability in disturbance patterns is of interest. The underlying philosophy is that maintaining an ecosystem within its range of natural variability provides an ecological template for sustaining species and ecosystem processes (Swanson et al., 1993). Second, because ecosystem management planning extends over long periods, the chance for natural disturbance events to occur is greater, so a land-use design must take the probability of these events into account.

These examples show that in the Pacific Northwest, long-term, landscape scale, integration (between scientific disciplines, between small and large scale, between short and long-term, between management and science), and disturbances are major concerns for research.

Ecosystem management is also gaining ground in Europe. We can document this with two examples. One is the International Seminar and Summer School, which was organised in Wageningen, The Netherlands, in 1997 (Nabuurs, 1998). The goal of the conference was to give a broad overview of current challenges, research interests and activities around ecosystem management. Another example is the special issue (Vol. 40, 1998) of the Elsevier journal «Landscape and Urban Planning», which was entirely dedicated to ecosystem management.

Forest ecosystems as part of a broader landscape

For hundreds of years, forests were studied and managed as such, and forest policy hardly went beyond forests. In other words, forests were considered for their own multiple values: «Traditionally the forester humbly stayed within the boundaries of the forest, diligently planted trees and tried to protect them from pests and thieves and finally felled them. But never did he really have a say in determining the share and place the forest should retain as the former matrix of the primeval landscape» (Anko, 1994). It is relatively recently that managers, scientists and policy makers have started to become aware that forests are not a closed system, but an element interacting with other elements in a broader world. For example, interactions can occur in the transition zones between forest and agricultural land, between forest and urban areas, or between managed forests and protected areas. We also know that functions of forest ecosystems are taking place at different spatial or temporal scales. Many processes, for example photosynthesis and biomass production, happen at the plant or site level. These levels have to be considered by science and management. But several phenomena related to forest ecosystems often happen at the landscape, regional or global level, such as the life of many animals, flows of matter and energy, climate change, air pollution, use of natural resources and land use changes.

This increasing awareness is the basis of several developments, including the taking into account of the landscape in ecosystem-based management. These developments are likely to be the basis for future forest and landscape management. Some of these developments are illustrated here.

Art.8.(e) of the Forest Principles formulated in Rio (Earth Summit, 1992) is a good illustration of our point:

> «Forest management should be integrated with management of adjacent areas so as to maintain ecological balance and sustainable productivity.»

This fundamental principle, rarely applied in current forest management, will be an important challenge in the future. It encourages managers to take account of the different interactions that can exist between forest ecosystems and neighbouring ecosystems or land-use types.

From a scientific point of view, the integration of forest management with the management of adjacent areas is closely related to landscape ecology. The main task of landscape ecology is, according to Zonnewald (1991), to gain knowledge about the relationships among the building blocks of the landscape and, from these, about the functioning of the landscape as a system. Zonnewald has also identified several areas of landscape study, each with its own specific set of methods:

- Morphology (description of the structure and its elements);
- Classification (systematic ordering);

- Chorology (study of spatial patterns and variation);
- Chronology (study of temporal variation) and
- Relationships (landscape ecology in the narrow sense), the study of the relationships as such between all land attributes as a scientific interest and as a basis for land evaluation.

The notion of landscape itself has different meanings, depending on the point of view. Some of them are presented in Farina (1998). For Forman and Godron (1986), a landscape is a heterogeneous land area composed of a cluster of interacting ecosystems that is repeated in similar form throughout. For Forman (1996), a landscape is a mosaic where the mix of local ecosystems or land uses is repeated in similar forms over a several kilometers wide area. In these definitions, we can easily recognise a forest as either an interacting local ecosystem or a land-use type, and we need to include this dimension in sustainable forest management. This was done by Franklin (1993), who proposed five elements essential to his concept of sustainable forestry and which are consistent with landscape ecology:

- Thinking holistically (focus on ecosystems rather than species or products);
- Planning at large spatial scales (landscape and region);
- Recognising the gradient of habitat maintenance (scales essential to species and processes range from square centimetres to thousands of hectares);
- Recognising the multiple roles of the matrix in conserving diversity. (Forman and Godron (1986) defined the matrix as the most extensive and most connected landscape element type present, which plays the dominant role in landscape functioning. Also a landscape element surrounding a patch) and
- Recognising that not all elements of diversity are equal and that more diversity is not necessarily better.

There are many good examples to illustrate Franklin's elements. Let us take the possible re-colonisation of Switzerland by the wolf, studied by Glenz (1999). The wolf is an animal with a home range of roughly 100 to 200 square km per family unit. The probability that a landscape is suitable as wolf habitat depends on many characteristics, among them the quality and quantity of forests in the landscape. Glenz, by using a logistic regression model developed for Italy by Massolo and Meriggi (1998) and including several habitat parameters, was able to give an estimation of the number of wolves which could live in the Canton of Valais. Another example is the relationship between forest fragmentation and biological diversity. This relationship is complex and depends on the landscape type and on the studied species. An interesting study on four species of North American tanagers (Piranga spp), based on 1107 sites including 1000-ha landscape blocks around census points, was performed by Rosenberg et al. (1999). The overall amount of fragmentation (a factor including several metrics), forest configuration (measured by the amount of edge in a landscape) and elevation affected tanager occurrence.

These two examples show that for managing biological diversity, decisions have sometimes to be taken at both the ecosystem and the landscape scale. In this respect, as pointed out by Bennett (1999), the maintenance or creation of linkages in the landscape, especially corridors and connectivity between forests plays an important role in wildlife conservation. However, any such assessment also needs to factor in a human element. In the wolf example cited above, the success of any wolves colonising the Canton of Valais will be dependent on a major shift in the attitudes of local farmers.

The landscape dimension is also an important theme in recent textbooks about forest ecology or forest management. For example, Waring and Running (1998) give special attention to spatial scaling methods for landscape and regional ecosystem analysis and the role of forests in global ecology. Nabuurs et al. (1998), recognising the increasing need to move towards ecosystem management at the landscape level, edited a book on forest scenario modelling for ecosystem management at the landscape level.

Conclusions

Forestry, like other domains, is following the rule that any new idea is first ignored, then questioned and passionately discussed, and finally, when scientifically sound, implemented or, when without scientific support, forgotten. The idea behind ecosystem-based management was first ignored. The emotionally intense discussion which followed is now calming down. We are slowly reaching the stage of implementation. This implementation, based on different instruments, could change forestry and foresters as we know them.

Implementation of ecosystem-based management

Introduction

Taking an ecological perspective in forest management involves decisions on how forest and landscape resources will be managed for sustainable development, which itself depends on our ability to:

- reach consensus among the different interests on the multiple objectives of management, including integration of ecological, economic and social considerations;
- define and assess progress towards sustainability; and
- improve scientific knowledge on the long-term processes, reactions, and interactions in ecosystems and landscapes.

The task of ecosystem-based forest management is formidable and could seem impossible, but, as in many other situations in life, it is a learning process. If we want to succeed, we have to go through all steps of the process, one step after the other. To do so, we need an understanding of management of natural resources for sustainable development and a vision of implementing it.

Principles, certification and questions

Progress can be made stepwise, based on the best available information and by always having in view the quality of life for future generations. Several instruments to improve management have been developed or are in development: the definition of sustainable forest management, the definition of criteria and indicators for evaluating the progress towards sustainability, ecosystem management and forest certification.

Several approaches are being developed to measure progress towards sustainable forest management. One is the notion of forest quality, which can be defined as the biological, social and economic components that together constitute the total significance and value of the forest (Dudley 1996, IUCN, WWF, EPFL, 1999). Forest quality can be assessed at different scales: stand, forest management unit, forest landscape, country, region or world. It can be measured using indicators. The challenge in evaluating forest quality is the combination of the different indicator values into an integrated assessment. Different approaches have been and are being developed. Bormann et al. (1994b) proposed the Pinchot Standard and the Pinchot Efficiency for assessing the condition and performance of the ecosystem relative to the demands made on it. The Pinchot Standard is based on the ecological properties that must be sustained for a given set of goods, services, and states which should be produced by the ecosystem and on what is actually sustained. The Pinchot Efficiency includes also the target values and the actually obtained values of the desired goods, services, and states. IUCN developed an approach to assessing progress towards sustainability in which people and ecosystem are equally treated. The method is based on measuring indicators, setting a performance scale and combining the indicator scores (Prescott-Allen, 1997). Prescott-Allen applied the general IUCN technique for the assessment of biodiversity (Prescott-Allen, 1998). Quality assessment methods are still not really operational and need more research, including field testing.

Principles of managing natural resources for sustainable development

Managing forest and landscape resources for sustainable development tends towards sustainable use of ecosystem resources

To move towards sustainable development, high priority should be given to the sustainable use of ecosystem resources. This objective is not trivial. Despite internationally accepted criteria, sustainable forest management has had severe limitations in its application, and many questions are still open. One is considering the quality of life of future generations, which means at least trying to avoid short-term solutions that aggravate long-term problems (Schulze et al., 1996). Management can often provide at least ecological improvements, rather than solutions for sustainability. Where management for sustainable development is truly possible, the cases should be described and used as models (Fig. 1).

Managing forest and landscape resources for sustainable development:

- **Tends towards sustainable use of ecosystem resources**
- **Is holistic**
- **Is ecosystem based**
- **Has a landscape perspective**
- **Fixes multiple objectives**
- **Is integrating**
- **Includes participation**
- **Is based on monitoring**
- **Is adaptive**
- **Is based on sound science and good judgement**
- **Takes cognitive, emotional and moral reactions into account**
- **Is based on the precautionary principle**

Figure 1. Principles of managing forest and landscape resources for sustainable development

Managing forest and landscape resources for sustainable development is holistic

People and their environment are one. This statement is true for the Earth, the ultimate ecosystem we want to sustain, as well as for ecosystems defined on smaller scales. Despite the partial decoupling that has occurred as the result of the increasing urbanisation of the population of the World, people, plants, animals, microorganisms, and the environment are all connected in a great and complex network. Closed systems that function independently from other systems do not exist. A holistic approach is therefore necessary. Savory (1988) called holistic resource management an approach that treats people and their environment as one. He derives his approach from a view of ourselves and our planet as one ecosystem functioning through four rudimentary processes: the development of living communities (succession), the cycling of mineral nutrients, the cycling of water, and the flow of solar energy. The approach also establishes clear goals in all situations, and the use of what he has termed a «thought model» to enable us to «see the whole» that we are managing.

An important management consequence of the holistic view is the need to include in the process, in so far as possible:

- ecological, economic, technological and social issues;
- the short-, mid- and long-term;
- different spatial scales: from the stand to the landscape;
- interactions between the different elements in the management unit and the interactions between the management unit and external systems and
- management experience and scientific information.

Managing forest and landscape resources for sustainable development is ecosystem based

From the following, we see that the terms *holistic* and *ecosystem-based* have some overlapping meanings:

> «Thinking holistically about sustainability means approaching it on an ecosystem rather than a species or product basis. It means thinking about the full array of species, processes, and structures and about their interrelationships, which underlies the productive potential of a given site or ecosystem.» (Franklin, 1993).

Because of the nature of ecosystems, human interventions usually affect not only one aspect of the ecosystem, but a series of components and processes. A timber harvest to obtain wood will not only decrease the standing wood volume of the forest but will also affect the soil, the water cycle, and the wildlife habitats. Similarly, the manipulation of a particular ecosystem may have important effects on the adjoining ecosystems. Typical examples are the effects of agricultural and forestry practices on streams and riparian areas. Ecosystem management implies processes in which the decisions are taken after an evaluation of the possible consequences for all key components and processes of the affected ecosystem and adjacent ecosystems at various scales.

Managing forest and landscape resources for sustainable development is done with a landscape perspective

Local ecosystems such as individual forest stands, riparian areas, and wetlands are not closed systems. They interact with other landscape elements. These interactions and the human manipulations done at a local level have cumulative effects at the landscape scale that can affect a whole region. According to Franklin (1993), planning at the landscape level is the only way that we are going to avoid undesirable, if not unacceptable, landscape dysfunction. As examples he describes the effects of large-scale clear-cutting on stream flow in the Pacific Northwest and the linkage between the extent of erosion processes, such as landslides, and the size of the road system, especially on steeper topography. Anko (1994) and Salwasser and Pfister (1994) also recognised the need to consider the forest not from the point of view of forest ecology alone, but also from the landscape perspective. This is applicable to all types of natural resources.

Managing forest and landscape resources for sustainable development has multiple criteria objectives

The nature of the objectives of managing natural resources for sustainable development is the result of many elements: all existing legal and political constraints, the holistic and ecosystem nature of such management, and the internationally defined criteria for sustainability, including biodiversity. Managing for a unique commodity is no longer an appropriate objective; objectives are based on a variety of issues corresponding to the identified ecological, economic, and social needs, or in other words, to the different functions of the ecosystems in the landscape. This situation suggests that an immediate priority is the need to strengthen multiple criteria decision making affecting land-use and therefore the need to manage with multiple objectives concerning ecological, economic, and social issues. These conclusions apply at all geographical and temporal scales and for different types of ecosystems. A vital question has to be answered: «What are the processes to be used for setting the objectives, and who should be involved?»

Related to this issue is the problem of land-use zoning. It is clearly impossible for every forest stand in a landscape to fulfil all possible functions, as some functions are directly incompatible. For example, the designation of an area as a natural preserve is incompatible with developing it for some recreational activities. Consequently, scale needs to be considered. It may be necessary to designate a priority objective for particular stands or areas, thereby enabling all functions to be fulfilled at the landscape scale or even over greater scales. An illustration is provided by the intensive management of pine plantations in New Zealand, which has reduced pressure to exploit the surviving native forests. In such circumstances, multiple criteria may still be important but tacit acknowledgement is made to certain priority uses.

Managing forest and landscape resources for sustainable development is integrating

Lang (1990) presented an excellent and short justification for integration:

> «Resource development today faces formidable problems. Demand for resources is accompanied by differing perceptions of their values, conflicts over their use, and concern about the natural and human environments affected. These problems are exacerbated by fragmented jurisdiction over the resource base, ambiguous government policies, lengthy review processes and weak regulations. Resource planners, developers and managers are responding to these problems by seeking more integrated approaches that will enable their projects and programs to deliver as many benefits as possible, within acceptable limits of social and environmental impact, and with minimum conflict and cost.»

The recognition of the need for integration in managing natural resources is not new. An overview of the evolution of the concept in the USA and Canada can be found in Mitchell (1990). Because land resources are used for a variety of purposes that interact and may compete with one another, the international community recognised at the Earth Summit the desirability of planning and managing all uses in an integrated manner (Agenda 21, chapter 10). Conferences with the theme of integration have been organised, for example, the *International Conference on Sustaining Ecosystems and People in Temperate Forests: Integrating Conservation of Biological Diversity with Social and Economic Goals*, British Columbia, 8 - 13 September 1996. Clark et al. (1996) explored the concept and meaning of integration with particular emphasis on fostering better understanding of social, cultural, and economic considerations. They summarised the literature and presented different perspectives on what integration is and is not. They examined some barriers and offered an approach for improving our capacity for effective integration, both in the short and long terms. They considered co-ordination, interdisciplinary teams, analytical science, modelling, and collaboration as means to achieve integrated resource management.

In our view, integration in managing forest and landscape resources should mean:

- Achieving agreement on the problems, issues, questions, and goals, including consideration of different ecological, economic, and social points of view and interests.
- Bringing together existing information about managed as well as unmanaged resources in the management unit, including the processes and interactions within and between the ecosystems.
- Linking together small scale and large scale issues.
- Deciding on a course of action only after having evaluated the possible ecological, economic, and social impacts and possible side effects of the foreseen manipulations on the considered ecosystem, on the adjoining ecosystems and on the whole managed unit over time.
- Recognising the middle and long-term consequences of planned short-term actions.
- Synthesising the existing scientific knowledge from the relevant disciplines.
- Linking managers, policy makers, scientists and the public.

Integration fosters appropriate choices and trade-offs, thus optimising sustainable productivity and use.

Managing forest and landscape resources for sustainable development includes participation of all stakeholders

How can the need to balance ecological, economic, and social issues be effectively satisfied? One approach is the participation of all the concerned parties in the management process. The stakeholders can be invited to help set goals, analyse the situation, and develop improvements and solutions. Shindler and Neburka (1996) identified attributes for the success of public participation in forest communities. They compared five different long-term planning processes that occurred between 1989 and 1995 in the Pacific Northwest. Eight attributes of success were identified by group participants. For example, it became evident that groups where members are selected by the agency for their understanding of the issues and their willingness to commit to a group process are more effective. Also, meetings that are structured to promote full group interactions, rather than simple information sharing and feedback, are much more productive. Management by participation is already practised with success in different cases, for example in the Elliott Forest in Oregon, or in the Canadian Model Forests Program (Science Branch Canadian Forest Service, 1996). It is clear that the participation process will not only depend on the main management issue, but also on the ownership situation: federal, state, private or common land.

An important case is community forestry, defined by Gilmour and Fisher (1991) as the control and management of forest resources by the rural people who use them especially for domestic purposes and as an integral part of their farming system.

Participatory management is required in some international conventions and regulations dealing with sustainable development and forest management (Muhlemann, 1997). But many obstacles are still existing for introducing participation in forest management. One of the greatest problems is the identification of the stakeholders themselves. Who are the stakeholders? Should all stakeholders carry the same weight? To what extent should the interests of local residents override the interests of occasional visitors? In cases of disagreement, who decides on the best course of action? We believe that research is needed to find out how these questions can be overcome.

Managing forest and landscape resources for sustainable development is based on monitoring

The term monitoring is used in two different ways. Firstly, it implies the measuring and assessment of data in relation to key variables to determine whether or not objectives or standards have been met. However, it can also signify the collection of data with the aim to detect trends and understand how a system is functioning (Canadian Forestry Terminology, 1988). Spellenberg (1991) summarised the value of biological and ecological monitoring by the following objectives:

- Monitoring as a basis for managing biological resources for sustainable development and resource assessment.
- Monitoring so that ecosystems and populations can be managed and conserved effectively.
- Monitoring land use and landscapes as a basis for better use of the land; that is, combining conservation with other uses.
- Monitoring the state of the environment - using organisms to monitor pollution and to indicate the quality of the environment.
- Monitoring as a way of advancing knowledge about the dynamics of ecosystems.
- Monitoring of insect pests of agriculture and forestry so as to establish effective means of control of those pests.

Management without monitoring is like playing football without recording the score. Effective monitoring helps managers to follow the consequences of their interventions, to be informed about unexpected events, and to measure deviations from the management objectives. In adaptive management, a monitoring philosophy is needed that fits with the broad learning goals (Bormann et al., 1966). It is particularly important that there is open communication between the managers and those doing the monitoring. This implies a degree of trust between the two, which is easier to achieve if the monitoring is aimed at determining the effectiveness of particular management activities rather than simply whether a manager is complying with specific standards. The former implies a partnership between management and sciences, with the ultimate beneficiary being the ecosystem.

Monitoring and evaluation processes must be scientifically and legally defensible, otherwise they are of doubtful value. The quality and relevance of monitoring depends on its objectives, its design, the methods used for measuring the variables, the variables and the processes chosen, the methods used for the analysis of the data, the interpretation of the results, and the way the results are disseminated.

Managing forest and landscape resources for sustainable development is adaptive

In managing biological resources we are confronted with the complexity of ecosystems, with long-term processes, with unpredictable natural disturbances as well as with human influences. Even with the best available science, complete answers to all management questions will not be attainable. We will always have to manage under conditions of uncertainty and will need to accept the inevitable mistakes as learning advances. These realities have been recognised by many authors and have led to the notion of adaptive management, defined by Holling (1978) as an interactive process using techniques that not only reduce uncertainty but also benefit from it. Walters (1986) asks how we should proceed to develop better understanding of managed system responses and potentials in a world of great uncertainty, limited research resources, and continuing pressure for more intense exploitation. His answer is to treat management as an adaptive learning process, where management activities themselves are viewed as the primary tools for experimentation.

Adaptive management is an important issue in the Pacific Northwest. The Forest Ecosystem Management Assessment Team (FEMAT) report (1993) proposed the creation of 10 adaptive management areas across Washington, Oregon, and northern California. These areas became realities in December 1994 with Judge Dwyer's decision affirming the legality of the Northwest Forest Plan (Stankey and Shindler, 1996). Bormann et al. (1994a) presented a systematic approach to adaptive management to simultaneously manage at the regional, provincial, and watershed scales and to reorganise the activity of agencies to better support the concepts of adaptive management. They proposed a strategy to improve management by making better decisions, improving public participation, developing science-based management, and using a systematic approach. They also recommended reorganising management activities into four groupings that support the desired function of the whole system: adjustment; linked, not single actions; feedback, including monitoring; and information synthesis.

Stankey and Shindler (1996), in a study of the FEMAT adaptive management areas, identified several barriers confronting effective implementation of the adaptive management. They presented a set of four propositions that address these potential barriers: (1) area boundaries must possess social meaning for stakeholders; (2) a focus on adaptive management areas will highlight limitations in scientific knowledge; (3) management of the adaptive management areas will highlight differences in how the world is perceived; and (4) effective management of the adaptive management areas will challenge existing institutional arrangements.

Adaptive management is intuitively extremely attractive, but its implementation in practice has met with little success (e.g. Lee, 1993). The barriers identified by Stankey and Shindler (1996) are only part of the problem. As explained by Bunnell (1998), adaptive management involves trying, noting the mistakes and trying again, but errors are made in all three steps. Goals are often inadequately defined, so there is nothing against which success or progress can be evaluated. Insufficient attention (and funding) is usually given to monitoring the effectiveness of the actions. And adequate funds have rarely, if ever, been allocated to implementing corrective actions. Even when funds are available, there may be an unwillingness to recognise that a mistake has been made and considerable inertia may be encountered when trying to bring about changes in management practices.

Bormann et al. (1996) explored new possibilities for achieving complex management goals on Federal lands by applying adaptive management. In their comprehensive paper, they used the term adaptive management to describe an approach to managing complex natural systems that builds on common sense and learning from experience to include experimenting, monitoring, and adjusting practices based on what was learned. In order to achieve balanced learning among citizens, scientists, and managers, they advocated adding learning and adapting as management goals. For speeding learning and to reduce the risk of biased or incorrect interpretations, they suggested incorporating scientific methods into everyday management actions whenever possible.

We have no doubt that adaptive management can be an efficient tool for managing forest and landscape resources for sustainable development. Adaptive management, as a combination of management and scientific methods, is a way to overcome the limitations of science as well as of management itself. Research is often in difficulty to answer questions at large spatial and temporal scales. Management, when based on only one course of action, has limited value as a learning process. Adaptive management can be considered as a common ground where managers, scientists, and citizens can try to learn to meet society's needs and wants while maintaining ecological capacity (Bormann et al. 1996). As an example let us mention what Bormann et al. (1996) called experimental management. This is a type of adaptive management in which suites of policies are directly compared in scientifically designed "management experiments" which become the focus of monitoring and evaluation. In Oregon, this technique is being tried at stand scales as well as at larger scales.

Managing forest and landscape resources for sustainable development is based on sound science and good judgement

It is generally accepted that scientific knowledge represents an important basis of managing natural resources for sustainable development. Thomas and Huke (1996) argue convincingly that:

> «The quality of policy and ecosystem management decisions is highly dependent on the quality and quantity of available information and science. Developing the foundation for ecosystem management will require not only sound science, but the right science - knowledge and understanding of how major ecosystems function; how they can support and tolerate human use; and how policies and management decisions affect resource use, the environment, and recovery.»

Ecosystem research is improving that knowledge through the publication and dissemination of its results.

Seeking to balance ecological, economic, and social considerations can induce many different, sometimes opposite views. To reach consensus in fixing the goals and the objectives for the society's use of natural resources, and to obtain the involvement, understanding, and support of all concerned groups in the implementation of management are among the most critical and difficult tasks of managers and policy makers. These challenges can be best met by basing decisions not only on sound science, but also on good judgement. We understand the term «good judgement» as does Yankelovich (1991), when he uses the term «public judgement» to mean a particular form of «public opinion» that exhibits more thoughtfulness, more weighing of alternatives, more genuine involvement with the issue, more taking into account a wide variety of factors than ordinary public opinion as measured in opinion polls, and that has more emphasis on the normative, valuing, ethical side of questions than on the factual, informational side. In managing forest and landscape resources, good judgement should also include more consideration of the long-term.

Managing forest and landscape resources for sustainable development takes cognitive, emotional, and moral reactions into account

In general, to obtain the desired goods, services, and states, ecosystem manipulations are needed, such as harvesting, harvesting exclusion, hunting, hunting exclusion, road construction, and many others. Often, required ecosystem manipulations or restrictions against manipulations are susceptible to cognitive, emotional, and moral obstacles that can seriously slow down the management process. The same is true for environmental issues like air pollution, climate change, or energy production. Yankelovich (1991) defined cognitive obstacles as the difficulties associated with perceiving, thinking, judging, connecting, sorting out, and absorbing information. For him, emotional obstacles involve coping with a wide swathe of feelings and the defences associated with them: hope, fear, anxiety, idealisation, anger, denial, avoidance, resignation, bitterness, self-aggrandisement, self-esteem, and so forth. The moral obstacles involve conflicts between personal desires and commitment to others. We can improve the effectiveness of management by identifying these obstacles and, if necessary, by taking them into account, for example by developing an adequate information strategy and including social scientists in the process.

Managing forest and landscape resources for sustainable development is based on the precautionary principle

Principle 15 of the Rio Declaration stipulates that

> «In order to protect the environment, the precautionary approach shall be widely applied by States according to their capabilities. Where there are threats of serious or irreversible damage, lack of full scientific certainty shall not be used as a reason for postponing cost-effective measures to prevent environmental degradation.»

The precautionary principle raises an extremely interesting, important, and complex issue: the role and the management of risks and uncertainties. In the long run, its application in managing natural resources should allow us to maximise the chance to sustain the different ecosystem functions.

Forest certification: a new policy instrument to promote sustainable forest management?

Introduction

Forest certification is a process which results in a written certificate being issued by an independent third-party, attesting to the location and management status of a forest which is producing timber (Baharuddin and Simula, 1994: 9-10, Upton and Bass, 1995).

There are normally two components:

- Forest management auditing involves inspection of forest management on the ground against specific standards, and a review of relevant documents such as management plans, working plans, inventories etc. Certification of forest management can be carried out at different levels ranging from the forest management unit, to the region or country. Existing certification programmes mostly work at the level of the management unit.
- Product labelling involves tracing round-wood and processed timber products through the «chain-of-custody» (or supply chain) from the forest floor to the final purchaser. This involves tracking the certified material through log transport and processing, shipping and further processing.

The result of these two processes is a label on a product that can be identified by a consumer.

The first certification was carried out in Indonesia in November 1990 by the Smartwood programme of the Rainforest Alliance, an NGO based in New York. Since then, certification has developed rapidly, and by June 1999 over 15 million hectares of forests in 26 countries had been certified (FSC 1999).

Goal and objectives of forest certification

Certification is an economic policy instrument with environmental and trade objectives. Economic instruments for environmental protection have been defined by the Organization for Economic Co-Operation and Development (OECD) as:

«Instruments that affect costs and benefits of alternative actions open to economic agents, with the effect of influencing behaviour in a way that is favourable to the environment.» (OECD 1991:10)

A basic objective of economic instruments is to ensure an appropriate pricing of natural resources in order to promote their efficient use and allocation. In a 1996 report on economic incentives for biodiversity conservation, over 40 types were identified, divided into four categories, as illustrated in Table 1.

	Category of Economic Instrument			
	Positive incentives	Disincentives	Indirect incentives	Removal of perverse incentives
Examples	Agricultural land set-aside schemes; Public land purchases; Incentive payments for organic farming.	User fees; Fines for damage; Performance bonds	Individual transferable fishing quotas; Eco-labelling (including certification).	Reform of tax structures; Full-cost pricing for water services; Reform of public forestry concession pricing.

Table 1. Categories and examples of economic instruments for biodiversity conservation. Source: OECD 1996:9

Certification is listed in Table 1 as an economic instrument, under the heading of ecolabelling which is defined as:

> «The provision of information about product characteristics, such as those that relate to the environment, to enable more informed consumer purchasing decisions and to differentiate products and create markets for the differentiated products.» (OECD 1996, p. 141)

Ecolabelling and certification are classified as «indirect incentives», defined as follows:

> «Any mechanism that creates or improves upon markets and price signals for biological resources, encouraging the conservation and sustainable use of biological diversity.» (OECD 1996, p. 141)

In line with this view of certification as an economic instrument, Elliott (1999) has defined the goal and objectives of certification as follows:

Goal: to improve the social, environmental and economic quality of forest management through the provision of incentives, including market access, to forest managers.

Objective 1: to promote and facilitate policy-orientated learning among actors in forest policy domains, such that acceptable standards of forest management, covering economic, social and environmental issues in a balanced manner, can be defined and used.

Objective 2: to provide incentives to forest managers to attain, or maintain, acceptable standards of forest management.

There is room for debate over what constitutes acceptable standards for forest management. However, these are being increasingly closely defined in the criteria and indicators that are being developed for sustainable forest management in different parts of the world. Although most certification organisations have their own standards, general standards for sustainable forest management are being increasingly set at the governmental and inter-governmental level, whereas compliance, and attestation of compliance, remains largely a voluntary issue. Eco-labelling and certification provide an economic incentive to achieve the necessary standards.

6.33 Certification in practice

Certification is a process which involves a number of different actors and steps. These are discussed below. The actors may include certifiers, accreditors, forest managers, wood buyers, stakeholders and government agencies. The roles of these actors are shown in Box 1.

The steps are normally as follows:

Step 1: The forest management decides to seek certification because it anticipates some benefits in terms of improved market access, image or forest management.

Step 2: Contact is made with a certifier which makes a visit to the operation and carries out a preliminary assessment of the feasibility and cost of certification. A preliminary report is given to the forest manager.

Step 3: If the forest manager decides to proceed, a certification contract is signed and a full assessment is carried out. In this assessment, a team of specialists engaged by the certifier checks forest management practices and procedures against a set of standards. These standards may be those of the certifier itself, be provided by the forest organization, or come from the accreditator. The result of this assessment is a certification report which is normally peer-reviewed.

The report will recommend one of the following to the certifier:

1. Unconditional certification
2. Certification subject to some preconditions
3. Certification followed by some corrective actions
4. No certification

Certifiers The certifier is a third-party which is independent from the forest manager. Certifiers may be non-profit or commercial enterprises. Their task is to assess the quality of the forest organization to be certified using a pre-established set of standards. This involves field visits and checks of the administrative procedures of the organization.
Accreditors The role of the accreditor is to ensure that the certifier is following reliable and transparent procedures in its assessments, and that the attributes of the certification programme (and accompanying label, if relevant) are clearly presented to consumers. Most countries have government-authorized accreditation bodies, but private accreditation is also possible.
Forest managers The forest manager is responsible for managing the forests of the organization that is to be certified. The organization may be a private forest owner, a community, a company or a government body. Similarly, the manager may be an individual or a group, with the mandate to manage the organization's forests.
Stakeholders A whole range of stakeholders from local communities to international NGOs may have views on what constitutes appropriate forest management for the organization to be certified. The certifier will normally have procedures for consulting these stakeholders on the standards to be used in the certification assessment, and on the perceptions of the overall performance of the organization.
Wood Buyers Companies which directly or indirectly buy wood from the forest organization may encourage the organization to seek certification. There have been a number of examples of this particularly from retailers in the UK.
Government Agencies Governments provide a framework for forest management through policy and legislation. In many countries, governments also own large areas of forests. Thus, governments may be able to encourage or discourage forest organizations from seeking certification.

Box 1. The Actors in Certification

The difference between 2 and 3 is that in the case of 2, the preconditions have to be fulfilled before certification can be granted whereas in the case of 3, corrective actions can be implemented within an agreed time frame after certification. A certifier may recommend both preconditions and corrective actions.

Step 4: The certifier analyses the certification report and if it is positive (i.e. not option 4) discusses a time frame for implementation of preconditions and/or corrective actions with the forest manager. If this discussion is successful, a certification contract is signed. This contract specifies the rights and responsibilities of both parties in terms of public communication etc. It also specifies the forest area that has been certified, and the length of time over which the certificate is valid.

Step 5: If product labelling is desired by the forest organization or its clients, the certifier must complete a chain of custody assessment. The purpose of this is to ensure that adequate and reliable mechanisms are in place to track wood from the certified forest through the processing and distribution chain, to the final consumer.

Step 6: Periodic repeat visits and checks are made by the certifier.

Costs and benefits of certification

As certification is a voluntary market instrument it will only be used if the benefits for the forest organizations involved exceed the costs. There is as yet insufficient data available to carry out a comprehensive cost-benefit analysis, partly because forest organizations generally consider this information to be proprietary. However it is possible to review the categories of costs and benefits involved in certification. Two factors have led some authors to cast doubts on the viability of certification as an economic instrument (eg. Berg and Olszewski, 1995; Wadsworth and Boateng, 1996:6-15; Sikod, 1996; Baharuddin and Simula, 1996:63-70; Brockman et.al., 1996; UNECE 1997:4-5; Côté, 1999). First, the costs and benefits are still subject to divergent estimates. Second, it is unclear whether certification will actually promote improved forest management, or simply recognise management which is already exemplary.

Costs

There are three categories of costs:

1. the cost of improving forest management to a level that will be adequate for certification;
2. the cost of forest management auditing (certification visit and repeat assessments); and
3. product certification costs associated with tracking the chain-of-custody.

The first category is likely to be the most significant and the least predictable. Again, it can be sub-divided into three components (Bach and Gram, 1993):

1.1 lower yields;

1.2 increased investment in planning; and

1.3 different distribution of costs and benefits over time.

Lower yields of timber may result from the need to adjust harvest levels to annual increment and to set aside forest areas for biodiversity and watershed protection. However, there are a number of means by which reduced timber yield might be offset, including the encouragement of more effective regeneration following harvesting, the use of improved seedlings, reduction in wastage during harvesting operations and reductions in the area lost to re-planting or regeneration. Certification requires detailed documentation, including management plans. If these are not available, an investment must be made in preparing them. This investment may be partially offset by improved operational efficiency resulting from better planning. Improved forest management may result in delaying harvesting of certain areas while making initial investments in planning. Depending on the interest rate and costs involved this different distribution of costs and benefits over time may become a significant cost component.

There are various estimates of the cost of certification assessments and of product certification. In both cases, there appear to be clear economies of scale which make certification relatively expensive for smaller scale operators. Table 2 presents some cost estimates from different countries.

Country	Cost Estimates for Certification Assessments
Brazil	So far, certification exercises have cost between US$ 20,000 and 100,000. In large operations the cost has been US$ 0.6 to 1.4 per ha.
Finland	Costs of certification assessments vary widely according to the scale of the operation. For a private holding of 30 ha., costs average US$ 24 per ha., whereas auditing a 50,000 ha. Forest Management Association would cost an average of US$ 0.4 per ha., and a 1.4 million ha. regional forestry centre US$ 0.02 per ha.
Indonesia	Certification assessments are estimated to cost between US$ 0.2 and 0.4 per cubic metre and chain of custody costs an additional US$ 0.3 to 1.3. (Bringing forest management standards up to the level required for certification could cost as much as US$ 13 per cubic metre.)
Switzerland	The cost estimate for the certification of the forests of Solothurn (400 ha) is US$ 3.0/ha/year, half of which are external.
Malaysia	Annual costs for the initial certification assessment and follow-up visits have been estimated at US$ 0.22 per hectare for a 100,000 ha concession.

Table 2. Estimates of the costs of certification assessments in selected countries. Note: costs are not directly comparable between countries as the certification standards are different in each case and the estimates should be considered preliminary. Sources: Baharuddin and Simula (1996), Simula (1998), Küchli (1998).

Benefits

The benefits of certification can be divided into two categories:

1. Market benefits; and
2. Non-market benefits.

The clearest market benefit is higher prices, but there are few examples of this. Collins Pine, a certified US company, has had some limited success in obtaining a price premium (Hansen and Punches 1998). In 1998, the Swedish Company AssiDomän was able to obtain a premium of 5% for certified sawn-wood and pulp. Even if the first companies to obtain certification can negotiate a price premium because of scarcity value, there is no indication that this will be maintained as more operations receive certification. A number of consumer surveys have been carried out in Europe and North America to try to determine consumers' «willingness to pay» for certified products. These are reviewed in Gale and Burda (1996) and Rametsteiner et. al. (1998). The surveys generally found that consumers and wood buyers were concerned about the impact of timber harvesting on forests. In several Northern European countries there was support for certification as well. However, there is little reliable evidence that consumers would be willing to pay more for certified products, and the results of surveys of the type described above are notoriously difficult to interpret, since what a consumer indicates as an acceptable premium in a survey is often very different from reality.

Other market benefits can include market access and «branding.» In an effort to strengthen the market support for certification and the Forest Stewardship Council (FSC), WWF and other NGOs have supported the establishment of «buyers groups.» By the end of 1997, such groups existed in 9 countries and were in preparation in 5 others (Ford, 1997). The groups include a variety of companies, mostly from the retail sector, which have made a commitment to selling certified wood and wood products. The first group to be established was the WWF UK «1995 Group» which in 1997 had 75 members covering 30% of the total UK retail market, and 15-20% of the wood products market (Lawton, 1997). Membership of a buyers group involves a commitment to tracing the origin of all wood products sold by the company and phasing in the supply of certified timber (WWF, 1996). Some buyers groups involve explicit recognition of the FSC as the only currently credible certification system, although this has been challenged in the UK as anti-competitive, and it is likely that this provision will gradually be dropped (Lawton, 1997).

In the Netherlands, a Dutch NGO campaign called «Heart for Wood» was started by Friends of the Earth and the development NGO Novib in 1992, with the aim of reducing Dutch tropical timber consumption to a level supplied only from sustainable sources (Murphy, 1996). It has involved 252 municipalities, 10 government departments, 72 real estate developers, 139 housing associations and the three largest do-it-yourself chains in Holland (Baharuddin and Simula, 1997, p. 23). The campaign was followed by the establishment of a buyers group in 1995.

The reasons given by companies for joining buyers groups vary but generally include a wish to avoid negative publicity on environmental issues, and a desire to seek competitive advantage in the market by demonstrating environmental commitment (Lawton, 1997; Bendall and Sullivan, 1995). Tetra Pak UK joined the WWF UK 1995 plus group because its Swedish suppliers were committed to obtaining FSC certification and its major UK retail clients were already members of the group, and because the company saw the group as a forum for policy learning about environmental issues (Gunn, 1997). Murphy (1996), who studied the UK buyers group and the Dutch «Heart for Wood» campaign, has suggested that international policy processes are no longer the exclusive domain of national governments but involve an increasing number of actors including international NGOs, local governments, trans-national corporations, grass-roots activists and small businesses. In this context, he argues that buyers groups are a manifestation of the emergence of new partnerships between business and NGOs, influenced by the concepts of corporate social responsibility and sustainable development (Murphy, 1996, p. 46).

Some of the members of buyers groups have undertaken exhaustive analyses of their wood suppliers to identify potentially problematic sources, such as old-growth forests, and in some cases supplies from these areas have been cancelled. This has created some difficulty, as definitions for old-growth forests differ between countries. In countries such as the USA and Canada, the criteria for recognising the old-growth stage in different forest types are being increasingly tightly defined (e.g. Wells et al., 1998; White and Lloyd, 1998), In the USA, old-growth has been recognised by the Forest Service as a distinct resource type and in addition to a generic definition, definitions are being developed for every recognised forest type in the country. However, it may take some time before such definitions become widely acknowledged outside the forestry community. A few retailers in the UK and Holland have been able to sell a limited number of product lines made from certified timber, although the demand for certified timber from buyers groups has exceeded the supply (Knight, 1995; Baharuddin and Simula, 1996, p. 42).

Buyers groups can thus influence market access by providing a growing market for certified products, and a shrinking market for uncertified products. However, the direct impact of the buyers groups on the market should not be overestimated, because of the limited volumes of wood that members purchase. Their indirect impact as trend setters is probably more important. Several certified forest organizations including AssiDomän in Sweden, Collins Pine and Seven Islands in the US (McNulty and Cashwell, 1995; Hansen and Punches, 1998) have reported that being certified has helped them obtain brand recognition and move towards distinguishing their products from others in what is basically a commodity market.

A number of non-market benefits have been mentioned by certified forest organisations. These include improved staff morale and operational efficiency, minimising the risk of being criticised by NGOs, and organisational image and identity in terms of good forest management. (Baharuddin and Simula, 1997; McNulty and Cashwell 1995). The process of preparing for, and going through, a forest management audit may help foresters.

Two approaches to certification

Two types of standards have been used for forest certification, one based on systems and the other on performance. Systems standards define the characteristics of an Environmental Management System (EMS). Certification involves assessing whether the EMS of the forest organisation in question is in place, and operating satisfactorily. The International Organisation for Standardisation (ISO) has played a leading role in the development of systems standards in quality and environmental management, specifically with the specification for environmental management systems (ISO, 14001) issued in September 1996. This is not specifically related to the forestry industry – rather it covers environmental management standards for all industries. It is unlikely that the ISO will develop a 14000 standard specifically for forestry, but there has been some interest in such an approach (Rotherham, 1996).

Performance standards define levels of performance to be reached by forest operations. Certification in this case involves assessing whether the forest operation meets these levels of performance in its activities. The Forest Stewardship Council (FSC) has been an important promoter of performance-based certification standards for forest certification. Performance standards can be seen as being constituted of various components organised in a hierarchical manner as shown in Figure 1. Both management systems and performance standards can be used as a basis for developing certification programmes.

Neither FSC nor ISO are certifiers. ISO's role is to promote and co-ordinate the development of international standards. FSC promotes and co-ordinates the development of performance based forest certification standards and accredits certifiers. The next two sections discuss the activities of FSC and ISO as they relate to forest certification, up until the end of 1997.

ISO is a world-wide federation of national standards bodies, established in 1947. Its secretariat is located in Geneva, Switzerland. In 1997, it had 120 members in the same number of countries. ISO's objectives are:

> «To promote the development of standardisation and related activities in the world with a view to facilitating international exchanges of goods and services, and to developing co-operation in the spheres of intellectual, scientific, technological and economic activity. The results of ISO technical work are published as International Standards.» (ISO,1997a, p. 3)

Most ISO member bodies are governmental institutions, and all have close links with governments in their countries. In addition to governments, ISO seeks to bring together producers, consumers and scientists in technical committees (TC) which prepare standards.

The FSC is a non-governmental association of individuals and organizations established in 1993. It has over 300 members from 37 countries (FSC, 1999). Its secretariat is located in Oaxaca, Mexico. The objectives of the FSC are to;

> «Promote environmentally appropriate, socially beneficial and economically viable management of the world's forests... through a voluntary accreditation programme for certifiers of forest management. The FSC shall evaluate and accredit certifiers based upon adherence to FSC Principles and adherence to FSC Guidelines for Certifiers.» (FSC, 1994:paragraphs 1 and 5).

Criteria for the assessment of certification programmes

Table 3 presents a number of criteria for the assessment of certification programmes (Elliott, 1999; Baharuddin and Simula ,1994).

Criterion	Necessary Characteristics for Certification Programmes to Meet This Criterion
Credible to consumers	Clear performance levels must be either included in the standard, or developed through a process involving the relevant actors and communicated in a reliable and transparent manner; NGOs and consumer groups must support the programme; A product label or other communication tool to identify certified products is needed.
Comprehensive to include all types of timber and timber products	The programme must not *a priori* exclude any forest types or regions.
Objective and measurable criteria	The standards in the programme must be clear and specific and the language used must be normative. A framework of principles, objectives, criteria, indicators and norms as proposed by van Bueren and Blom (1997) would be useful here.
Reliable in assessment results	The framework mentioned above would be helpful here. The certifier carrying out the assessment needs to have clear and rigorous assessment procedures.
Independence from parties with vested interests	The certification must be carried out by an independent, third-party, accredited certifier.
Voluntary in participation	Certification programmes cannot be required by law.
Equal treatment, non-discriminatory in trade impact	The programmes must be independent of government and conform with World Trade Organization agreements and rules.
Acceptable to the involved parties	There must be an open and transparent process for standards and programme development, conducted according to professional norms, which takes into account the views of all the actors in the policy domain, and allows for policy-orientated learning.
Institutionally adapted to local conditions	See above. The process must take into account any special local conditions.
Cost-effective	The programme must be designed to be as efficient as possible in reaching its objectives.
Transparent to allow external judgement	Actors involvement must not be limited to standards development, but should continue through implementation and evaluation.
Goal orientated and effective	The structure of principles, criteria, indicators and norms will help here.
Practical and operational	The programme should be as simple as possible while complying with the other criteria, and mechanisms for feedback and policy-orientated learning should be included to promote improvements.
Applicable to all scales of operation	Special measures should be taken where necessary to avoid discrimination against small-scale forest organizations.

Table 3. Criteria for assessing certification programmes (Elliott, 1999; Baharuddin and Simula, 1994).

Future developments: what can certification contribute to ecosystem-based forest management?

If we define «impact» as the broad economic, technical, social, political and environmental consequences of a policy instrument, it is too early to assess the impacts of forest certification, or indeed its effectiveness and efficiency.

However, a number of recent studies (e.g. Elliott, 1999) suggest that the market benefits of certification may have been overestimated in some cases, and the non-market benefits underestimated. Market benefits should not be neglected, and in some cases (such as the company AssiDomän in Sweden) these have included both better market access and higher prices for certified products . However these market benefits have accrued so far mainly to large companies exporting to «ecosensitive» markets in Northern Europe.

What is interesting, and to some extent surprising, is the extent and variety of non-market benefits, including: increased motivation of staff in certified operations, increased operational efficiency, improved image of company or indigenous group managing the forest, and learning and exchanges between stakeholders in developing the certification standards.

If certification lives up to the expectations of its proponents and provides a process for policy-oriented learning and exchange between stakeholders as well as a market-based incentive for improved forest management, it may prove to be a useful tool for implementing ecosystem-based management.

Other tools for implementing ecosystem-based management

It is beyond the scope of this paper to provide a comprehensive overview of the various instruments that could be used to promote ecosystem-based management. There range from community forestry in Nepal, through participatory land-use planning, to economic incentives. What can be stated is that instruments that provide incentives, develop capacity and encourage learning will be particularly important, whereas the use of «command and control» instruments should be limited. A number of economic incentives exist (Pearse, 1998) that could be used to promote ecosystem-based management, including appropriate property rights, pricing mechanisms, taxes and subsidies, deposit-refund systems and compliance incentives. A full analysis of these in relation to ecosystem-based forest management still needs to be done.

Basic questions for implementation of ecosystem-based management

The main principles and instruments presented above are the basis for implementation. Following this is an overview of a possible approach for management, a systematic and iterative process based on a series of relevant management questions.

- What are the guiding principles and what is the vision for management?
- What are the goals and general objectives?

These questions allow us to become aware of the reasons for managing and the most logical lines to follow.

- How intense should the planning process be?

  This step allows us to adapt the intensity of the planning process to the general objectives. Often, extensive planning, based on a minimum of information and actions, can be sufficient.

- Who should belong to the planning team?

  A good way to assure participation of the stakeholders is to invite them to participate at the beginning of the planning process. To avoid frustration, the role of the planning team and the rules of the game should be clearly defined.

- What are the available sources of information?

  There is a huge volume of information available, and this is becoming increasingly available through resources such as the internet. Access to information is likely to improve steadily in the coming years, and the principal problem is going to be the location of the most appropriate information.

- What is the current situation and what are the key issues in this management unit?

  The depth of the analysis depends on how intense a planning process is chosen. The current status and the likely changes in the following items should be considered:

- Geographical limits and landscape characteristics of the management unit. What are the specific management objectives?

  The specific objectives should be formulated in terms of the criteria and indicators defining and quantifying sustainable development, and should include ecological, economic, and social components. We can follow the three steps proposed by Bormann et al. (1994), for calculating sustainability: select candidate goods, services, and states desired by society; determine ecosystem patterns and processes which are needed for the desired goods, services, and states; and jointly evaluate and set priorities among societal demands and ecosystem patterns and processes. Learning objectives should also be included.

- What alternative courses of action could be used?

  To determine the optimal solution, generating different management alternatives that should allow us to reach the formulated objectives, is useful, perhaps essential. In tradional forestry, this is often done through linear programming (e.g. David and Johnson 1987) and this technique is still very much applicable today. In addition, the development of suitable decision support systems for managers seems to be an urgent priority (Rauscher, 1999).

- What are the mid- and long-term effects on the defined criteria and indicators of each alternative?

  On the basis of given assumptions, the effects of the different courses of action on the ecological, economic, and social criteria and indicators are simulated. In particular, we evaluate the possible effects of the proposed manipulations on the elements, processes, and interactions in this ecosystem, as well as on adjacent ecosystems.

- Which management alternatives should we choose and how should they be designed?

  This step is based on comparing the elaborated alternatives, including their cost/benefit ratios. Bormann et al. (1996) advocated comparing more than one alternative at the same time to increase the rate of learning.

- Which are the priorities?

  Knowing that everything cannot be done immediately, priorities have to be defined and weighted, based on an analysis of budget and timelines. This step is a basis for the implementation.

- Plan the implementation

  The plan should answer the questions: What has to be done? Why? Who has to do what? When? How? and How much does it costs?

- Plan the monitoring and the assessment

  Monitoring is a learning process and informs us about the effects of our management. Therefore, monitoring and assessment are based on the criteria and indicators used to define the management objectives, and what will be monitored, how, and at what intervals is contained in the original design.

- Inform the concerned organisations and persons

  Relevant and truthful information is the basis of good communication between people. Who should be informed, when, and about what should also be described in the planning process.

- Implementation, monitoring and assessment

  Essentials in this step are the choice of the most competent people and a clear mission statement, based on the planning document.

- Adapt the management

Results of management rarely correspond precisely and entirely to the planning. Assessment informs us about deviations from what was expected and their causes. The management strategy may then be adapted (Noss, 1993; Bormann et al., 1994).

- **What are the guiding principles and the vision for management?**
- **What are the goals and general objectives?**
- **How intense should the planning process be?**
- **Who should belong to the planning team?**
- **What are the available sources of information?**
- **What is the current situation and what are the key issues in the management unit?**
- **What are the specific management objectives?**
- **What alternative courses of action could be used?**
- **What are the mid- and long-term effects on the defined criteria and indicators of each alternative?**
- **Which management alternatives should we choose and how should they be designed?**
- **Which are the priorities?**
- **Plan the implementation.**
- **Plan the monitoring and the assessment.**
- **Inform the concerned organizations and persons.**
- **Execution, monitoring and assessment.**
- **Adapt the management.**

Figure 2. Questions to be answered for the implementation of ecosystem-based management

Scientific challenges for the implementation of ecosystem-based management

The contributions of science are numerous and valuable but many uncertainties remain about the ecosystem-based management of natural resources. Baker et al. (1995) identified the following major research needs: considering multiple endpoints, multiple stressors, and their interactions and trade-offs at multiple spatial and temporal scales; synthesizing the relevant information from many scientific disciplines; and presenting scientific information in a user-friendly format useful for policy makers and understandable to the wide array of stakeholders.

The following list is an attempt to illustrate some general questions about ecological, social, and technical issues, and that could be used to define ecosystem-based management research priorities.

Ecological questions

- How can we assess the quality of forest ecosystem and landscape?
- How do adjoining ecosystems interact and what are the effects of these interactions at the landscape scale?
- How do human activities, including management and engineering, affect the ecological integrity and the health of manipulated ecosystems, e.g. their composition, structure, processes, interactions, state and dynamics, as well as their ability to satisfy the economic and social needs of the society, in both the short and long-term ?
- How do human manipulations of a particular ecosystem affect the adjoining ecosystems as well as the landscape as a whole, in both the short and long-term?
- How should the role of disturbances in ecosystems at all scales be integrated with management?
- How can we monitor the ecological state and dynamics of ecosystems and landscapes and their actual and potential capacity to fulfil the economic and social needs of society?
- How can we describe and identify the risks of degradation to which ecosystems and landscapes can be exposed? How can we take account of these risks in management?

Social questions

- How do changing socio-economic conditions like population growth and increasing energy consumption influence the social choices related to use of forest and landscape resources?
- Given the different interests and points of view, how can we reach agreement in defining the ecological, economic, and social objectives for managing forest and landscape resources? Who has the authority to resolve conflicting points of view?
- How can we assure the involvement of all concerned groups in implementing the management of forest and landscape resources?

Questions about sustainability and certification

- Given the changing economic conditions and available scientific knowledge, how do we understand sustainable use of forest and landscape resources? In particular, what should we sustain and at what spatial and time scales?
- Which criteria and indicators are feasible to measure progress towards sustainability? How can these criteria and indicators be used in ecosystem-based management?
- What are the impacts of certification on forest management and on the quality of forest ecosystems and landscapes?
- Would a landscape certification be useful for sustainable development? If yes, how should this be done?
- How can we best use the available scientific knowledge (both biological and social) and engineering technology for the multiple and sustainable use, protection, restoration, and rehabilitation of all types of ecosystems to serve as habitat for wildlife (especially for endangered species), to produce commodities, and to satisfy social needs?
- How can we implement adaptive management in order to take account of the existing uncertainties and of the evolution of knowledge?
- What does ecosystem-based management cost? Who gains and who loses?

Impacts associated with the introduction of ecosystem-based management

It is not possible to advocate a major change in the current approach to forest management without looking at some of the potential implications of such a change. Any change will be associated with costs, and these costs will not be equitably distributed across all countries and all individuals. This may pose a major barrier to the introduction of the management changes. For example, a reduction in the volume of old-growth cutting would have little impact on European countries, but would have major implications for countries such as Canada and Russia (Sedjo et al., 1998). Restrictions on the use of exotic species would have a greater impact on countries such as the UK, South Africa and New Zealand than on Canada or the USA.

For forest management to be effective, the costs must not outweigh the benefits. Unfortunately, the economic valuation of the non-timber benefits remains extremely difficult (Stewart Roper and Park, 1999). As a result, many studies have emphasised the costs to the timber industry of changing forestry practices. For example, the annual costs of fully implementing the Forest Practices Code in British Columbia have been estimated at between $220 million (van Kooten 1994) and $2.1 billion (Haley, 1996). These costs extend throughout the economy, with Binkley et al. (1994) estimating the annual losses to the GNP of British Columbia at $1 billion annually. The costs can be broken down into a number of different aspects, including implementation costs, costs associated with reduced timber supply, costs of retraining and redeploying labour displaced by the reduced economic activity, public sector costs associated with administration, monitoring and enforcement and costs associated with generating the necessary information and its extension (Haley, 1996).

While it is important to consider the costs of any change, it is equally important to consider the costs of not making the changes. Acccording to Salwasser (1999), these include:

> «Continued loss of soil, water, and biotic resources; expansion of globally irresponsible local behaviors; deterioration of parks, refuges, and wilderness areas; disruption of human communities; and disintegration of the common ground that people worked so hard for over the past century in conservation.»

In moving from a philosophical approach to an operational one, it is important to recognise that there are many problems facing ecosystem-based management other than the political and economic ones described above. Rauscher (1999) lists these as:

- there are many societal goals, preferences and values, which are often poorly defined and in conflict;

- legal mandates may be complex and unclear and may even be contradictory;
- in some countries, policy direction is missing, ambiguous or incomplete and may change rapidly;
- there is no well-defined and widely accepted decision making process;
- participating decision makers and stakeholders have varying degrees of interest and may vary the amount of effort they put into issues that are not of direct concern to them;
- there is no widely accepted method to achieve consensus among stakeholders, with legal processes induced by a minority in disagreement sometimes having the power to block further progress; and
- decisions need to be made in the face of uncertainty over the consequences.

Such problems are not insurmountable, and they are likely to have more than one solution. Rauscher argues that such problems need to be handled using «knowledge, organization, judicious simplification and inspired leadership», a combination that is sometimes in short supply, particularly in view of the restricted training still provided by some schools to forestry graduates.

Conclusions

People, world-wide, are becoming aware of the need to use forest and landscape resources with the quality of life of current and future generations in mind. Policy makers, managers, scientists, governments, private industry, and non-governmental organisations are developing, sometimes together, new visions for their management. These new visions require a shift from mainly commodity-production oriented management towards a management addressing simultaneously ecological, economic, and social issues. In other words, a paradigm shift from «sustainable forest management» to «ecosystem-based management.»

This change of paradigms is not without costs, contradictions, controversy, and tensions. For example, the concept of ecosystem management developed in the United States is still defined and interpreted in many different ways. However, the concept also includes enough useful principles that if applied in practice, it could allow society to take some steps in the direction of sustainable development. Foresters are increasingly managing with the whole ecosystem in mind, and are also using the ecosystem concept to allocate attention across both internal interactions and between systems, though we cannot expect to manage everything in the whole ecosystem to produce natural resources. Increasingly, the landscape scale (thousands to hundreds of thousands of hectares) is seen as the scale over which planning, managing and monitoring biological phenomena can best be undertaken (Noss, 1998), and forest managers need to adjust their horizons to this scale.

The key issue in managing forest and landscape resources for sustainable development is to find the balance between production and the productive capacity of the ecosystems – locally, regionally, and globally, and in both the short and long-terms. One major challenge, is how to simultaneously use and conserve ecosystems and landscapes and to integrate the ecological, economic, and social needs of society, while considering the interactions and processes within and between the different ecosystems, and taking into account the natural and human factors affecting the management unit. There are successes, but there are also emerging problems that must be addressed. A basic condition for progressing is, of course, that managers, scientists, policy makers, teachers, and all concerned people are recognising the need to act with a view towards future generations and are accordingly willing to implement the principles of ecosystem-based management.

We believe that forest management can still improve its contribution to sustainable development. The quality of this contribution depends on several factors. One of them is the ability to evaluate the quality of the forest landscape and to reach multiple objectives concerning not only the forest ecosystem itself, but also the processes and the interactions taking place in the surrounding landscape. This ability can be trained by introducing in the curriculum of forestry schools basic education in landscape ecology and its application in forestry, and by developing feasible techniques to evaluate forest and landscape quality. Such courses are already available in some forestry schools, but many schools remain strongly attached to methods and techniques that are no longer appropriate in the current and future socio-economic context (Clark et al., 1999). A second factor is the attitude towards the needs of other domains. It could be helpful not only to defend our internal forest concepts but also to include in our thinking the philosophy and vision existing in other areas, such as biological conservation, urban planning, agriculture, industry or social needs. This would mean that foresters are on the way to realise Anko's vision (1994):

> «In the future forester will not be just a «lumberjack with a university degree», as seen in many environments, but rather a fine interpreter of the share, place and of the role forest should have in the landscape tomorrow.»

We began this chapter with a provocative question «is this the end of forestry?» If Francis Fukuyama thought that the fall of the Berlin wall and the collapse of communism meant the end of history, he was wrong, as recent events in Iraq and the Balkans have shown. Let us learn from his mistake and not assume that the advent of ecosystem-based management means the end of forestry. However it does mean significant changes for both foresters and forestry. Foresters will have to learn to deal with other professionals sûch as biologists, anthropologists and economists on an equal footing. They will also have to adapt to more public participation from local communities, urban dwellers and other «non professionals.» If foresters can adapt to this, they will be able to use their undoubted skills at a landscape level beyond the confines of the forests. If they cannot, they will continue to find their role, even in the forests, being dramatically reduced.

Acknowlegements

The authors thank John Innes, University of British Columbia, Vancouver, Canada, and anonymous reviewers for the critical comments and useful suggestions. Thanks are also due to Sonja Desploz, Laboratory of Ecosystem Management, Swiss Federal Institute of Technology, Lausanne, for her support and excellent page-setting of the document.

References

Anko, B., 1994: Application of landscape ecology in forestry. Landscape ecology - Ecologia del paesagio, a cura di D. Cattaneo & P. Semenzato, Atti del XXXI Corso di Cultura in Ecologia: 129-143.

Aplet, G. H., Johnson N., Olson, J. T., Sample, V. A., (ed.) , 1993: Defining Sustainable Forestry. Island Press, Wasfington D.C., Covelo, California. 328 pp.

Bach, C. F. and Gram, S., 1993: *The Tropical Timber Triangle*. Department of Economics and Natural Resources, Royal Veterinary and Agricultural University, Copenhagen, Denmark.

Baharuddin, H. G. and Simula, M., 1994: Certification Schemes for all Timber and Timber Products. Report for the International Tropical Timber Organization, Yokohama, Japan.

Baharuddin, H. J and Simula, M., 1996: Timber Certification in Transition: Study on the Development in the Formulation and Implementation of Certification Schemes for all Internationally Traded Timber and Timber Products. Report for the International Tropical Timber Organization, Yokohama, Japan.

Baharuddin, H. G. and Simula, M., 1997: Timber Certification: Progress and Issues. Report for ITTO, Yokohama, Japan.

Baker, J. P., Landers, D. H., Lee, II P., Ringold, P. L., Sumner, R. R., Wigington, P. J., Bennett, R. S., Preston, E. M., Frick, W. E., Sigleo, A. C., Specht, D. T. and Young, D. R., 1995: Ecosystem management research in the Pacific Northwest: Five-year research strategy. Washington D.C.: US Environmental Protection Agency, EPA/600/R-95/069.

Bendall, J. and Sullivan, F., 1996: Sleeping with the Enemy? Business-Environmentalist Partnerships for Sustainable Development. The Case of the WWF 1995 Group, p. 3-33 in Aspinall, R. and J. Smith (ed.) *Environmentalist and Business Partnerships: A Sustainable Model? A Critical Assessment of the WWF-UK 1995 Group*, Cambridge Environmental Initiative Professional Seminar Series, The White Horse Press, Cambridge, UK.

Bennett, A. F., 1999: Linkages in the Landscape: The Role of Corridors and Connectivity in Wildlife Conservation. IUCN, The World Conservation Union, Gland, Switzerland and Cambridge, UK.

Bennett, M., 1996: Ecosystem management: Opportunities and implications for woodland owners. Corvallis, OR, U.S.: Oregon State University Extension Service, EC 1469.

Berg, S. and Olszewski, R., 1995: Certification and Labelling; A Forest Industry Perspective *Journal of Forestry* 93: 30-32.

Binkley, C. S., Perky, M., Thompson, W. A. and Vertinsky, I. B., 1994: A general equilibrium analysis of the economic impact of a reduction in harvest levels in British Columbia. *The Forestry Chronicle* 70: 449-454.

Biolley, H. , 1920: L'aménagement des forêts par la méthode experimentale et spécialement la méthode du contrôle. Dans Œuvre écrite. 1980. Supplément no 66 aux organes de la Société forestière suisse, p. 51-135.

Bormann, B. T., Cunningham, P. G., Brookes, M. H., Manning, Van W. and Collopy, M. W., 1994: Adaptive ecosystem management in the Pacific Northwest. Gen. Tech. Rep. PNW-GTR-341, Portland, OR, U.S.: Department of Agriculture, Forest Service, Pacific Northwest Research Station, 22 pp.

Bormann, B. T., Brookes, M. H., Ford, E. D., Kiester, A. R., Oliver, C. D. and Weigand, J. F., 1994: Volume V: A framework for sustainable ecosystem management. Gen. Tech. Rep. PNW-331. Portland, OR: U.S. Department of Agriculture, Forest Service, Pacific Northwest Research Station, 61 pp.

Bormann, B. T., Martin, J. R., Wagner, F. H., Wood, G., Alegria, J., Cunningham, P.G., Brookes, M. H., Friesema, P. and Henshaw, J., 1996: *Adaptive management: Common ground where managers, scientists, and citizens can try to learn to meet society's needs and wants while maintaining ecological capacity*. Draft chapter T-27 (Science and management) Ecological Stewardship book.

Bouman, O. T. and Brand, D. G. , (ed.), 1997: *Sustainable Forest: Global challenges and local solutions*. Food Products Press, an imprint of The Haworth Press, Inc. New York, London.

Brooks, D.J., 1996: Compiling and interpreting criteria and indicators for sustainable forest management. Draft May, 1996. Corvallis, OR: U.S. Depart. of Agriculture, Forest Service, Pacific Northwest Research Station.

Brown, K. and Pearce, D. W., (ed.),: *The causes of tropical deforestation. The economic and statistical analysis of factors giving rise to the loss of tropical forests.* UBC Press, Vancouver.

Bunnell, F. L., 1976: The myth of the omniscient forester. *The Forestry Chronicle* 52: 150-152.

Bunnell, F. L., 1998: Setting goals for biodiversity in managed forests. In: Bunnell, F.L. and Johnson, J.F. (ed.) The living dance. Policy and practices for biodiversity in managed forests. UBC Press, Vancouver, 117-153.

Caldwell, L. K., Wilkinson, C. F. and Shannon, M. A., 1994: Making ecosystem policy: Three decades of change. *Journal of Forestry* 92: 7-10.

Christensen, N. L., Bartuska, A. M., Brown, J. H., Carpenter, S., D'Antonio, C., Francis, R., Franklin, J. F., MacMahon, J. A., Noss, R. F., Parsons, D. J., Peterson, C. H., Turner, M. G. and Woodmansee, R. G., 1996: The Report of the Ecological Society of America Committee on the Scientific Basis for Ecosystem Management, Ecological Applications, 6(3): p. 665-691.

Clark, R.N., Stankey, G.H., Brown, P.J., Burchfield, J.A., Haynes, R.W. and McCool, S.F., 1996: Towards an ecological approach: Bringing together social, economic, cultural, biological, and physical considerations. Draft April 29, 1996. The paper is based on a presentation by Roger N. Clark at the Ecological Stewardship Workshop, Tucson, Arizona, December 4-14, 1995.

Clark, R. N., Stankey, G. H. and Kruger, L. E., 1999: From New Perspectives to ecosystem management: A social science perspective on forest management. In: Aley, J., Burch, W.R., Conover, B. and Field, D. (ed.) Ecosystem management. Adaptive strategies for natural resources organizations in the 21st century. Taylor & Francis, Philadelphia, p. 73-84.

Clutter, J. L., Fortson, J. C., Pienaar, L. V., Brister, G. H. and Bailey, R. L., 1983: *Timber management: a quantitative approach.* John Wiley and Sons.

Cortner, H. J., Shannon, M. A., Wallace, M. G. and Burke, S. M., 1996: Institutional barriers and incentives for ecosystem management. Gen. Tech. Rep. PNW-GTR-354, Portland, OR: U.S. Department of Agriculture, Forest Service, Pacific Northwest Research Station. 35 pp.

Côté, M.-A., 1999: Possible impact of forest product certification on the worldwide forest environment. *The Forestry Chronicle* 75: 208-212.

Cotta, H., 1821: Tables de cubage et d'expériences sur l'accroissement des bois et instruction pour le calcul de la valeur des bois. Traduit en français par de Salomon, M. 1837. Dans le Traité de l'aménagement des forêts, tome second. Bureau de l'Almanach du Commerce, Paris.

Davis, L. S. and Johnson, K. N., 1987: *Forest management.* 3rd edition. McGraw-Hill, Inc., New York.

Dixon, J. A. and Sherman, P. B., 1991: *Economics of Protected Areas: A New Look at Costs and Benefits.* Earthscan, London.

Dubourdieu, J., 1997: *Manuel d'aménagement forestier.* Office national des forêts, Technique et documentation Lavoisier, Paris.

Dudley, N., 1996: Authenticity as a means of measuring forest quality. *Global Ecology and Biodiversity Letters* 3: 6–9.

Dudley, N., 1999: Evaluation of Forest Quality-Towards a Landscape Scale Assessment. Part of an IUCN/WWF/ EPFL Project: Forest Conservation for the 21st Century, Gland/Lausanne, Switzerland. Not published.

Earth Summit, 1992: Agenda 21: The United Nations programme of action from Rio. The final text of agreements negotiated at the United Nations Conference on Environment and Development (UNCED), 3-14 June 1992, Rio de Janeiro, Brazil. New York: United Nations Department of Public Information. 294 pp.

Elliott, C.; Hackman, A., 1996: Current issues in forest certification in Canada. A WWF Canada discussion paper. Toronto: WWF Canada, 63 pp.

Elliott, C., 1999: Forest Certification: Analysis from A Policy Network Perspective, PhD Dissertation, Chair of Ecosystem Management, Swiss Federal Institute of Technology, Lausanne.

Farina, A., 1998: *Principles and Methods in Landscape Ecology*. Chapman and Hall, London.

Ferguson, I. S., 1996: *Sustainable forest management*. Oxford University Press.

Fitzsimmons, A. K., 1996: Sound policy or smoke and mirrors: Does ecosystem management make sense? *Water Resources Bulletin* 32(2): 217-227.

Ford, D., 1997: The US Forest Products Buyers Group. Paper presented at the International Conference on Certification, Criteria and Indicators; Global Approaches to Sustainable Forest Management, 21-26 September 1997, Prince George, British Columbia. Canadian Institute of Forestry.

Forest Ecosystem Management Assessment Team (FEMAT), 1993: Forest ecosystem management: An ecological, economic, and social assessment. Portland, OR: U.S. Dept. of Agriculture: Forest Service, U.S. Dept. of Commerce: National Oceanic and Atmospheric Administration, National Marine Fisheries Service, U.S. Dept. of the Interior: Bureau of Land Management, U.S. Dept. of the Interior: Fish and Wildlife Service, U.S. Dept. of the Interior: National Park Service, and U.S. Environmental Protection Agency. Variable pagination.

Forman, R. T. T. and Godron, M., 1986: *Landscape Ecology*. John Wiley and Sons. New York.

Forman, R. T. T. , 1995: *Land Mosaics: The Ecology of Landscapes and Regions*. Cambridge University Press, Cambridge, 632 pp.

Franklin, J. F., 1993: The Fundamentals of Ecosystem Management with Applications in the Pacific Northwest. In Aplet, Gregory H., Johnson, Nels, Olson, Jeffrey T., Sample, V. Alaric (ed.). Defining Sustainable Forestry. Island Press, Washington D.C., USA: 127-145.

FSC (Forest Stewardship Council), 1994: FSC Statutes, FSC, Oaxaca, Mexico.

FSC, 1999: FSC Members Short Report, FSC Document 4.1, FSC, Oaxaca, Mexico.

Fukuyama, F., 1993: *The End of History? The Last Man*. Avon Books, New York.

Gale, F. and Burda C., 1996: *Attitudes Towards Eco-Certification in the BC Forest Products Industry*. Eco-Research Chair, Faculty of Law and Environmental Studies Programme, University of Victoria, Canada.

Garman, S. L., Spies, T. A., Cohen, W. B., Bradshaw, G. A. and Dippon, D., 1995: Research problem analysis - modeling, monitoring and displaying ecological change at watershed to landscape scales: Tools for ecosystem management, final report. Submitted to: Forest and Rangeland Ecosystem Science Center, USDI/NBS, Corvallis, OR 97331.

Gilmour, D. A., Fisher, R. J.,: *Villagers, Forest and Foresters: The Philosophy, Process and Practice of Community Forestry in Nepal*. Sahayogi Press, Kathmandu, Nepal.

Glenz, C., 1999: Lebensraumanalyse des Wolfes (Canis lupus) für das Wallis mit Hilfe des GIS. Diploma thesis at the Laboratory of Ecosystem Management, Swiss Federal Institute of Technology, Lausanne, Switzerland.

Glück, P.,1987; Social values in forestry. Ambio 16 (2-3): 158-160

Gordon, J. C., 1993: Ecosystem management: An idiosyncratic overview. In: Defining Sustainable Forestry, Aplet et al. ed. Washington D.C.: Island Press. p. 240-247.

Granholm, H., Vähänen, T. and Sahlberg, S., 1996: Background Document for the Intergovernmantal Seminar on Criteria and Indicators for Sustainable Forest Management (ISCI). Ministry of Agriculture and Forestry, Helsinki.

Graybeal, N., 1999: The Northwest Forest Plan. Statement made on 21 June 1999 before the Subcommittee on Department Operations, Oversight, Nutrition and Forestry, Committee on Agriculture, United States House of Representatives. Available at: http://www.fs.fed.us/intro/testimony/19990621b.html

Grumbine, R.E., 1994: What is ecosystem management? Conserv. Bio. 8: 27-38.

Gunn, G., 1997: Tetra Pak UK and the WWF 1995 plus Group. Paper presented at Tetra Pak Europe Environment Seminar, 10 September 1997, Pully, Lausanne.

Haley, D., 1996: Paying the piper. The cost of the British Columbia Forest Practices Code. Paper prepared for Working with the B.C. Forest Practices Code, an Insight Information Inc. Conference, sponsored by the Globe and Mail, April 15-17, 1996. Vancouver, BC.

Hampton, J. C., 1994: A problem, not a solution. *Journal of Forestry* 92(4): p. 25.

Hansen, E. and Punches, J., 1998: Collins Pine: Lessons from a Pioneer, Case study from The Business of Sustainable Forestry, The John D. and Catherine T. MacArthur Foundation, Chicago, USA.

Henne, A., 1973: Forsteinrichtung als betriebswirtschaftliche Planung und Kontrolle. Mitteilung der Hessischen Landesforstverwaltung, Band 11, Sauerländer's Verlag, Frankfurt am Main.

Heywood, V. H. and Watson, R. T. (ed.), 1995: *Global biodiversity assessment*. Published for the United Nations Environment Programme. Cambridge University Press, 1140 pp.

Holling, C.S., 1978: *Adaptive environmental assessment and management*. New York: Jon Wiley & Sons, 377 pp.

ISO, 1997: Memento 1997, ISO, Geneva, Switzerland.

IUFRO,1983: Forest management in various countries of the world. Institutul de cercertari si amenajari silvice, Bucharest, Romania.

Jowett, B. (ed), 1892: *The Dialogues of Plato*. Volume 3, Oxford University Press, London.

Kauppi, P., 1996: Global Change. In: Caring for the forest: Research in a changing world, Congress Report, Volume II, IUFRO XX World Congress, 6-12 August 1995, Tampere, Finland: 29-34.

Kennedy, J. J. and Quigley, T. M., 1993: Evolution of Forest Service organizational culture and adaptation issues in embracing ecosystem management. In Jensen, M.E and Bourgeron, P.S. (eds.) Eastside Forest Ecosystem Health Assessment: vol. II. Ecosystem Management: Principles and applications. USDA Forest Service General Technical Report PNW-GTR 324. Pacific Northwest research Station, Portland, Oregon, p. 19-29.

Kimmins, J. P., 1987: *Forest ecology*. New York: Macmillan and London: Collier Macmillan. 531 pp.

Knight, A. P., 1996: A Report on B&Q's 1995 Timber Target, pp 34-44 in Aspinall, R. and J. Smith (eds) □Environmentalist and Business Partnerships: A Sustainable Model? A Critical Assessment of the WWF UK 1995 Group□, Cambridge Environmental Initiative Professional Seminar Series, The White Horse Press, Cambridge, UK.

Kramer, H., 1976: Begriffe der Forsteinrichtung. Schriftenreihe der Forstliche Fakultät der Universität Göttingen und Mitteilung der Niedersächsischen Forstlichen Versuchsanstalt, Band 48, Sauerländer's Verlag, Frankfurt am Main.

Kramer, H., 1982. *Nutzungsplanung in der Forsteinrichtung*. Sauerländer's Verlag, Frankfurt am Main.

Küchli, C., 1998: Comparaison de systèmes de certification actuels. In: Küchli, Christian and Zanetti, Waldzertifizierung und Holzlabelling, Materalien der Tagung vom 13.Februar 1998, ETH-Zürich. BUWAL/Eidgenössische Forstdirektion. Bern, Switzerland.

Kurth, H., 1987: Forsteinrichtung: Nachhaltigkeit. Hochschulstudium Fostingenieurwesen, Technische Universität Dresden, Sektion Forstwirtschaft, Tharandt.

Lammerts van Bueren, E. M., 1997: Hierarchical framework for the formulation of sustainable forest management standards. The Tropenboos Foundation.

Lang, R., (ed.), 1990: Integrated approaches to resource planning and management. The Banff Centre for Continuing Education. Calgary, Alberta, Canada: The University of Calgary Press.

Lawton, S., 1997: Markets for Certified Products. Paper presented at the International Conference on Certification, Criteria and Indicators; Global Approaches to Sustainable Forest Management, 21-26 September 1997, Prince George, British Columbia. Canadian Institute of Forestry.

Lee, K. N., 1993: *Compass and gyroscope: Integrating science and politics for the environment*. Island Press, Washington DC.

Leopold, A., 1949: **A Sand County almanach**. Oxford University Press. Available by Ballantine Books, New York

Leuschner, W. A., 1992: *Introduction to forest resource management*. Wiley, New-York.

Mantel, W., 1959: *Forsteinrichtung*. Sauerländer's Verlag, Frankfurt am Main.

Massolo, A. and Meriggi, A., 1998: Factors affecting habitat occupancy by wolves in Northern Apennines (Northern Italy): A model of habitat suitability. Ecography 21: 97-102

Mather, A. S., 1990: *Global Forest Resources*. Belhaven Press, London.

McGarigal, K. and McComb, W. C. , 1995: Relationship between landscape structure and breeding birds in the Oregon Coast Range. Ecological Monographs, 65(3): 235-260.

McKillop, W.,1994: Critique of economic aspects. *Journal of Forestry 92(4)*: p. 37.

McNulty, J. W. and Cashwell, J. H., 1995: The Land Manager's Perspective on Certification. *Journal of Forestry* 93: 22-25.

Meffe, G. K. and Carroll, C. R., 1994: *Principles of Conservation Biology*. Sunderland, Massachusets: Sinauer Associates, Inc., 600 pp.

Meyer, H. A. . 1934: Die rechnerischen Grundlagen der Kontrollmethoden: Genauigkeit der stehenden Vorrats- und Zuwachskontrolle. Supplément no 13 aux organes de la Société forestière suisse.

Ministerial Conference on the Protection of Forests in Europe, 1993: Resolution H1: General guidelines for the sustainable management of forests in Europe. Helsinki: Ministry of Agriculture and Forestry.

Mitchell, B., 1990: The evolution of integrated resource management. In: Integrated approaches to resource planning and management. Lang, R. ed. The Banff Centre for Continuing Education. Calgary, Alberta, Canada: The University of Calgary Press.

Montreal Process (The), 1995: Criteria and indicators for the conservation and sustainable management of temperate und boreal forests. Hull, Québec: Canadian Forest Service.

Muhlemann, P., 1997: Participatory Forest Management. In: People, Forests and Sustainability. Report of the FAO/ECE/ILO Team of Specialists on Social Aspects of Sustainable Forest Management. International Labour Organisation (ILO), Geneva, Switzerland.

Murphy, D. F., 1996: Business, NGOs and Sustainable Development: Towards a Global Perspective p. 45-72 in Aspinall R. and J.Smith (eds) *Environmentalist and Business Partnerships: A Sustainable Model? A Critical Assessment of the WWF-UK 1995 Group*, Cambridge Environmental Initiative Professional Seminar Series, The White Horse Press, Cambridge, UK.

Nabuurs, G.-J., Nuutinen, T., Bartelink, H. and Korhonen, M. (ed.), 1998: Forest Scenarion Modelling for Ecosystem management at Landscape Level. EFI Proceedings No. 19. European Forest Institute, Joensuu, Finland.

Navey, Z. and Lieberman, A., 1994: *Landscape ecology theory and applications*. New York: Springer Verlag. 360 pp.

Noss, R. F., 1993: Sustainable forestry or sustainable forests? In: Defining sustainable forestry. Aplet et al., ed. Washington D.C.: Island Press. p. 17-43.

Noss, R.F., 1998: At what scale should we manage biodiversity? In: Bunnell, F. L. and Johnson, J. F. (ed.) *The living dance. Policy and practices for biodiversity in managed forests*. UBC Press, Vancouver, p. 96-116.

OECD (Organisation for Economic Co-operation and Development), 1991: *Environmental Policy: How to Apply Economic Instruments*, OECD, Paris, France.

OECD, 1996: *Saving Biological Diversity; Economic Incentives*. OECD, Paris, France.

Overbay, J. C., 1992: Ecosystem management. *Proceedings of the National Workshop: Taking an ecological approach to management, 27-30 April 1992, Salt Lake City, UT.* USDA Forest Service, Watershed and Air Management Branch, WO-WSA-3. Washington DC. p. 3-15.

Palo, M. and Uusivuori, J., 1999: *World forests, society and environment.* Kluwer Academic Publishers.

Pearse, P. H., 1998: Economic instruments for promoting sustainable forestry: opportunities and constraints. In: Tollefson, C. (ed.) *The wealth of forests. Markets, regulation, and sustainable forestry.* UBC Press, Vancouver, p. 19-41.

Poore, D., 1995: Forestry and Nature Conservation: Changing Perspectives. *Commonwealth Forestry Review* 74: 5-19.

Rackham, O., 1986: *The History of the Countryside.* J.M. Dent, London.

Rametsteiner, E., Schwartzbauer, P., Juslin, H., K□rn□, J., Cooper, R., Samuel, J., Becker, M. and K□hn, T., 1998: Potential Markets for Certified Forest Products in Europe. Discussion Paper 2, European Forest Institute, Joensuu, Finland.

Rauscher, H.M. , 1999: Ecosystem management decision support for federal forests in the United States: A review. *Forest Ecology and Management* 114: 173-197.

Relisch, B., 1972: Planung in Forstwirtschaftsbetrieben. Duncker und Humblot, Berlin. Salomon (de), M. 1837. Traité de l'aménagement des forêts, tome premier. Bureau de l'Almanach du Commerce, Paris.

Ripple,W. J., Bradshow, G. A. and Spies, T. A., 1991: Measuring forest landscape patterns in the Cascade Range of Oregon, USA. Biological Conservation 57: 73-88.

Rosenberg, K. V., Lowe, J. D. and Dhondt, A. A., 1999. Effects of Forest Fragmentation on Breeding Tanagers: A Continental Perpective. *Conservation Biology*, 13(3): p. 568-583.

Rotherham, T., 1996: Forest management certification: objectives, international background and the Canadian program. *The Forestry Chronicle* 72: 247-252.

Rotherham, T.,1997: Chain of Custody. Paper presented at the International Conference on Certification, Criteria and Indicators; Global Approaches to Sustainable Forest Management, 21-26 September 1997, Prince George, British Columbia. Canadian Institute of Forestry.

Salwasser, H., MacCleery, G. W. and Snellgrove, T. A., 1993: An ecosystem perspective on sustainable forestry and new directions for the U.S. National Forest System. In: Defining sustainable forestry. Aplet et al., ed. Washington D.C.: Island Press, p. 44-90.

Salwasser, H. and Pfister, R. D., 1994: Ecosystem management: from theory to practice. In: Covington, W.W.; DeBabo, L.F. Technical coordinators. Sustainable ecological systems: Implementing an ecological approach to land management. 1993 July 12-15; Flagstaff, Arzona. Fort Collins, CO: Gen. Tech. Rep. RM-247. U.S. Department of Agriculture, Forest Service, Rocky Mountain Forest and Range Experiment Station, 363 pp.

Salwasser, H., 1999: Ecosystem management: A new perspective for National Forests and Grasslands. In; Aley, J., Burch, W. R., Conover, B. and Field, D. (ed.) Ecosystem management: Adaptive strategies for natural resources organizations in the 21st century. Taylor & Francis, Philadelphia, p. 85-96.

Sample, V., Johnson, N., Aplet, G. and Olson, J., 1993: *Defining Sustainable Forestry*, Island Press, Washington DC.

Savory, A., 1988: *Holistic resource management.* Washington, D.C.: Island Press, 564 pp.

Schaeffer, A., Gazin, A. and D'Alverny, A., 1930: *Sapinières: Le jardinage par contenance (méthode du contrôle par les courbes).* Les presses universitaires de France, Paris.

Schlaepfer, R., 1987: L'aménagement des forêts et la planification intégrée de l'entreprise. Schweiz. Z. Forstwes., 118 (1): 21-38.

Schlaepfer, R., 1997: Ecosystem based management of natural resources: a step towards sustainable development. IUFRO Occasional paper no 6. Vienna.

Schulze, P. C., (ed.), 1996: *Engineering within ecological constraints*. Washington, D.C.: National Academy Press, 213 pp.

Science Branch Canadian Forest Service, 1996: Model forest network, year in review 1994-1995. Ottawa: Natural Resources Canada.

Sedjo, R. A., Goetzl, A. and Moffat, S. O., 1998: *Sustainability of temperate forests*. Resources for the Future, Washington DC.

Shands, W. E., 1994: National forests and the human legacy: Some history. In Foley, L. H. (compiler) Silviculture: From the cradle of forestry to ecosystem management. Proceddings of the National Silviculture Workshop, 1-4 November 1993, Hendersonville, NC. USDA Forest Service, General technical Report SE-88, p. 3-11. Southern Forest Experimental Station, Asheville, North Carolina.

Shindler, B. and Neburka, J., 1966: Public participation in forest communities: Identifying attributes for success. To be published in the Journal of Forestry.

Sikod, F., 1996: Certification Processes in Sustainable Forest Management; Economic Concepts and Indicators. p. 125-146 in *Proceedings of the UBC-UPM Conference on the Ecological, Social and Political Issues of the Certification of Forest Management, Faculty of Forestry, UBC, Vancouver, Canada.*

Simula, M., 1998: *Testing of the Finnish Certification Standard*. Indufor, Helsinki, Finland.

Simula, M., 1996: International and Institutional Arrangements for Certification of Forest Management and Eco-labelling of Forest Products, p. 170-193 in *Proceedings of the UBC-UPM Conference on the Ecological, Social and Political Issues of the Certification of Forest Management*, Faculty of Forestry, UBC, Vancouver, Canada.

Solberg, B., Brooks, D., Pajuoja, H., Peck, T. J. and Wardle, P. A., 1996: Long Term Trends in World Supply and Demand for Wood and Implications for Sustainable Forest Management: A contribution to the CSD Ad Hoc Intergovernmental Panel on Forests (IPF), Draft paper. Joensuu, Finland: European Forest Institute and As, Norway: Norwegian Forest Research Institute.

Speidel, G., 1957: Die rechnerischen Grundlagen der Leistungskontrolle und ihre praktische Durchführung in der Forsteinrichtung. Schriftenreihe der Forstliche Fakultät der Universität Göttingen und Mitteilung der Niedersächsischen Forstlichen Versuchsanstalt, Band 19, Sauerländer's Verlag, Frankfurt am Main.

Spellenberg, I. F. , 1991: *Monitoring ecological change*. Cambridge University Press, 334 pp.

Spiecker, H., Mielekäinen, K., Köhl, M. and Skovsgaard, J. (ed.), 1996: Growth Trends in European Forests. European Forest Institute Research Report No. 5, Springer-Verlag, 371 pp.

Spurr, S. H. and Barnes, B. V., 1980: *Forest Ecology*. New York: Wiley, 687 pp.

Stankey, G. H., 1995: The pursuit of sustainability: Joining science and public choice. *The George Wright Forum* 12(3): 11-18.

Stankey, G. H. and Shindler, B., 199X: Adaptive management areas: achieving the promise, avoiding the peril. Gen. Tech. Rep. PNW-GTR-XXX. Portland, OR: U.S. Department of Agriculture, Forest Service, Pacific Northwest Research Station. XX pp.

Steinitz, C. (ed.), 1996: Biodiversity and landscape planning: Alternative futures for the region of Camp Pendleton, California. Harvard University, Graduate School of Design.

Stewart Roper, C. and Park, A., 1999: *The living forest: Non-market benefits of forestry*. The Stationary Office, London.

Swanson, F. J., Jones, J. A., Wallin, D. O. and Cissel, J. H., 1993: Natural variability - Implications for ecosystem management, p. 89 - 103. In: Eastside forest ecosystem health assessment - Volume II: Ecosystem management: Principles and applications.Jensen, M.E.; Bourgeron, P.S.ed. Portland, OR: U.S. Dept. of Agriculture, Forest Service, Pacific Northwest Exp. p. 89-103.

Tassy, L., 1858: Etudes sur l'aménagement des forêts. Bureau des annales forestières, Paris.

Thomas, J. W. and Huke, S., 1996: The Forest Service approach to healthy ecosystems. *Journal of Forestry* 94(8): 14-18.

Upton, C. and Bass, S., 1995: *The Forest Certification Handbook*. Earthscan Publications Ltd., London.

Van Dyne, G. M., (ed.), 1969: *The ecosystem concept in natural resource management*. New York and London: Academic Press.

Van Kooten, G. C., 1994: *Cost-benefit analysis of BC's proposed forest practices code*. Forest Economics and Policy Analysis Research Unit of British Columbia, Vancouver.

Viana, V., Ervin, J., Donovan, R. Z., Elliott, C. and Gholz, H.,1996: *Certification of Forest Products: Issues and Opportunities*. Island Press, Washington D.C.

Wadsworth, J. and Boateng, P.,: 1996: Study on Markets and Market Segments for Certified Timber and Timber Products. Report for the International Tropical Timber Organization, Yokohama, Japan.

Waley, A., 1939: *Three ways of Thought in Ancient China*. George Allen and Unwin, China.

Walters, C. J., 1986: *Adaptive management of renewable resources*. New York: Macmillan Publishing Company, 373 pp.

Waring, R. H. and Running, S. W., 1998: *Forest Ecosystems: Analysis at Multiple Scales*. Academic Press.

Wells, R. W., Lertzman, K. P. and Saunders, S. C., 1998: Old-growth definitions for the forests of British Columbia, Canada. *Natural Areas Journal* 18: 279-292.

White, D. L. and Lloyd, F. T. , 1998: An old-growth definition for dry and dry-mesic oak-pine forests. US Department of Agriculture, Forest Service. General Technical Report SRS-23. Southern Research Station, Asheville.

Wiersum, K. F., 1995: 200 Years of Sustainability in Forestry: Lessons From History. Environmental Management 19(3): 321-329.

Wood, C. A., 1994: Ecosystem management: Achieving the new land ethic. *Renewable Natural Resources Journal* 12: 6-12.

Yankelovich, D., 1991: *Coming to public judgment*. The Syracuse University Press.

Zonneveld, I.. S., 1991: Scope and Concepts of Landscape Ecology as an Emerging Science. In Zonneveld, Isaak.S. and Forman, Richard T.T. (ed.): Changing Landscapes: An Ecological Perspective. Springer, New York: 3-20.

Sustainability in Forestry: Theory and a historical case study

M. Hauhs and H. Lange[1]

Abstract

The concept of sustainable development contains two independent goals: The connotation of unlimited economic growth in the term "development" and the perpetual conservation of natural resources implied by the concept of sustainable management. These two goals have been difficult to reconcile. Traditionally, sustainability is connected to the search for truly renewable resources in the realm of natural sciences and to the search for truly human needs in the realm of social sciences. Here we argue that in order to operationalize the concept of sustainable development it is necessary to distinguish two fundamentally different types of technologies by which humans develop and utilise resources. The proposed dualism of technologies is based on the respective historical contexts of their origin. One type of technology, exemplified by European forestry, depends on systems derived from natural history, but empirically by trial and error changed into an ahistoric, i.e. sustainable *management system that may be extended infinitely. The second type of technology depends on scientific understanding of ahistoric material building blocks that can be recombined and* developed *into a virtually infinite technical potential through (open-ended) cultural history. The current dilemma of ecological problems becomes one of learning which of the technical potentials and traditions belongs to which category and how to organise the interface between them. We present and discuss a historic example were the interdependent use of the two types of technologies is exceptionally well-documented: The Rammelsberg mine at Goslar, Germany. Delineating a technical culture into realms organised by one or the other type of technology remains a non-trivial problem, but the continuous updating and learning process can be organised in a straight-forward manner. The proposed concept generalises lessons from the example of European forest history and suggests an operational implementation of sustainable ecosystem utilisation.*

Keywords: *ecosystem utilisation; Rammelsberg mine; exploratory and emancipatory technology; Harz mountains.*

---

[1] Bayreuth Institute of Terrestrial Ecosystem Research (*BITÖK*), University of Bayreuth, D-95440 Bayreuth, Germany

K. von Gadow et al. (eds.), Sustainable Forest Management, 69–98.

Introduction: Two related problems

Sustainability and sustainable development have been proposed as general organising principles with respect to ecosystem utilisation, and political goals that are required to meet with a globally broad consensus. Within the political arena these terms are widely used today, but are notoriously difficult to operationalise when it comes to practical management schemes (Jamieson 1998). No theoretical or pragmatic framework embraces the technical, social, political, economic and ecological aspects of this vision of management principles. The term „sustainable development" may even be regarded as an oxymoron (Gunderson et al. 1995): something needs to be kept unchanged (sustained) while at the same time it needs to be changed (developed).

The broad acceptance of the term „sustainable development" across vastly different cultural and political camps has caused doubts that it is non-operational (Gatto 1995). Is it a mere illusion that has become popular at times of rapid social, technical and environmental change or can it be given an operational basis in management and an observational basis in science? How can the two be related? Specifically, what is the role of science and modeling for assessing the sustainability of ecosystem *management schemes* (opposed to sustainability of natural resources)?

The pace and global scales of changes induced by human civilisation have reached such an extent that it has become a matter of concern to decide how and what must/should be kept unchanged. Humans have become major participators of biogeochemical cycles not only at local ecosystem scales, but at a global scale. Here we will focus at the technical questions of sustainability only: what is it that *can* be kept unchanged, and by which technical means? How can the effect and the efficiency of these technical means be controlled and improved? The scope of sustainability will be addressed in the next section.

While sustainability as a management constraint has a long tradition within the forestry profession, it had little impact on the public debate of what is meant by sustainable development and how sustainable development should be organised outside forestry. Here we will use specific examples from forest history in Germany to demonstrate how the installation of sustainability in the narrow realm of forestry can be generalised as a feasible approach to the vision of sustainability of ecosystem utilisation in general. The implementation of sustainability within forestry was based on a comparison of expected growth rates with actual cutting rates. The discussion of the proposed generalisation of the sustainability concept will also show why the actual boundary conditions of forestry in Central Europe strongly require a revised version of sustainability even within forestry itself.

The actual threats to European forestry ultimately result from the changing civilisation into which it is embedded and which i) change site conditions for forest productivity, e.g. by atmospheric deposition of toxic or nutritional pollutants, climate change (Pretzsch 1999) and ii) the value system for forest products (Oesten 1995). These trends have exposed and opened the forestry sector to public discussion with other interest groups (Baskerville 1995) for which it was ill prepared.

Silvicultural management schemes from forestry in Central Europe have been proposed by its proponents as successful practical implementations of sustainability. According to a recent poll more than 95% of German foresters are convinced that their silvicultural management is sustainable (Schanz 1994). However, environmental interest groups view forestry as an outdated economic sector in which the focus on timber production jeopardises more urgent concerns of nature conservation. The most recent arena for settling such issues of ecosystem management is the debate about certification of forest products in Europe. By certification environmental interest groups try to regulate and restrict the allowed set of silvicultural techniques (a ban on clearcuts, on application of chemicals, on import of new species, etc.) in order to achieve what is really a shift in the ultimate goal in ecosystem management: conserving „nature" and „biodiversity[2]" versus producing valuable timber. For German foresters it has been difficult to understand how in this public controversy about sustainability in ecosystem management forestry could have been forced into a defensive role.

Science has been considered the authority capable of solving the technical problems of sustainable ecosystem management[3]. However, in examples of conflicting management options in forestry, science and especially modelling has played only a minor role in bridging across this controversy (Baskerville 1995). The German poll mentioned above (Schanz 1994) showed that science and research are currently ranked as *least* important among the twenty external factors that foresters might regard as essential for achieving sustainability. This representative opinion of ecosystem managers sheds light on the practical relevance of the scientific results from a decade with the most extensive (and expensive) forest ecosystems research program that was ever funded in Germany!

---

[2] Lets assume for this discussion that these goals can be given a specific meaning.

[3] The authors are actually funded under a research framework entitled: Foundations for Sustainable Development of ecosystems in a changing environment (BMBF Nr. PT BEO 0339476 A-C).

Here we propose that the two problems described above are related: The difficulties to implement sustainability in a broader context of ecosystem management and the difficulties in science to generalise and transcendent historic examples of such management outside their specific realm of e.g. forestry result from the same conceptual flaw: Neglect of the evolutionary, adaptive changes in the studied systems. In the following section we will propose a general concept that allows us to investigate the implications of sustainability in various contexts and demonstrate how historical developments may serve as a examples, helping us to see actual environmental problems in a new light. We will come back to the disappointing role of science in this controversy below.

A theoretical framework

The search for a proper distinction of changes relating to an unchanged background, or the identification of lasting systems in a sea of change is an old one. This selfreferential dichotomy of permanence and change can even be regarded as one of the most fundamental ones (Ruhnau 1997). Western philosophy has wrestled for a long time whether this complementary relation between change and permanence can be turned into a hierarchical one: e.g. material permanence has been proposed as dominant. Natural sciences have tried to identify the unchanging laws and material components that are responsible for the permanence of a given system.

The recent interest in the study of complex systems within many fields of science is closely related to a fresh look at the role of change and its predictability. In the disciplines dealing with ecosystems, economics, social or immune systems change has been identified as a much more pervasive and characteristic feature than in former equilibrium-oriented views. All these systems suddenly appear now as complex, evolving and adaptive systems[4], which are hard, if not impossible to understand and predict from the external viewpoint of a detached observer (Fontana and Ballati 1999, Gell-Mann 1994, Morowitz 1998). „The way key subdisciplines in the natural and social sciences view the world is converging on these theories of changes" (Gunderson et al. 1995, page 4). This ambitious research program[5] (re-) addresses the relation between the law-like character of certain aspects of the world and the fact-like character of many phenomena, but without yet being able to unify or solve it (Casti

[4] In modern philospophy the notion of becoming has become already as important as the notion of being, in the natural sciences, which are still dominated by Cartesian dualism, the division of reality into res extensa and res cognitans, the genuine creativity of evolving systems still needs to be fully appreciated (Kampis 1991).

[5] See http://www.santafe.edu

1997; Horgan 1995). It allows, however, to address questions about the limits of scientific knowledge when dealing with complex systems, i.e. systems in whose history a lot of frozen incidents have been involved and which therefore represent a major part of the fact-like aspects of the world. We are searching the history of ecosystem utilisation for a paradigmatic case of implementing sustainable management of complex adaptive systems. We are convinced that the success in doing so critically depends on a proper interface between science and other forms of knowledge, derived from the mutual insights into their respective potentials, limits and constraints.

Modelling of complex systems has to respect a fundamental distinction between i) state systems that are predictable to an outside observer and ii) systems that remain contingent on their history (Rapp et al. 1999). Ecosystems belong to the second category (Lange 1999). That is why the question about sustainable utilisation of ecosystems can be regarded as a fundamental one and not only one of the applied sciences. Modelling concepts that work well for the first category may fail entirely for the second class of systems. In complex systems, historical changes are not introduced externally into the models, e.g. from their boundary conditions or from an unbounded free „spirit“ (a „global subject“, see Schellnhuber 1999), but the emergence of new degrees of freedom is an intrinsic property of such systems.

What is the scope of sustainability?

Human civilizations that up to now depend essentially on the functioning of biological entities from which they evolved may wish that the *potential* of these providers should be sustained. Civilizations cannot transcend their own history, e.g. by becoming fully autonomous, which means without constraint in rearranging their interactions with the environment. It is assumed as politically wise to sustain any „natural“ potential that human civilisation may depend on and which is historically contingent and cannot be recreated, e.g. after extinction of biological species that carried this unique potential.

Human culture evolved from the biological realm. The question of sustainable development becomes one of interfacing the remaining dependencies on the biological realm with the acquired technical advancement in the cultural realm. In the subsequent sections we discuss the difficult *technical* question of how such biological potential, on which human culture remains dependent, can be identified or how to discover the (im)possibility of substituting them. At this stage, it is unnecessary to further consider the ethical implications of the sustainability concept, as the issue is reduced to a technical one. However, the technical aspect cannot be restricted solely to the material domain, because we define the scope of sustainability within the domain of biological potential. The discovery of these genotypic potentials, inferred from the actualised phenotypic behavior, requires continued learning.

What is the scope of development?

Development is the unfolding and testing of a novel, culturally acquired potential. Apart from the technicalities of its realisation it includes the difficult problem of *valuating* it afterwards, as opposed to the task of sustaining a potential, where values are given a priori and the potential preservation is left as a technical issue.

In this universe of discourse the apparently technical question about the nature of potentials is the fundamental one: Firstly, humans may discover a biological potential. They may strive to "tame" and "freeze" this potential in order to achieve reliable utilization, which means repetitive actualisation becomes possible (as in agriculture and forestry). This process represents the onset of civilization. However, they may not be capable of substituting it once it is lost. Moreover, civilizations may also be able to invent (*create*) novel potentials within their reach, whose realisation, test, and valuation depends on interaction with the external world.

In this approach sustainability and development are not defined relative to a given state of the (earth-)system and its future path in state space (Schellnhuber[6] and Wenzel 1998), but to the history of biological entities and the origin of their potential behaviour. Change is a genuine characteristic of evolving systems themselves and not a predictable or manageable external intervention.

Two types of technologies

In the proposed concept, sustainable development is the framework by which subjects may classify and organise biological and technical potential. Before we apply this new perspective to typical examples from forestry (section four) and then generally to environmental problems (section five) we will oppose it with the traditional way of accessing sustainability.

Traditional approaches to sustainability

Prior to scientific investigations of the term, sustainability was related to the heuristic search for truly renewable resources in the realm of the natural sciences and to the search for truly

[6] Earth system analysis as proposed by Schellnhuber (1999) represents the most rigorous application of the Cartesian dualism with its detached human observer vs. a material state system to environmental problems.

human needs in the realm of the social sciences. The difficulties to accommodate and operationalise sustainable development by scientific methods can be traced back to a conceptual flaw: The idea of a "detached observer" (Gell-Mann 1994) independent from his environment and resembling Cartesian dualism, i.e. one who does not share a common (evolutionary) history with the system he manages. As our example shows, the above question of identification of biological potentials is extremely difficult to answer and the appropriate methodologies remain dependent on the historic situation. In fact the concept of splitting the problem into one of the natural and one of the social/cultural sciences beraves the problem of its historic perspectives, on which we base our distinction of technologies.

The image of a detached observer confronting a stationary world may become a metaphor for the whole civilisation: The spaceship earth metaphor does just this: It conceptualises a boundary based on stationary fluxes (solar energy), and seeks to sustain human potential relative to that boundary. Imperatives that the anthroposphere must be based entirely on renewable resources have their origin in this metaphor.

Human agents need resources such as water or food. However, the task of identifying environmental boundaries for which resource fluxes can be considered stationary has remained elusive (Hauhs and Lange 1996). There have been two major attempts to introduce the notion of equilibrium or stationary states into natural forest ecosystems. The first was based on vegetational dynamics resulting in the concept of climax vegetation (Clements 1936). The second one was based on budgets of matter and energy (Odum 1953). Today both approaches are considered as unsuccessful (Gunderson et al. 1995). Undisturbed, natural ecosystems are first of all characterised by changes, i.e. by natural evolution at all scales considered. This view has been supported by detailed studies of vegetational dynamics, e.g. the Holocene redistribution of forest species across North America (Huntley and Webb III 1988), and studies of nutrient budgets in relatively undisturbed ecosystems. The former shows that actual species composition may depend on the initial conditions in a contingent way and on the history of recolonisation; the latter shows that nutrient budgets are transient and dependent on external input fluxes at all time scales considered (Chadwick et al. 1999, Liski et al. 1998, Wardle et al. 1997).

The failure of ecology to derive an empirical or theoretical framework that allows to predict stationary states of ecosystems under a given stationary environment has left the public controversy about sustainable ecosystem management open for ideological and philosophical debates.

As long as these debates are based on the non-evolutionary view of Cartesian dualism they remain unresolvable in principal. One camp still regards undisturbed nature as the ultimate source of stability, whereas the other camp regards the human technological potential in achieving stability as unlimited. We call the former camp as believers in a „lost paradise" and the latter as believers in an „achievable paradise". Both sides in this debate are characterised by the assumption that the true nature of the problem is solvable by basic sciences: „Which ecosystems are naturally stable and to what extent?", „How can organisms be engineered to make up stable ecosystems?" etc. Technical questions of learning, acting, testing and valuating are considered as secondary: „How do ecosystem managers learn, act and evaluate?", etc. These topics will be first considered in section 3.2.

The above two opposing perspectives are therefore associated with typical assumptions about the „true" nature of the problem:

A) Believers in a lost paradise prefer an equilibrium view of ecosystems: Nature „knows" best for itself what to do in order to achieve stability. Left on its own and undisturbed by humans the biosphere is expected to be stationary. Through their advanced learning capabilities humans can gain insight into this true picture of nature and are obliged to act accordingly: Nature is treated as a social partner and the ethics towards nature closely resemble a social ethic, where intentions are decisive for morality. Environmental conflicts require that human wants and needs must respect the domain of its social partner „Nature" and are settled such that humans strive to distinguish „true needs" from secondary and derived wants that can be abandoned. Environmental policy hence becomes primarily a moral issue.

B) Believers in an achievable paradise adhere to a technocractic view of ecosystems. With the help of modern biotechnology it will be possible to reengineer ecosystems towards any conditions needed. Complex ecosystems are regarded as similar to other complex technical devices such as cars, air planes, etc. Environmental conflicts require that the needed ecosystem functions can be intentionally and rationally redesigned. Environmental policy hence becomes primarily a technological issue.

The first view completely trusts nature's wisdom, while the second completely trusts cultural i.e. scientific wisdom. It is easy to find proponents of such views among interest groups in environmental policy ranging from deep ecology to genetic engineering. In each camp sustainability is easy to define, but remains totally incompatible with that of the opposing view. However it is remarkably obvious that both positions are derived from a static world view of the Cartesian dualism in which the ability to learn and act (the realm of res cognitans) remains unrelated to the material world (of res extensa).

The two camps do not share a joint basis to settle this controversy and their views are both founded in untestable assumptions. Science has in the past done little to settle the controversy between these two views. It seems that membership to one of the camps is basically a question of belief.

Apparently the problems of implementing and organising sustainability under real world conditions cut right across the traditional definitions and work sharing among scientific disciplines. That is why we will not pursue this traditional concept further. The results can be regarded as an indication that the old controversy about the role of humans in nature is beset by a conceptual misunderstanding of the issue. In the following section we will turn to the proposed alternative concept which regards the distinction between two types of fundamentally different technologies in an evolving system as the key issue. Implementation of sustainable development becomes a process of how to organise learning rather than a decisive category for a detached observer.

A new perspective at sustainability

If potentials are chosen as the scope of consideration, sustainable development becomes a technical question of how the natural and the cultural history of mankind have been and can be linked. Natural history prior to man is loaded with frozen accidents (it could have been different, but did not happen to occur ..). As discussed above, ecology has found little evidence of law-like behaviour in ecosystems, each of which may be marveled at as being unique. The formation of civilizations presupposes reproducible and reliable patterns of biotic entities. Technologies aiming at this behavior led to a (partial) *emancipation* of man from nature; we thus term these management schemes (including silviculture) *emancipatory technologies*. Before human civilisations emancipated themselves from ecosystems, humans took part in an unpredictable open-ended *natural* history (Gould 1989). In a similar manner *cultural* history can also be regarded as loaded with frozen accidents, and even less law-like behaviour is expected by social scientists (Ingold 1986). The concept of development extends this notion of an open-ended cultural history into the future. The appropriate developmental technologies will be termed *exploratory*.

The human techniques which have transformed natural history into cultural history are a complicated field of research. Even though anthropology classifies the earliest human societies and ages by the presence or absence of techniques for food collection among hunters, gatherers, pastoralists and farmers, „techniques are excluded from the prevailing vision of societies" (Sigaut 1994[7]). Technology as the science that studies lines of technical evolution has generally not been considered to be a fundamental issue. Here we will argue that some of the confusion about the term „sustainable development" may be due to a failure of drawing principle distinctions in technology. We use the term technology in a very general sense as „the way people do things" (White Jr. 1940). That is why silvicultural techniques are classified as technology as well as information technology, though they may be fundamentally different.

Reasons for the difficulty of separately assessing natural and cultural histories have been addressed above (Ingold 1994, Gunderson et al. 1995). Attempts to identify laws, stationary or even equilibrium states have been unsuccessful, either in respect of the nature around us (i.e. the search by ecology) or regarding the nature within us (i.e. the search in anthropology for „true" human needs versus culturally determined wants). Why should then coupling of natural and cultural history contain fundamental issues and resulting in a new simplicity? It is this hope that motivates us for the proposed concept of sustainability.

We will consider the two types of technologies as being fundamentally different, because of the historic and gradual character of human emancipation from natural history. In type one (exploratory technologies) the discovery of persistence (laws, conserved matter) marks the initial point. In the other type (emancipatory technologies) persistence (sustainable application) marks the ultimate goal. The distinction goes back to two kinds of technical practice: one in which knowledge and understanding is subordinated to material action and the other in which action is subordinated to knowledge (Sigaut 1994). In both cases the concept of ahistoric temporal persistence of patterns is an abstract idealisation of real world history. This will be exemplified in the discussion of historic examples.

In the case of exploratory technology this abstraction consists of an scientific understanding of building blocks which can be isolated from external forces (conceptually or experimentally) and which are governed by simple laws of nature. Once humans gain insight into such laws, they can experiment with and reconfigure the simple building blocks in order to achieve complicated devices that serve human purposes. This simplified scheme is exemplified by most modern technologies ranging from cars to computers. Explorative technology describes how simple, well understood modules can be reconfigured into devices performing potentially useful action under guidance of human choice.

---

[7] „Yet the whole edifice of social sciences was built as if technics did not exist or were of only minor

In the case of emancipatory technology this abstraction consists of a regulative success while interacting with a complex historic system. The practical skill that exploits and generates patterns in nature is in this case the starting point. Once humans have acquired regulative skills and competence over natural systems, they can experiment with and reinitialise some of its components to achieve simple interference and utilisation schemes that serve human needs. Here the interference is devoted to repetition of the discovered pattern, and if this becomes successful the resulting management systems may even approach an ahistoric state, i.e. the interference scheme yields a „frozen history" and may then be termed sustainable.

A historical case study

In this example three natural resources, forests, polymetal ores, and water, that happened to co-occur in the same geographically distinct area of the Harz mountains (Northern Germany) became closely related and combined through a utilisation history lasting more than 1000 years. We will focus on the exploitation of timber and ores through the better documented part of this history starting in the 16th century when Welfian dukes gained full control over these resources.

The *Welfian* dukedom which had been in control of the area had no access to major rivers or to coastal harbors. That is why the natural resources within the Harz mountains could only be developed locally and in a highly interdependent manner resulting in the export of valuable goods such as copper, lead or silver. We start our summary at a time when the ownership rights were forcefully and finally settled (1530). One of the oldest maps of Northern Germany originates from that conflict between the city of Goslar, the former owner of the Rammelsberg mine, and the Welfian Duke at Wolfenbüttel. This map delineates the forest resources within the Northern Harz claimed by the city of Goslar (Bauer 1981). Timber was already appreciated as a scarce resource. During the conflict the duke was able to halt mining operations at Goslar`s Rammelsberg by an embargo on timber trading from forests outside this area.

After the shift in ownership the utilisation of water, forest and ores was reorganised. Initially expertise, man-power, and financial investments were needed to exploit these resources. In the 16th century expertise on mining and smelter technology started to become disseminated through printed books. A number of rights were granted to attract people for

importance" Sigaut 1994, p. 452.

settlement and capital for investment into the Harz mountains. These rights included free use of Harzian woods for mines, smelters, and the corresponding workforce. As exemplified in other cases (North 1973) such warranted rights proved decisive for the initialisation of economic development. The right for free timber supply was regulated and supervised in detail by a local administration of the forests and the mines (Henschke 1974).

The goal of a persistent (sustainable) utilization of the natural resources in the Harz can clearly be traced to the self-interest of the governing Welfians: Several documents from that time explicitly mention the "posterity" (of the dynasty). The Welfian dukes wished to sustain the substantial contribution of income from the mining activities in the Harz. For example in 1579/80 more than 40% of the total income in the dukedom was derived from Harzian mining operations (Bartels 1992, p.47). In addition to the Rammelsberg, the duke possessed the right to several mines in the Upper Harz. As timber was scarce already at the beginning of this reorganisation, the Harzian forests were included into the planning efforts of the centralised mining administration.

We have chosen this case to discuss the distinction between the two types of technologies as described below, because at the onset of written records the distinction between the different nature of resources was generally unknown to practitioners of mining and forestry. In the 16th century ores, timber and water were all regarded by most scholars as *renewable* resources. Over the subsequent centuries leading-edge technologies were applied and developed for the interdependent use of these resources. There is hardly any other region in Germany where forests and water were to such an extent exclusively devoted to one single use, namely the production of metals (Cu, Ag, Pb, Zn) over such a long period. All other utilizations were indirectly motivated and prioritized relative to the main goal[8].

Of the many centers of activity in the Harz we examine the Rammelsberg as an important, particularly well-documented and thoroughly researched mine (Bornhardt 1931, Bartels 1997), and the Bramke valley, as a typical and well-studied forested catchment area (Hauhs 1989) that served over centuries as a timber resource. It belonged to the so-called Communion forest (1635-1788), then to the Hannoverian Harz (1788-1866), the Prussian state forest (after 1887) and is today state forest of Lower Saxony. The Lange Bramke

[8] For example at the Upper Harz farming and even the possession of a plow were banned by law. The food supply for inhabitants of the Upper Harz was organised through the mining administration. These laws stem from the 17th century when food was scarce after the thirty year war (Bartels 1992, p. 21).

catchment is situated 3-5 km south of the Rammelsberg mine and belongs to the Upper Harzian forests. It extends in height from 500 to 700 m above sea level.

The Rammelsberg mine, an exploratory technology example

The *Rammelsberg* near the town of Goslar is situated at the northern edge of the Harz mountains. The oldest traces of the history of mines and smelters date back to Roman times (Klappauf 1996). In historical documents, mining in the Harz mountains extends over more than 1000 years. The exceptional character of the Rammelsberg mine, now included into the UNESCO cultural world heritage sites, has been documented and discussed elsewhere (Bornhardt 1931, Bartels 1988, Bartels 1997, Ließmann 1997). Its compact ore deposit of about 26 Mill. tons was one of the largest of this type in the world. Since the 15th century written records allow us to quantify the total results of the mines and smelter operations. The first quantitative estimates from the state of the forests in this area date back to the 16th century (Steinsiek 1999).

The distinction between renewable and non-renewable resources was a result of a learning process mainly by trial and error. The changes in the idea of sustainability are reflecting the contemporary technical knowledge and the scientific concepts available at that time. Böhme (1988) pointed out that a comprehensive set of the corresponding ecological arguments was already discussed by one of the first "modern engineers" Agricola (1556) in his famous "*De re metallica libri*". Without explicitly using these terms, Agricola takes up the criticism against an extensive use of non-renewable resources. Confronted with the expectation that mining, unlike agriculture, does not provide continuous long-term economic results, Agricola (1556, p.3) pointed to the Rammelsberg mine as a counter example lasting to his knowledge already for about 600 years. He nevertheless already acknowledged a principal difference between agriculture and mining pointing out that mining resources may not last forever (Agricola 1556, p.3).

Agricolas view was exceptionally modern. During the 16th century there was still a widespread belief that both wood and ores grow at comparable time scales. Hieronymus Cardano asked suggestively: "What is a mine more than a plant covered with earth?" or "Metals in the mountains have roots, stem and leaves like a tree."[9] The first printed books on mining proposed that ores grow in a mountain (Pieper 1955) and sometimes mines were temporarily closed to allow for "regrowth". Close observation of secondary minerals in mines even seemed to prove this point which was, however, reflecting the technical progress in smelter technologies (Bartels 1992). After some decades additional ores could usually be processed after reopening the mines. This produced an illusion of "regrowth" of ores. At the Rammelsberg it occurred at least twice that material originally used to backfill emptied parts of the mine was later regarded as valuable ore.

Initially the Rammelsberg contained in total about 26 million tons of rich pyrite ores (Cu, Ag/Pb, Zn[10]) concentrated in two massive pockets, one of which intercepted the surface over a length of ca. 500 m (Mohr 1993). The metal-containing minerals within these ores were intensively mixed and until well into the 20th century these ores could not be separated prior to the smelting process. Thus Rammelsberg ores, unlike vein ores from the Upper Harz, were initially easy to access but diffcult to process (Bartels 1988, 1997).

The ore resources of most Upper Harzian mines showed a vertical gradient in metal content. Silver ores were enriched at the surface due to weathering processes. That is why after an initial boom based on such accessible rich ores, the mines were facing substantial technological difficulties as their resource deteriorated in metal content with depth, and the shafts became technologically more challenging due to drainage problems. A major transition in mining technology occurred between 1580 and 1620, when the use of gun powder was introduced at the Upper Harz and production was shifted towards mass production of poorer ores (Bartels 1992).

The transition period is characterised by contrasting views about the nature of the resource. Löhneyß, director of the mining administration at the time, was a dedicated alchemist. He rejected the possibility of patterns and regularities in the distribution of ores. Ores were God given gifts and more related to the consience of the finder than to a predictable pattern in nature. Afterwards he was blamed by his followers to have ruined mining by this attitude. Another document[11] from 1584, however, argues against the alchemist view and claims that poorer ores require an application of the most modern technologies in order to be economically viable.

For the Rammelsberg the quantitative record of ore production has been documented from 1460 when mining commenced after a standstill. From an estimate of the missing ores at

---

[9] Cited after M. Eliade (1992).

[10] In the historic sequence by which they dominated the economy of the mines, due to progress in smelter technologies.

[11] OBA Cl, F 462/7 cited in Bartels 1992, p. 150

that time, which is possible due to the unique compact character of the deposit, the total production until 1360 when the mine was temporarily closed for a century amounts to about 3 mill. tons. The production lasted almost continuously from 1460 until 1988 when the deposit was exhausted with only 0.4 million tons left. We use ten-year averages of cumulative production figures to reconstruct the exploitation history (Fig. 1). Only two major phases can be identified, even though this development is stretching over more than five centuries.

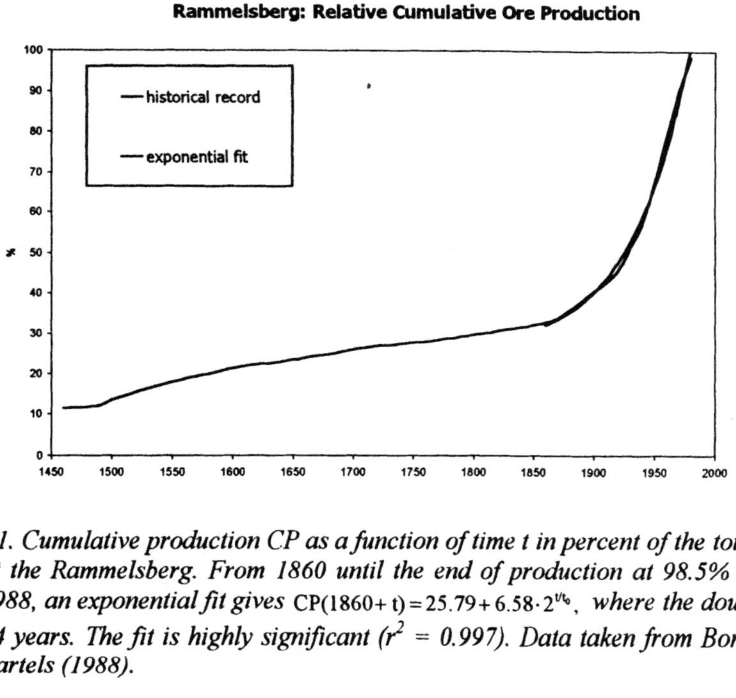

Figure 1. Cumulative production CP as a function of time t in percent of the total ore deposits at the Rammelsberg. From 1860 until the end of production at 98.5% exploitation in 1988, an exponential fit gives $CP(1860+t) = 25.79 + 6.58 \cdot 2^{t/t_0}$, *where the doubling time t_0 is 34 years. The fit is highly significant (r^2 = 0.997). Data taken from Bornhardt (1931), Bartels (1988).*

The first phase appears as one of the world's longest records of the sustainable utilisation of a natural resource, though in the end it clearly turned out not to be a renewable one. In fact, between the 16th and early 18th century, it was often regarded by its managers as a sustainable resource, as measurements of its size or the rate of "ore regeneration" were impossible at that time. A transition between these two phases occured in the 1860ies when industrialisation took off in the Harz region. Before this time the curve is roughly linear, indicating a relative constant production rate; after 1860 the accumulated production began to turn into an exponential trajectory.

The second phase in figure 1 starts after several major changes cooccurred at the Rammelsberg: the second part of the deposit had been discovered and opened, the smelters processing these ores were connected to the German railway net and got access to cheap coal, ownership and administration changed, investments into steam-powered engines for pumping and lifting became economically worthwhile.

Water-powered drainage machines had been in operation at the Harz for centuries and their efficiency had been state of the art as the increasing depth of the mines continuously increased the technological challenges. Water-powered lifts for miners were invented in the Harz region. At the end of the 18th century the most important mines in the Upper Harz had reached depths between 300-450 m below the surface, the Rammelsberg having reached about 300 m. Energy in form of water power was limiting[12] the water pumps (Bartels 1992, p.360) in the mines and energy in the form of expensive charcoal limited the smelters operation. Whenever timber was temporarily available in excess, as for example due to an insect calamity during the late 18th century, the ore production was rapidly adjusted to the increase in wood supply.

Over the last thirty years prior to the end of operations in 1988, the production again remained relatively constant, and a linear fit would do even better than the one shown in Fig. 1. At these times the market prizes for the produced metals (then mainly zinc) limited the production rates. Continuing technical development was invested into higher efficiency only, as indicated by the number of employees dropping from about 1100 in 1958 to about 300 in 1988 (Bartels 1988).

The technical improvements based on trial and error were able to uphold diminishing returns in the mines such that the overall production rate remained relatively constant (figure 1, see Bartels 1992 for many examples). From the 1860ies onwards[13], technical innovations became more and more science based. Before that time the innovators responsible for the technical innovations between 1750 and 1860 were mainly practical people and engineers.

The history of the mines and smelters contains many examples in which conservative attitudes resisted technical innovations. In the Harz this became especially well-documented as mining administration between 1635 and 1788 was shared between two dukedoms, with yearly alternating directorship. Over that time usually one administration played the conservative and the other one the innovative role. Repeated testing of the economic outcome

[12] Water power remained limiting despite a very sophisticated system of ditches (600 km), tunnels (30 km) and reservoirs (120) sustaining water supply to the mines through drought or frost periods (Schmidt 1989). The main part of this system was completed by the 18th century.

[13] The two phases of figure 1 are very similar to the historic development of the gross national product in Western Europe, with a linear phase before and an exponential phase after 1860.

of innovations allowed the Harzian mines to stay at the forefront of technological development and to abandon alchemistic attitudes entirely.

That is why we regard this as a well-suited example of an exploratory technology in which information about technological potentials is relatively easy to transfer, but valuation of its realisation is the limiting step. During the two linear phases, production was either limited on the input, energy supply up to 1860, or on the output side, market prizes after 1958, respectively. During the transitional phase the exponential growth, lasting for more than three doubling times, implies that further production was proportional to the degree of exploitation already achieved! No external limiting factors were effective and the learning process about the best suited technology became apparently self-enforcing: Increased production meant an increased competence and an increased ability for even further investments into new production capacities.

The fact that these resources were used at relatively constant rates over an extended period of time obviously does not prove sustainability of the management or renewability of the resources. The nature of the potential has to be understood, and if it is a potential originating from an exploratory technology *development* becomes the organising principle (i.e. extension of the technological potential available). Sustainability as defined above does not apply to this type of technology. Also in forestry, as shown in the next section, an extended utilisation at relatively constant rates may mask the diminishing supply of a non-renewable nutrient such as nitrogen or magnesium. After several rotation periods or increased export due to leaching such nutrients may become depleted and finally lead to non-sustainable growth rates. Therefore we suggest to make distinctions about aspects that are instantaneously testable and aspects which are able to support continuous updating and learning during management. The distinction between renewable and non-renewable resources does not match this criterium, as it is only helpful for an all-knowing, *detached* observer or if a resource has indeed been proved as non-renewable. The proposed distinction among two types of technologies is, however, testable for *participating observers*, i.e. resource managers.

Regarding operations in the mining/smelting and forestry sector as independent was justified for about 100 years only. About a century after the rate limiting effect of the charcoal-based energy supply was removed, forests and the mines had again to be regarded as a coupled system. By that time the emissions caused by the widespread use of fossile energy after the industrial revolution had largely changed the chemical climate of Central Europe and contributed to the forest decline in the Harz forests (Hauhs 1989). The question reappeared:

how can development of exploratory technologies be linked with sustainable management of emancipatory technologies?

The Harzian forests, an emancipatory technology example

The Harz mountains, of which the Rammelsberg forms the northern edge, are climatically unsuited for agriculture. Since about 5000 years B.P. beech (*Fagus sylvatica* L.) had become the dominating tree in this area. In the higher locations (>650 m) it was mixed with Norway spruce (*Picea abies* K.). With the onset of human utilisation, this mixture was more and more changed into pure spruce stands due to the difference in natural regrowth after clearcuts (Schubart 1978, Hillebrecht 1982). In the area neighboring the Rammelsberg, which will be discussed here, this transition to almost pure spruce forests was nearly completed by the 16th century (Schubart 1978).

During the period between the 16th and the 19th centuries, about 70-80% of the total wood consumption of the upper Harz, about 40 000 ha of forests, was converted into heath under supervision of the mining administration (Baumgarten 1933). Most cut timber was processed to charcoal and then transported to the smelter facilities, e.g. in the vicinity of the Rammelsberg. Forests at the Harz needed about 25-30 years for a regrowth to the typical tree size used for charcoal production and 80-120 years for a regrowth to the size needed as construction wood for the mines. These long rotation periods required a difficult and painful learning process to reach an adequately organized utilisation, i.e. one that approached a continuous wood supply. The longer rotation periods needed to be organised in the vicinity of the mines due to high transportation costs (e.g. in the Bramke valley).

Forests were quite early recognized as a finite resource the oldest evidence of regulations endorsed by foresters dating back to 1271. Prior to that the only competing utilisation was royal hunting. No permanent settlement occurred in the Harz before the development as a mining area. The wood supply was limiting throughout most of the mining history and thus methods were needed for a quantitative assessment of this resource. In the 16th century estimates were based on visual inspections only ("Forstbereitungen"). These estimates suffered from lack of quantitative knowledge about the spatial reference area. The actual growth rates were unknown, but the exploitation rates could be compared with the age of the standing volume. Hence warnings about excessively high cutting rates and thus neglect of the needs of posterity appeared already in the inventory of 1580.

In the 17th century the forests were accurately mapped and hence the quantitative estimates of standing volume improved. The forest units inside the Bramke valley with a typical size of about 100 ha each were covered by the first true map from 1680. Compared to the correct size of these units as measured today, these first areal estimates are usually below 5% in error. In the 18th century estimates and measurements of stand growth were introduced and modern forestry developed.

The historical estimates and records of the total cut over the last 150 years indicate an average growth rate of about 6 m³/ha/a (Figure 2). A yearly record is available since 1871, when Prussian forest administration was installed and the Harzian forestry became independent of the mining administration. Around the second World War, the principle of sustainability (that had been adopted since the mid-19th century) was twice deliberately abandoned. For war preparations and during the war, the Nazi regime ordered increased cutting rates. After the war an unsustainable cutting regime was ordered for the British zone. After these short external influences, the cut was rapidly adjusted to compensate for these losses (Figure 2).

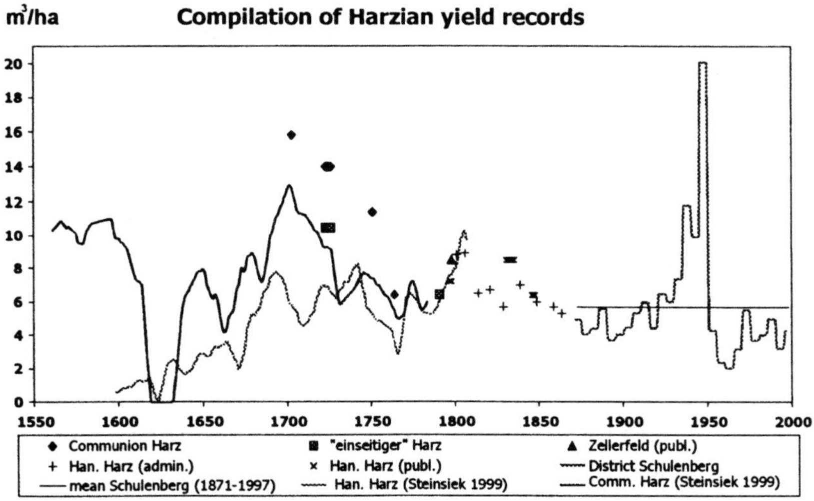

Figure 2. A compilation of yield data for various parts of the Harz. Over time this record zooms in on the forest district Schulenberg, which can be regarded as a representative example of spruce forests in the Upper Harz, containing the Bramke valley. The reference areas for the data change between 3.000 and 54.000 ha. The straight line is an estimate of the long-term timber production *for the area. Data from Baumgarten (1933, corrected after Riehl, 1968), Steinsiek (1999) and Lück (pers. comm.).*

The message of Figure 2 is not to suggest a gradual decrease from an overcutting regime to the level of sustainable yield, but the improvement in the density and reliability of the monitoring process itself, hence the organisation of institutional learning. In the early 18th century non-sustainable cuts were also a result of the ignorance *within* forestry about the long-term growth rates. This period is characterized by relatively large fluctuations in the wood supply and a sparse record thereof. Much planning had to be done in organizing the supply of mines and smelters, but little control was actually applied to assess forest growth rates. Thus no short term corrective responses as shown after the second world war were possible in the 18th century. Sustainability was learned "the hard way". Hence it is not the overall extent of deviations from sustainable harvests that changed much over time in this example, but their length of duration, their reasons (internal vs. external) and the availability of knowledge to diagnose and compensate them.

Throughout the history of forestry, sustainability evolved from an externally imposed constraint in the interest of the sovereign and organised by a central administration into a local management criterium for individual silvicultural decisions. The spatial and temporal scales at which deviations from sustainable use could be recognised and compensated were greatly refined throughout this history. The use of timber and ores, even though originating from a common starting point and remaining coupled through most of the history, resulted in entirely separate administrations.

From 1871 on, when regular checks of sustainability had been institutionally installed, it became necessary to improve the models by which local and actual growth rates could be estimated and compared to the actual cutting rates. The standard instruments for these predictions were yield tables. The first yield table for Norway spruce (Picea abies *Karst.*) in Germany, Schwappach (1912), was based on a new network of 472 forest growth monitoring plots most of which were started from the 1880ies onwards. One of these plots had been installed in 1889 at an altitude of 700 m within the Lange Bramke valley[14]. Almost 99 years later in the wake of the forest decline debate three similar plots were established at Lange Bramke within a distance of 1 km from the historical one, but without any knowledge of the old experiments. With both data sets it is possible to directly compare growth rates of subsequent spruce stands (Hauhs et al., in prep.).

---

[14] Niedersächsisches Staatsarchiv Hannover 182 Schulenberg acc 51/71 No. 18

Several similar comparisons of this type have shown that today's growth rates are generally higher than that of preceding stands at the same site (Schübeler, 1997; Pretzsch, 1999). At Lange Bramke recent growth rates are about doubled relative to the former inventories. The 19th century stands at Lange Bramke had among the lowest yields ever observed in German Norway spruce stands (Schwappach 1912). The Bramke stand had reached only a height of 13m at an age of 71 years. On the basis of recent nutrient inventories and budgets it can be inferred that nitrogen had been dramatically limiting after the repeated biomass exports over the preceding centuries. Today, however, with ample nitrogen deposition, Mg has become the limiting nutrient. The few stands at the Lange Bramke that were experimentally kept unlimed when the whole Upper Harz was treated with dolomitic lime still show symptoms of severe Mg-deficiency.

Implementing sustainability in forestry on the basis of renewable soil resources has thus at best been only temporarily successful. It needs a two step procedure of which only the first step, organising a system of monitoring and documenting the actual growth rates, has been completed. The second necessary step, explaining the observed and inferred growth potential in terms of causal and modular factors such as gene pool, soil nutrient stores, climate and management methods has not yet been completed. All these factors were considered or imposed as static by the forestry administration and hence the transient validity of yield tables was overlooked until drastic deviations in growth ocurred.

Forestry at the Harz can no longer rely on stationary boundary conditions as shown by ecosystems research and empirical data. Similar trends in site conditions may have happened in many forests of Central Europe. In such a situation the distinction between renewable and non-renewable resources provides little guidance for practical management. Errors in the basic assumptions about the character of the resource (e.g. replenishing rates of the available soil nutrients) have been detected relatively late. Under such circumstances it is mandatory to determine already incipient changes in growth potentials. Appropriate mitigation and inference techniques must be learned in order to find out what can be done in time to keep a potential at a sustainable level.

The traditional categories for organising sustainable management either by the objective character of the resources (renewability?) or by the moral character of human intentions (basic needs?) are no longer appropriate under these circumstances. We suggest that the differences between the history of mining and forestry in the Harz region are more efficiently described by the different learning requirements. The way in which the process of learning and updating knowledge about the available technological potentials and skills had been organised led to the different courses of forestry and mining, from a common starting point. This argument takes up our initial hypothesis that the different origins of the technological potentials mark the most critical distinction among utilisation of natural resources.

General lessons from the case study

Organising the development of exploratory technologies

One of the most striking differences that divides the production period discussed here from times prior to the 16th century is simply the amount of written material that was generated. From early modern times onwards, not only the exploitation of the resources changed but clearly also the information management that occurred during this time. During the 16th century mining and smelting technologies became disseminated in printed books. The special form of shared ownership of Harzian forests and mines between 1635 and 1788 (Communion forests and mines) led to a well-documented history of its administration (Baumgarten 1933, Riehl 1968, Bartels 1992), only part of which has yet been compiled and edited. At several instances during this mining history, external experts and scientists were contacted and contracted to investigate and solve specific technological problems, including famous personalities such as G.W.Leibniz and C.Polhem (Bartels 1992).

Over time the mining enterprises were gradually liberated from the alchemistic myths that characterized their origins. At the end, modern mining and smelter industries proved to what extent it pays to apply scientific methods for improving and transferring knowledge and engineering competence about these resources. Throughout history the Upper Harzian mines and the Rammelsberg formed a landmark in the development of both geology and mining technology (Bartels 1992, Page 30). In mining the economic value of applying the available technology was decisive. The transfer and updating of the respective technological potential available at the time was apparently not critical. Hence we regard this as a typical exploratory technology, where entirely novel technical possibilities may become available at any time, but need to be evaluated. This evaluation process has to include the interface towards the other (emancipatory) technology such as forestry or nature conservation. Thus testing and evaluation of technical potentials need to be organised.

Knowledge about exploratory technologies can be regarded in essence as knowledge about *ahistoric objects* and modules that can be transformed into useful devices such as arrow heads, axes, plows, cars and computers that almost *define* ages in cultural history. Time remains an external parameter to such technologies, as long as ageing phenomena of man-made crafts or artefacts are not an issue. There is no intrinsic timescale and ideally no change at all. When external constraints are missing the realisation of accumulated technological potentials can be reinforcing and trigger exponential growth as exemplified by the Rammelsberg mining activities between 1860 and 1960, or industrial revolution in general. In other words, the speed with which these technologies can be implemented and the respective goals can be materialized is subject to economic reasoning alone. What remains of the original alchemistic idea is the concept that man can master time and accelerate "natural progress" in achieving his goals (Eliade 1992). This view of technology has thus inherited from alchemy the vision of an achievable paradise (Binswanger 1994).

Organising the sustainability of emancipatory technologies

We suggest that the introduction and development of modern forestry necessarily resulted in a different relation between technology and science compared to the mining example. In forestry the most crucial lesson is to appreciate the overall temporal growth pattern that can by no means be accelerated and extends well beyond the life span of any single forester, i.e. the rotation period. Thus in this type of technology, in which the biological potential must be discovered in natural history, time remains an intrinsic parameter. The time pattern of growth as imposed by the chosen rotation period is an invariant for all respective silvicultural techniques. In the Harzian example the ability of recognising such patterns was facilitated by two conditions: The extended life-time and the relative constant demand of the mines such as at the Rammelsberg and the relatively stationary framework in terms of ownership, administration and site conditions. The latter would not irreversibly deteriorate after the first exploitation of stand biomass.

During the "classical" period of forestry, the first half of the18th century, practical and theoretical knowledge about forestry were merged by experienced *practitioners* (Rubner 1967), still similar to the history of mining technologies (see above). From the beginning of organized forestry, the problem of practical training and education was considered as the most critical issue. Modern forestry in which sustainability is a firmly established principle was first made possible after such a regular practical training had been established within the entire hierarchy of forest administration. In the Harz this happened through the first half of the 19th century. The forestry faculty at the University of Göttingen is a successor of the practical forest school at the Harz that was started as part of the mining administration in the 18th century and subsequently moved to Hann. Münden and then to Göttingen.

In essence the technical knowledge in forestry is about the temporal patterns that can be imposed by human interference at historical objects (tree populations) under a given management regime. The technical competence refers to a holistic perspective of the forest ecosystem. That is why for foresters it is easy to accomodate an ecosystem view, even when the ecosystem concept remains vaguely defined. The difficulty and empirical expertship in silviculture lies in recognising the *potential* behavior in the *actualised* states of the managed system. For example in a thinning operation the growth potential of individual trees over the remaining rotation period needs to be inferred on the basis of growth realised in the past, i.e. on the basis of the actual phenotypes. In an exploratory technology the inference works in the reverse direction.

It is a difficult scientific task after the temporal growth patterns have been discovered by silvicultural experience to assign this utilised/realised (growth) potential to individual factors such as soil nutrient stores and dynamics, climatic conditions, genetic potential of the managed tree population, and the interference scheme itself. On this basis, the growth potential becomes modular and further refinements of the management, e.g. experimenting with planting and thinning regimes or breeding and fertilisation become possible.

Science has the role of decomposing a "holistic potential" inferred from practical ecosystem management into simpler modules. This does not (yet) imply the technical capacity to reconstruct (de novo) the respective potential from those modules, but it facilitates the learning and updating processes about the extent of the biological potential used. However, decomposition of the technical skills into simple modules is extremely difficult for silviculture itself, e.g. for thinning regimes. Forestry text books dealing with this matter seem to have inherited much of the "darkness" typical of alchemy writings (see Leibundgut 1978). They may appear obscure being meaningful only to those already initiated. For example the definition and description of growth phases over a typical rotation period in forestry seems to resemble analogues in the opus alchymicum[15].

The above mentioned similarities in the written description of silvicultural techniques with alchemistic texts were not mentioned to ridicule foresters, but to demonstrate a genuine difficulty in documenting and teaching such technologies. We regard these as typical features of an emancipatory technology, i.e. one that deals with potentials discovered in nature. The technical checks whether such a potential is indeed sustained require a comparison with its past realisations and hence require a long-term *memory* for past management actions. Central European forestry can be regarded as an instutionalised memory of this sort. However, key elements of the silvicultural experience and expertise can still only be described and taught in practical training and not through books. Scientific abstractions have been of relatively little help in facilitating this learning process. This situation is no longer adequate to the present challenges of forest management.

This silvicultural experience derived in German forestry appears today as highly endangered. Firstly, it is devalued by changing site conditions, as demonstrated by unreliable yield tables; secondly, it is devaluated due to the economic crisis of forestry that leads to an extensivation of silvicultural practice and training; and thirdly, it is at risk because it is not highly respected by sciences and must remain an expertship that can only be judged and appreciated by "insiders". These trends are equivalent to a loss of memory about the past realisations of biological potentials which will thus diminish the possibility of testing the sustainability of today's management. In this perspective, the de facto loss of sustainability occurs regardless whether actual growth rates have increased or decreased relative to the tabulated experience contained in the yield tables.

The difficulty of dealing with changes in forest ecosystem management under changing boundary conditions has also been reflected by the forest decline debate in the 1980ies. It was difficult to isolate the effects at the material level of forest ecosystems, whereas it has been easy to identify the effects on forest ecosystem management (Pretzsch 1999). Founding the monitoring of sustainability on silvicultural management, i.e. testing whether potentials that could be attained in the past can still be realised, may be the only practical implementation that matches historical experience *and* the results of ecosystems research. We expect that in the future the debate on forest decline in Germany might be followed by a forester decline debate.

The real obstacle in implementing sustainable development is to be found in organising the interface between the two types of technologies within our civilisation, especially as it is no longer possible to ignore or abstract from the mutual interferences of these technologies at all scales including global biogeochemical cycles. To this end, practical experiences from ecosystem management need to be updated, documented, and disseminated outside the inner

---

[15] Compare for example the phrases and descriptions proposed by Leibundgut (1977) with summaries of the alchemistic opus (Atmanspacher 1992, White 1997).

circle of experts in a much more efficient way. The last section will shortly demonstrate the technical means by which this could be achieved.

Outlook on new technologies: A flight simulator for foresters

We proposed that the continuous accumulation of practical silvicultural competence in forestry is the best operational indicator of sustainability. This criterium, however, requires efficient techniques for testing, updating, communicating and training of such expertship. This condition is no longer fulfilled by the traditional forms of testing sustainability in forestry which are only applicable under the elusive condition of stationary boundary conditions and a stationary society. In our real world of changing demand and changing site conditions we propose a more general criterium that contains the old one as a special case: Whenever it can be shown that experiences (about a biological potential) that were derived in the past are no longer valid in the current situation, one can suspect that sustainability is no longer achieved. Sustainability becomes therefore restricted to those potentials that *can* demonstratively and by historical experience be transferred into repetitive management schemes and only as a secondary step to modular ecosystem properties that can be scientifically tested for rates of change. This task of documenting and checking present against past management competence can be largely supported by modern technologies.

Interactive simulators are already used in areas where non-verbal pratical competence is difficult to train under real-world scenarios, e.g. flight simulators in pilot training. The example of complicated modern planes can serve as an analogue to forests in an area where the know-how of constructive design became increasingly irrelevant for the practical use of the system. On the contrary, the degree to which the user-interface can be shielded against the constructive background is itself a key criterium for technical progress.

As an analogy to pilots that are able to utilize *augmented vision* in their training and flying we suggest that emancipatory technologies such as forestry would benefit from *augmented memory*. The concept of a "flight-simulator for foresters" has been implemented into the model TRAGIC++ (Hauhs et. al 1999). It uses basic structural and functional aspects of forest ecosystems as the building blocks from which experienced growth patterns can be interactively reconstructed (Figure 3). The rationale of such a model is similar to that of "ghost-images" that help to communicate the experience of an eyewitness of a crime scene. In the case of ecosystems, communication of past utilization potentials becomes the bottleneck for assessing sustainability. Here science is able to provide the correct set of parts (corresponding to shapes, eyes, ears, noses, etc. in the case of ghost images, and the witness (i.e. forester) adds the knowledge about the overall pattern. Similarly, in forestry practitioners have difficulties to communicate their acquired competence about temporal growth patterns in written form, while scientific growth models are notoriously underdetermined through scientific observation and often employ a circular reasoning called calibration (Hauhs et al. 1999).

Development, testing and implementation of such a scheme will result in a different organization of the interface between practical ecosystem utilization and science. In such a model, information flows in *both* directions. Similarly, it opens the possibility of a rationally organised interface between the two types of technologies. Thus it may help to overcome ideological discussions about sustainable development and lead to an operational technique of implementing it not only in forestry.

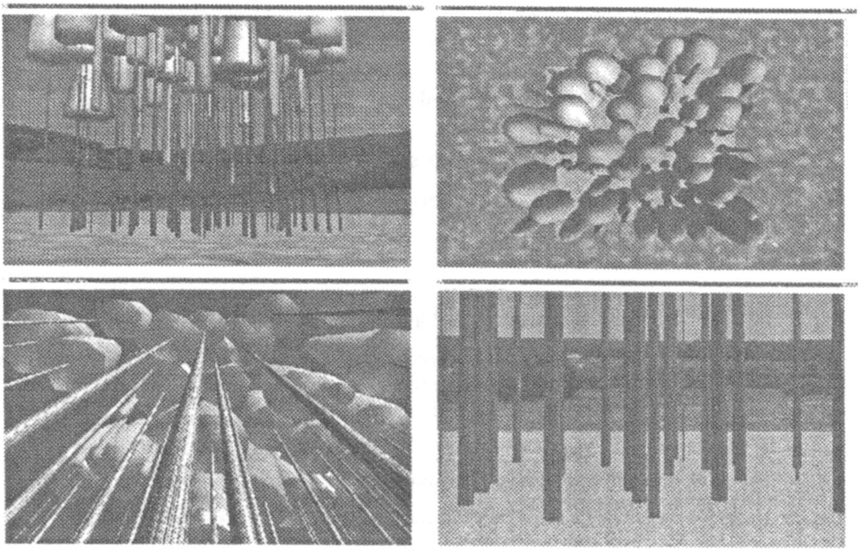

Figure 3: Typical interface for a growing spruce stand in TRAGIC++. The thinning and planting of trees is interactively controlled during the simulation. The range of potential growth behaviors from which the forester selects the appropriate ones can be "bred" from evolutionary runs of the model. For details see Hauhs et al. (1999), Lange et al. (1998).

Acknowledgements

We thank F. Lück for the growth data from the district Schulenberg and P.M. Steinsiek for the historic growth data from the Harz. Ch. Bartels, O. Froer, A. Kastner-Maresch and G. Müller-Christ are thanked for comments and help. This work was founded by BMBF Nr. PT BEO 0339476 A-C.

Literature

Agricola, G., 1556: De re metallica Libri, German translation reprinted 1978, VDI Verlag, Düsseldorf.

Atmanspacher, H., 1992: Wolfgang Pauli und die Alchemie Teil II: Das opus alchymicum. Zeitschrift für Parapsychologie und Grenzgebiete der Psychologie 34, 33-63.

Bartels, Ch., 1988: Das Erzbergwerk Rammelsberg. Preussag AG Metall (ed.), Goslar.

Bartels, Ch., 1992: Vom frühneuzeitlichen Montangewerbe zur Bergbauindustrie. Veröffentl. Deutsches Bergbau-Museum Bochum Vol. 54.

Bartels, Ch., 1997: Strukturwandel in Montanbetrieben des Mittelalters und der frühen Neuzeit in Abhängigkeit von Lagerstättenstrukturen und Technologie. In H-J. Gerhard (ed.): Struktur und Dimension Bd. I: 25-70. Franz-Steiner Verlag, Stuttgart.

Baskerville, G.L., 1995: The Forestry Problem: Adaptive Lurches of Renewal, in Gunderson, L.H., Holling, C.S., and Light, St. S. (eds.): Barriers and Bridges to the Renewal of Ecosystems and Institutions, Columbia University Press, New York, 37-102.

Bauer, H.: Die älteste Karte des nördlichen Harzes bei Goslar. Harz-Zeitschrift 33, 45-77.

Baumgarten, W., 1933: Beziehung zwischen Forstwirtschaft und Berg- und Hüttenwesen im Kommunionharz, Diss. Univ. of Munich.

Binswanger, H.C., 1994: Money and Magic: A Critique of the Modern Economy in the Light of Goethe`s Faust. Univ. Of Chicago Press.

Böhme, H., 1988: Natur und Subjekt. Suhrkamp, Frankfurt a.M.

Bornhardt, W., 1931: History of the mining at the Rammelsberg from its beginnings until modern times (in German), Archiv für Lagerstättenforschung 52, 1-366.

Casti, J.L., 1997: The Borderline. On the limits of scientific knowledge. Complexity 3(1), 5-7.

Chadwick, O.A., Derry, L.A., Vitousek, P.M., Huebert, B.J. and Hedin, L.O., 1999: Changing sources of nutrients during four million years of ecosystem development. Nature 397, 491-497.

Clements, F.E., 1936: Nature and structure of the climax. J. Ecology 24, 252-284.

Eliade, M., 1992: Schmiede und Alchemisten. Herder Spektrum, Freiburg. (German translation from: Forgerons et Alchemistes in 1956).

Fontana, W. and Ballati, S., 1999: Complexity – Why the sudden fuss? Complexity 4(3), 14-16.

Gatto, M., 1995: Sustainability: Is it a well defined concept? Ecological applications 5, 1181-1183.

Gell-Mann, M., 1994 The Quark and the Jaguar. W.H. Freeman, New York.

Gould, St.J., 1989 Wonderful Life. W.W. Norton, New York.

Gunderson, L.H., Holling, C.S. and Light, S.S. (eds.), 1995: Barriers and Bridges to the Renewal of Ecosystems and Institutions. Columbia Univ. Press, New York.

Hauhs, M., 1989: Lange Bramke: An Ecosystem Study of a Forested Catchment. In D.C. Adriano and M. Havas (eds.), Acidic Precipitation, Vol. I: 275-305. Springer Verlag, New York.

Hauhs, M. and Lange, H., 1996: Das Problem der Prozeßidentifikation in Waldökosystemen am Beispiel Wassertransport. IHI-Schriften Zittau 2, 212-222.

Hauhs, M., Dörwald, W., Kastner-Maresch, A. and Lange, H. , 1999: The Role of Visualization in Forest Growth Modelling. In: Amaro, A. und Tomé, M. (eds.), Empirical and process-based models for forest tree and stand growth simulation (EDIÇAOES SALAMANDRE, Lissabon), 403-418.

Henschke, E. , 1974: Landesherrschaft und Bergbauwirtschaft, Schriften zur Wirtschafts- und Sozialgeschichte Bd. 23, Duncker und Humbolt, Berlin.

Hillebrecht, M.L. , 1982: Die Relikte der Holzkohlewirtschaft als Indikatoren für Waldnutzung und Waldentwicklung, Göttinger Geographische Abhandlungen, Vol. 79. Erich Goltze Verlag, Göttingen.

Horgan, J. , 1995: The End of Science. Broadway Books, New York.

Huntley, B. and Webb III, T. , 1988: Vegetation history. Kluwer, Dordrecht.

Ingold, T. , 1986: Evolution and Social Life. Cambridge Univ. Press, Cambridge.

Ingold, T. , 1994: Companion Encyclopedia of Anthropology. Routledge, London.

Jamieson, D. , 1998: Sustainability and beyond. Ecological Economics 24, 183-192.

Kampis, G. , 1991: Self-modifying systems in biology and cognitive science. Pergamon Press, Oxford.

Klappauf, L. , 1996: Montanarchäologie im Harz. In: Jockenhövel, A. (ed.), Vierteljahreszeitschrift für Sozial- und Wirtschaftsgeschichte, Beiheft 121, 93-111.

Lange, H., Thies, B., Kastner-Maresch, A., Dörwald, W., Kim, J.T. and Hauhs, M. , 1998: Investigating forest growth model results on evolutionary time scales. In: Adami, C., Belew, R.K., Kitano, H. and Taylor, C.E. (eds.), Artificial Life VI , pp. 418-422. MIT Press.

Lange, H. , 1999: Are Ecosystems Dynamical Systems? *International Journal of Computing Anticipatory Systems* 3, 169-186.

Leibundgut, C. , 1978: Die Waldpflege. Haupt Verlag, Bern.

Ließmann, W. , 1997: Historical mining in the Harz (In German). Springer, Berlin.

Liski, J., Ilvesniemi, H., Mäkelä, A., Starr, M. , 1998: Model analysis of the efffects of soil age, fires, and harvesting on the carbon storage of boreal forest soils. European Journal of Soil Science 49, 407-416.

Mohr, K. , 1993: Geologie und Minerallagerstätten des Harzes. Schweizerbart'sche Verlagsbuchhandlung Stuttgart.

Morowitz, H. , 1998: Evolutionary Complexity. Complexity 3(6), 12-14.

North, D.C. , 1973: The rise of the Western world. A new economic history. Cambridge University Press, Cambridge.

Odum, E.P. , 1953: Fundamentals of ecology. Philadelphia.

Oesten, G. , 1995: Zur forstökonomischen Diskussion über das Leitbild einer nachhaltigen Waldwirtschaft. Forst und Holzwirt 6, 171-175.

Pieper, W. , 1955: Ulrich Rülein von Calw und sein Bergbüchlein. Freiburger Forschungshefte D7, Akademie Verlag, Berlin.

Pretzsch, H. , 1999: Waldwachstum im Wandel – Konsequenzen für Forstwissenschaften und Forstwirtschaft. Forstwissenschaftliches Centralblatt (in print).

Rapp, P.E., Schmah, T.I., and Mees, A.I. , 1999: Models of knowing and the investigation of dynamical systems. Physica D 132, 133-149.

Riehl, G. , 1968: Die Forstwirtschaft im Oberharzer Bergbaugebiet von der Mitte des 17. bis zum Ausgang des 19. Jahrhunderts. Aus dem Walde, Vol. 15. Lower Saxony forest administration, Hannover.

Rubner, H. , 1967: Forstgeschichte im Zeitalter der industriellen Revolution. Schriften zur Wirtschafts- und Sozialgeschichte Bd. 8. Duncker und Humbolt, Berlin.

Ruhnau, E , 1997: The Deconstruction of Time and the Emergence of Temporality. In: Atmanspacher, H. and Ruhnau, E. (eds.), Time, Temporality, Now: Experiencing Time and Concepts of Time in an Interdisciplinary Perspective. Springer, Berlin, 53-70.

Schanz, H., 1994: Forstliche Nachhaltigkeit aus der Sicht von Forstleuten in der Bundesrepublik Deutschland. Arbeitspapier 1994, Institut für Forsteinrichtung und Forstliche Betriebslehre, Univ. Freiburg.

Schellnhuber, H.J. and Wenzel, V. (eds.) , 1998: Earth System Analysis. Integrating Science for Sustainability. Springer-Verlag, Berlin.

Schellnhuber, H.J. , 1999: Earth system analysis and the second Copernican revolution. Suppl. Nature 402 (6761), C19-C23.

Schmidt, M. , 1989: Die Wasserwirtschaft des Oberharzer Bergbaues. Schriftenreihe der Frontinus Gesellschaft, Bd. 13, Bergisch–Gladbach.

Sigaut, F. , 1994: Technology. In: Ingold, T. (ed.), Companion Encyclopedia of Anthropology. Routledge, London, 420-459.

Schubart, W. , 1978: Die Verbreitung der Fichte im und am Harz vom hohen Mittelalter bis in die Neuzeit. Aus dem Walde, Vol. 28. Lower Saxony forest administration, Hannover.

Schübeler, D. , 1997: Untersuchungen zur standortabhängigen Wachstumsmodellierung bei der Fichte. Diss., Institute of Forest Management and Growth research, Georg-August-Univ. of Göttingen.

Schwappach, 1912: Ertragstafeln der wichtigeren Holzarten. J. Neumann Verlag, Neudamm.

Steinsiek, P.M., 1999: Zur Geschichte staatlicher Maßnahmen des Ressourcenschutzes am Beispiel der nutzungsbedingten Veränderungen von Waldökosystemen des Westharzes zwischen 1550 und 1810. Diss., Univ. Göttingen.

Wardle, D.A., Zackrisson, O., Hörnberg, G. and Gallet, C. , 1997: The Influence of Island Area on Ecosystem Properties. Science 277, 1296-1299.

White, L. Jr , 1940: Technology and invention in the Middle Ages, Speculum 15, 141-59.

White, M. , 1997: Isaac Newton the last sorcerer. Fourth Estate, London

Sustained water yield in afforested catchments - the South African experience

P.J. Dye and J.M. Bosch[1]

Abstract

South Africa is a relatively dry country, and is particularly dependent on limited areas of high-rainfall mountain catchments as reliable sources of streams and rivers. The natural vegetation within these catchments is mostly seasonally-dormant grassland or fynbos (sclerophyllous shrubland), but has largely been replaced by evergreen forest plantations during the course of this century. This change of land use has brought about declines in streamflow over large parts of the country, to the detriment of downstream users. Sustaining water yields from these catchments is a development imperative, in light of increasing demand for both water and forest products. This paper traces over six decades of research information, and the parallel development of legislation and policies, which has brought South Africa within sight of an effective integrated water resource management system. This experience may offer useful insights to the problem of sustaining water yields from afforested catchments in other water-scarce countries.

Keywords: *afforested catchments, South Africa, catchment yields, streamflow, forestry impacts*

---

[1] Environmentek, CSIR, C/O Agrometeorology. Faculty of Science and Agriculture, UNP, Private Bag X01, 3209 Scottsville, South Africa

K. von Gadow et al. (eds.), Sustainable Forest Management, 99–120.

Introduction

South Africa is a relatively dry country with a mean annual rainfall of only 475 mm (Department of Water Affairs, 1986). Limited areas receiving higher rainfall occur mainly as narrow bands along the seaward slopes of mountains and escarpments, as well as certain coastal belts along the eastern seaboard. The natural vegetation in most of these high rainfall areas comprises low, seasonally-dormant grassland or *fynbos* (sclerophyllous shrubland). The ratio of runoff to rainfall in virgin catchments is relatively high. Versfeld (1994) reported a mean ratio of 0.36 (with a range of 0.15 to 0.54) for nine research catchments situated in such areas.

Natural forests are of very limited extent in South Africa, and therefore the country relies heavily on plantations of pines, eucalypts and wattle for its supply of wood. Demand for wood products has grown consistently since the first forest plantation was established in 1876, and plantation forestry now covers nearly 1.5 million hectares of land. These plantations are confined to high rainfall areas. Thus, 87% receive more than 800 mm / year, while 29% receive more than 1000 mm (Le Maitre *et al.*, 1997). Scott *et al.* (1999) showed that catchments in which some degree of afforestation has occurred comprise only 14% of the country, yet produce 53% of the mean annual streamflow and 70% of the mean annual low flows of South African rivers. Plantation forests are characterised by tall, dense, evergreen canopies, and deep root systems, and therefore contrast strongly with the typically short, seasonally-dormant pre-afforestation vegetation with shallow root systems. There is thus high potential for negative impacts by forest plantations on catchment water yields.

Complaints about decreased runoff from afforested catchments began to be voiced by 1915 (Malherbe, 1968), and in subsequent years have been particularly strident during times of drought. A steady rise in demand for water by all sectors of the economy has been fuelled by population increase, rising living standards, and industrial and agricultural development. Despite the importance of the forestry industry to the national economy, plantation forests continue to attract severe criticism for alleged excessive use of the precious water resources of the country. Sustaining the water yield of afforested catchments is a vital necessity for the country. Can this be achieved in a way to minimize checks on the growth of the equally vital impacts, and the introduction of systems to manage forests for sustainable water yield. Although these systems contributed greatly to the conservation of water in many catchments, they had severe shortcomings. One of the most important of these was the lack of an integrated water resource management approach. Water management in forestry was treated in a piecemeal fashion, without regard to economic, social, and alternative water uses.

An integrated water resource management (IWRM) system now seems, for the first time, within reach. In large measure this is due to political developments which led to a new constitution in 1996, and to new water laws and policies for environmental management. New legislation now recognises

- the need for legislative instruments to integrate water management
- the principle of equity
- the need for participation and negotiation by all affected parties in decision making
- the principle of sustainability of natural resources
- the need for institutional capacity to implement IWRM

In this paper we trace over six decades of research into the hydrological impacts of plantation forests in South Africa, as well as the evolution of policies and legislation designed to regulate these impacts and ultimately introduce an equitable system for integrated water resource management. We believe that the South African experience provides some valuable pointers to the necessary requirements for a system of sustainable management of water resources in the forestry regions of other semi-arid countries.

Historical overview of research

Research catchment studies

Early criticisms of the effects of forest plantations on water supply led to a decision by Government to initiate a research programme to address this question. A number of long-term paired catchment experiments (Hewlett and Pienaar, 1973) were set up at Jonkershoek (close to Stellenbosch in the Western Cape Province, in 1935) and at Cathedral Peak (1945) in the mountains along the western boundary of KwaZulu-Natal. Subsequently, additional research catchments were set up further north at three locations (Mokobulaan, Witklip and Westfalia) along the Drakensberg escarpment in Mpumalanga and Northern Province.

Long term impacts of afforestation on streamflow

Streamflow and precipitation data from seven such catchment studies (Bosch and Gadow 1990 ; Smith and Scott 1992 ; Versfeld 1994) have been used to illustrate the post-afforestation trends in evapotranspiration (ET). A brief description of these catchments is provided in Table 1, while the trends are illustrated graphically in Figure 1. ET was calculated for each year as the difference between annual precipitation and annual streamflow, assuming that catchment leakage and year-to-year changes in soil storage were insignificant.

In summary, the impacts of forest plantations on annual catchment water yields were found to be high, reducing annual streamflow by 200-500 mm. This is in accord with more general relationships between deforestation and streamflow enhancement which have been described by Hibbert (1967) and Bosch and Hewlett (1982).

The trends in figure 1 all show marked increases in ET as the trees mature through the rotation. Most of this increase occurs in the earlier years as the tree canopies close. There are obvious species differences, with eucalypts showing an earlier rise in ET than in the pines. This is attributed to the faster development of eucalypt canopies.

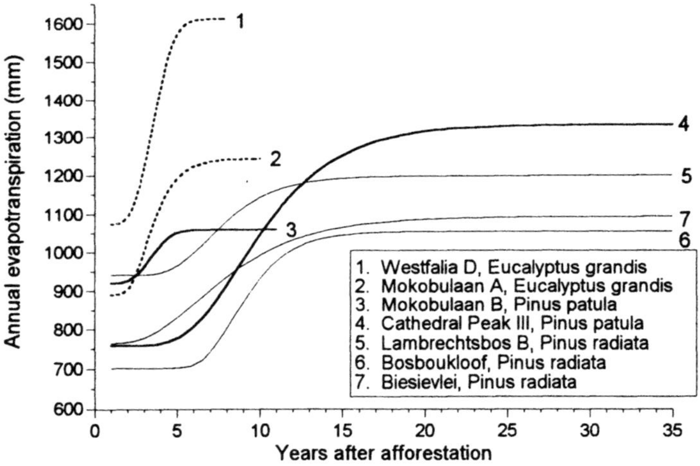

Figure 1. Trends in estimated annual evapotranspiration calculated from streamflow data recorded in seven research catchments

Catchment	Westfalia D	Mokobulaan A	Mokobulaan B	Lambrechtsbos B	Cathedral Peak III	Bosboukloof	Biesievlei
Province	Northern	Mpumalanga	Mpumalanga	Western Cape	kwaZulu - Natal	Western Cape	Western Cape
Area (ha)	40	26	35	65	142	210	27
mean altitude (m)	1225	1354	1396	660	2019	543	396
MAP (mm)	1611	1135	1135	1473	1587	1296	1427
Virgin MAR (mm)	548	244	217	531	760	593	663
Runoff:rain ratio	0.34	0.21	0.19	0.36	0.48	0.46	0.46
Pre-afforestation vegetation	Scrub forest	Grassland	Grassland	Fynbos shrubland	Grassland	Fynbos shrubland	Fynbos shrubland
Plantation species	*Eucalyptus grandis*	*Eucalyptus grandis*	*Pinus patula*	*Pinus radiata*	*Pinus patula*	*Pinus radiata*	*Pinus radiata*
Afforestation extent (%)	83	100	100	84	86	57	98
References	Bosch & Smith, 1989; Smith & Bosch, 1989	Van Lill *et al*, 1980	Van Lill *et al*, 1980	Van Wyk, 1987	Nänni, 1956; Bosch, 1979	Van Wyk, 1987	Van Wyk, 1987

Table 1. A description of experimental catchments used to illustrate trends in annual evapotranspiration recorded at a selection of afforested catchments (MAP=mean annual precipitation; MAR=mean annual runoff).

The estimated peak ET is clearly limited by the annual rainfall occurring within the catchment. Highest ET occurs in the Westfalia catchment, where high rainfall and deep soils ensure minimal constraint on ET rates. Lags in this limitation to ET have been shown to occur on sites with deep, permeable soils, where large soil storage capacity may introduce several years of delay in the response of streamflow to altered ET caused by land-use changes (Scott and Lesch, 1997).

The impacts of riparian trees on streamflow

The effect on streamflow of clearing woody riparian vegetation has been investigated in three research catchments in South Africa (Scott and Lesch, 1995). Table 2 summarizes streamflow increases recorded in the first year following treatments. Supportive evidence for gains in streamflow following removal of both indigenous trees and exotic plantation trees species in riparian zones has been reported by Rycroft (1955), Banks (1961), Nänni (1972), Dye and Poulter (1995) and Prinsloo and Scott (1999).

Catchment	Area (ha)	MAP (mm)	Experimental treatment	First year increase in runoff	
				(m^3 ha^{-1})	(% per 10%)
Westfalia D	40	1611	Clearfelling indigenous riparian forest	5 445	9
			Clearfelling non-riparian indigenous forest	2 700	3.5
Witklip 2	136	996	Cut and poison riparian scrub	7 965	44
			Clearfelling non-riparian pines	4 045	37
Biesievlei	27	1427	Clearfelling riparian pines	11 505	44
			Clearfelling non-riparian pines	3 430	14

Table 2. A description of experimental treatments and first-year runoff responses recorded in three experimental catchments. Response is expressed in terms of volume increase per unit area of catchment treated, and as a percentage flow increase per 10% of the catchment cleared.

Early regulation

The Department of Forestry adopted a policy in 1932 of discontinuing afforestation within 20 m of all perennial streams and other surface water to protect water yields from forestry catchments (Görgens and Lee, 1992). This policy was entrenched in the 1969 Soil Conservation Act.

Two committees were appointed in 1966 to investigate afforestation and water supplies (Van der Zel, 1995). Both recommended restricting afforestation where water resources are already committed, and where development could be threatened through shortage of water. These recommendations led to an amendment of the Forestry Act (No 72 of 1968) whereby Government was empowered to restrict new afforestation where the national water resources required protection.

These policies were enforced through the Afforestation Permit System (APS) under the 1972 amendment (Act 46 of 1972) of the Forest Act of 1968 (Bosch and Gadow, 1990; Van der Zel, 1995). Major catchments in the country were classified into three categories of allowable reductions in streamflow based on downstream demand. In each category of catchment, afforestation was allowed to the point where the 1972 mean annual runoff (MAR) was reduced by a specific percentage. In Class 1 catchments, no further reduction was allowed. In class 2 catchments, a 5% reduction was allowed, and in class 3 catchments, a 10% reduction was allowed. A stipulation in planting permits was that no afforestation was permitted within 20 m of any perennial stream or water body.

While the APS served a valuable purpose by limiting afforestation in areas where catchment water resources were close to fully committed, several shortcomings were apparent which led to unhappiness in the forestry industry.

- Water demand within catchments could only be determined subjectively
- The method of calculating streamflow reductions was based on limited data involving one species (*Pinus patula*) from one locality (Cathedral Peak). The „Nänni curves" used for this purpose are illustrated in figure 2. Extrapolation of these results to other areas was therefore open to serious doubt.
- Afforestation effects could be severe on a local scale, yet substantially less at a larger scale of catchment.
- The APS restricted afforestation, while no other dryland crops were regulated in a similar

way.

- Restrictions on afforestation did not take account of the value of the industry to the economy and society, and the effectiveness of forest plantations in using water.
- The 20 m delimitation of riparian zones to be left unplanted, was seen as too arbitrary. With the revision of the Forest Act in 1984, the 20 metres was changed to a variable width dependent on ministerial discretion. Bosch *et al.* (1995) developed a model to determine the widths of riparian zones, based on estimates of water use differences of different riparian vegetation types, and on the downstream demand for the water.
- The fact that the impact of forests on low flows (which were deemed the important dry season component) was not considered.
- The lack of an integrated water resource management approach.

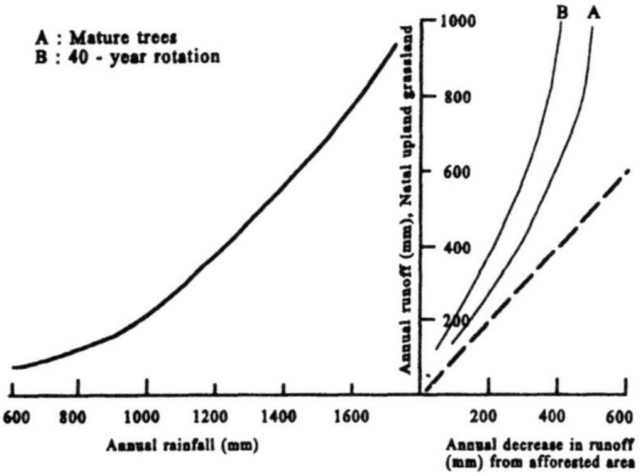

Figure 2. The relationship between mean annual runoff and the expected reduction in runoff as a result of afforestation of a grassland catchment (after Nänni, 1970).

Improved empirical models based on research

The shortcomings listed above led to renewed research efforts to improve both the available empirical models and our understanding of the hydrological processes that are affected by forestry. This research also resulted in a range of new models for determining the hydrological impacts of forestry. Further analyses of the catchment experiments mentioned in Table 1 resulted in a set of „flow reduction curves" that applied to „optimum" and „sub-optimum" sites, to low flows as well as total annual flows, and to eucalypt and pine plantations. The refined set of curves were adjusted according to the assumption that the riparian zones were not afforested. These „flow reduction curves (Scott *et al.*, 1998) are shown in figures 3 and 4, and reflect expected reductions in MAR relative to the pre-afforestation natural vegetation.

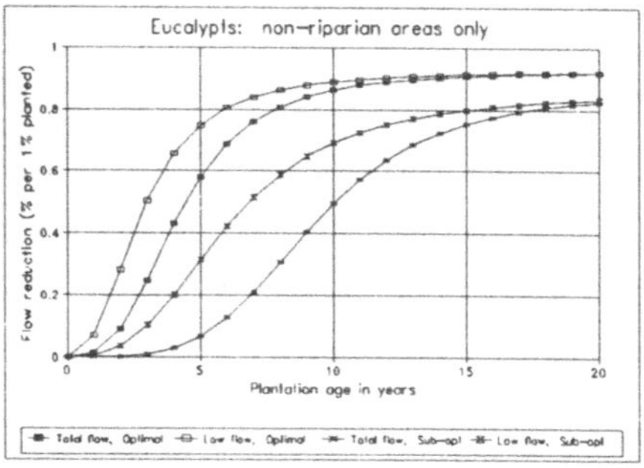

Figure 3. The flow reduction curves for eucalypts modified from Scott and Smith (1997) to take account of riparian zones being left unplanted (after Scott et al., 1998).

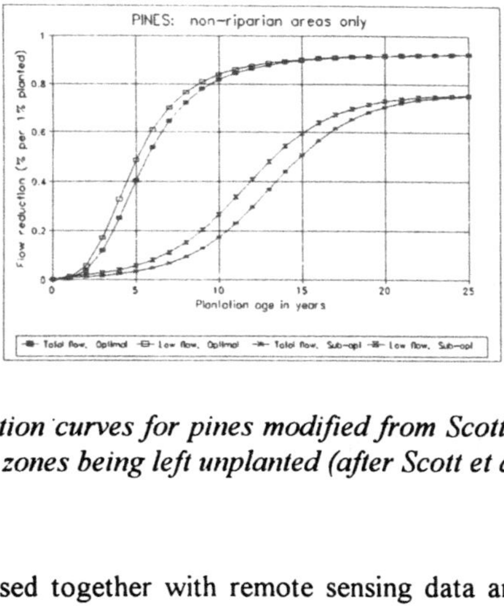

Figure 4. The flow reduction curves for pines modified from Scott and Smith (1997) to take account of riparian zones being left unplanted (after Scott et al., 1998).

Such trends have been used together with remote sensing data and spatial climatic data to estimate impacts of forest plantations over the entire national forestry estate (Scott *et al.*, 1998). This study has provided a most useful overview of forestry impacts, providing much needed perspective and helping to defuse some of the wilder claims of detrimental impacts by forest plantations. Thus, an annual average reduction in river flow of 99 mm per unit of planted area was calculated for the whole country. This is equivalent to a reduction in mean annual streamflow of 3.2%, and of low flows by 7.8%. At the provincial scale, greatest impacts were estimated in Mpumalanga, where the highest concentration of forestry (7.2% of land area) reduced mean annual streamflow by 10%, and low flows by 18%. At the secondary catchment scale, the greatest flow reduction was estimated in the Sabie/Sand catchment (12.5% afforested), where total runoff reduction equalled 18.9% and low flow reduction equalled 36.8%.

Hydrological process studies

By the 1980s, there was increasing interest in measuring hydrological processes in South African forest plantations. This was partly fuelled by frequent extravagant claims of excessive water use by forest trees, but also by the development and increasing availability of sophisticated electronic equipment and sensors. Such hardware permitted direct measurement of site energy balance and canopy microclimates (Bowen ratio, Eddy correlation), sap flow in trees (Heat pulse velocity), and hydrologically important physiological features of trees such as stomatal conductance (porometer), leaf area index (canopy analyzer) and xylem water potential (thermocouple psychrometers). Data loggers allowed continuous measurement in the field. Such studies led to significant advances in understanding of rainfall interception by tree canopies (Schulze, *et al.*, 1978; Versfeld, 1988; Dye and Versfeld, 1992a; 1992b) and litter layers (Dye and Mostert, 1994), soil water uptake patterns and rooting depths (Boden, 1992; Dye *et al.*, 1997a), stress development and transpiration rates in trees (Dye *et al.*, 1997a; Dye, *et al.*, 1997b; Olbrich, 1994), and evapotranspiration from multi-species communities and soil moisture dynamics (Everson, *et al.*, 1998). Contrasting differences in annual water use between eucalypt plantation and mountain grassland (Fig. 5 and 6) have been shown to result largely from differences in green leaf area dynamics, canopy height, rooting depths and duration of favourable growth conditions.

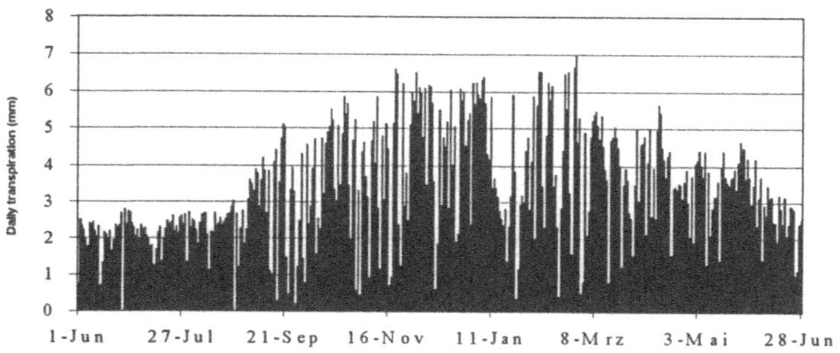

Figure 5. The pattern of daily transpiration recorded in a 3-year-old stand of Eucalyptus grandis in the Sabie district.

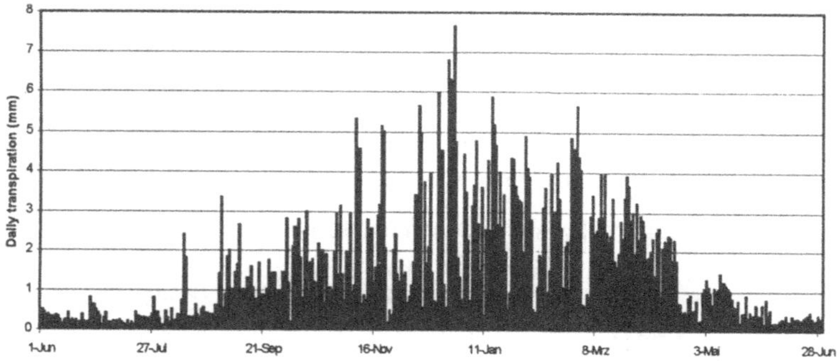

Figure 6. The pattern of daily evapotranspiration recorded above montane grassland in the Cathedral Peak district.

Process models of forest hydrological impacts

Empirical streamflow data from afforested research catchments are effective in illustrating long-term trends in afforestation effects, but their usefulness is limited by the following short-comings:

- The forestry regions encompass great site diversity. This diversity is likely to be poorly represented by the limited number of research catchments.
- Research catchments represent high rainfall forestry sites, yet 71 % of all forestry plantations occur in areas with a mean annual precipitation of less than 1000 mm (Le Maitre *et al.*, 1997).
- Forestry management practices are changing, and the original choice of genotype, site preparation, spacing and rotation length are often no longer representative.
- Trends mask the important year-to-year variation in rainfall and streamflow, which has major water supply implications.
- Forestry impacts are known primarily from small, first order watershed catchments, but increasingly need to be assessed within larger and more complicated catchments, where a mix of land use activities and water demand conditions apply.
- There is increasing need for a wider perspective on forest water use, where the impacts of forestry can be compared to the impacts of alternative crops and land use activities.

The hydrological model preferred by South African engineers working within very large catchments is the Pitman model, which has become a standard for water resource studies in South Africa (Görgens and Lee, 1992). The model was designed in the early 1970s as a robust method for solving water resource problems using ideas from the well-known Stanford Watershed Model. The Pitman model is strictly a calibration model in which reference to measured data is required. A later version of the Pitman model, called WRSM90 (Pitman and Kakebeeke, 1991) has four modules (runoff, reservoir, irrigation and channel). Attention is given, within the runoff module, to broad impacts of forests on water yields. However, no attempt is made to simulate forest hydrological processes. Görgens and Lee (1992) advise that this model is not suitable for hydrological simulations at small scales or within regions where there is heterogeneous topography, soil and land use activity.

Estimates of the impact of land-use changes on water yields are more commonly undertaken using the ACRU model (Schulze, 1989; Schulze, 1995). ACRU (Agricultural Catchments Research Unit) is a daily time step, soil water budgeting model which is designed to simulate physical hydrological processes. Forest hydrology has received special attention within the constraints of the ACRU model, and is believed to be better positioned than the Pitman model for simulating hydrological impacts of forests (Görgens and Lee, 1992). The wider agricultural focus permits a comprehensive range of crop comparisons to forestry. An additional advantage is the interface to a national hydrological database which can supply required climatic input data for any region of South Africa. Crop factors form the basis on which ACRU determines water use of plantations. The model employs several routines to moderate these factors, depending on prevailing local conditions that affect soil water availability, leaf area index and climate. Successful simulations have been performed on both large catchments (Tarboton and Schulze, 1991) and small catchments (Schulze and George,1987).

The Penman-Monteith equation has proved to be a useful model for predicting evapotranspiration from forest and grassland canopies (Dye, 1987; Everson *et al.*, 1998; Burger, 1999) in a variety of intensive studies where the necessary inputs are available. It has been used over a range of time scales, varying from the typical 20 minute or hourly time interval of intensive process studies, to monthly time intervals in the case of the 3-PG forest growth and water use model which runs over an entire rotation. This model is believed to be especially useful in promoting understanding of the year-to-year variation in water use and growth caused by rainfall variability (Dye, 1999).

Improved plantation management for sustainable water supply

In recent years, the forest industry has demonstrated a greatly increased concern over the impacts of forest plantations on the sustainability of catchment water yields, and possible ways in which these impacts can be reduced. Forest owners have shown an increased willingness to withdraw plantations from riparian zones, and to encourage the development of indigenous grasslands and forests in these zones, in the belief that these are more water-conservative than forest plantations. Forestry companies have been heavily involved in supporting the „Working for Water" scheme, where self-established, invasive exotic tree species are cleared by previously unemployed people. The industry has also embraced the concept of environmental auditing on forest plantations, and some of the standards set by ISO14000 and the Forest Stewardship Council relate to practices which influence catchment hydrology.

There are a number of other management options which could potentially help to sustain water yields from afforested catchments, but these are largely constrained by economic factors. Both pine and eucalypt plantations are grown over short (pulpwood) and long (saw-timber) rotations, according to the desired product. Wicht (1967) first hypothesized that long-rotation plantations have a lower long-term impact on catchment water yields than short rotation plantations. Recent research results generally supports this contention. In a study of an age-sequence of *Eucalyptus grandis*, Olbrich (1994) reported a decline in transpiration per unit leaf area with tree age. In a similar age sequence study of this species, Van Lochem (1986) recorded a decline in leaf area index with tree age. Measurements of sap flow recorded in a 3-year-old stand, and a thinned nine-year-old stand of *E. grandis* revealed a reduced stand water use by the older trees (Dye *et al.*, 1997a). The declines in leaf area index (LAI) and sap flow are attributed largely to thinnings performed in the stands selected in the two latter studies. Plantation management regime is currently determined by market forces, proximity of saw mills, and site quality, but could in future become increasingly influenced by water use considerations where water supply is critically short.

Whereas past tree breeding efforts were devoted largely to improving the yield and quality of timber, some recognition is now being given to the efficiency of water use and resistance to drought. These criteria are assuming more importance as forestry expansion increasingly takes place in the dry end of the potential range of forestry sites. In recent years, South African eucalypt growers have developed and planted *Eucalyptus* clones selected for high growth rates. Hybrids have been produced which combine the growth potential of *E. grandis* with the drought resistance of species such as *E. camaldulensis*, *E. tereticornis* and *E. urophylla* (Denison and Kietzka, 1993). These new clones exhibit considerable physiological variability, and results from comparative studies have revealed large differences in sap flow rates (Dye, *et al.*, 1997b). Fig. 7 illustrates the magnitude and consistency of differences in annual sap flow recorded in three eucalypt hybrid clones in Zululand (a region of the province of KwaZulu-Natal).

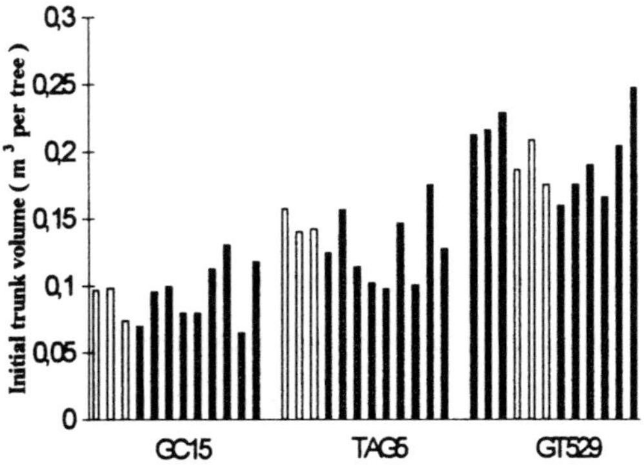

Figure 7. Total annual sap flow (l) recorded from sap flow measurements in 10 sample trees in each of three Eucalyptus clones in Zululand. GC15 is a grandis/camaldulensis hybrid clone, TAG5 is an E. grandis clone while GT529 is a grandis/tereticornis hybrid clone. Unfilled bars identify sample trees where sap flow was continuously monitored over an entire year. Filled bars identify trees where sap flow measurements were performed intermittently.

In contrast to eucalypt stands, evidence from South African catchments planted to pines suggests that thinning and pruning have little lasting effect in promoting streamflow. Thus, streamflow analyses by Van Wyk (1977) for the Biesievlei catchment in Western Cape, and for other catchments, could not demonstrate significant water yield increases. It is believed that growth in these catchments was so dynamic that the treatments had little more than a holding effect on the trend towards decreasing water yield (Versfeld and Bosch, 1982; Van der Zel 1970; Lesch and Scott, 1997).

Site preparation and weed control can significantly improve survival and early growth, providing easier access by tree roots to soil water. Such practices result in larger trees, more rapid depletion of soil water reserves, and drier profiles (Boden, 1992; Jewitt, 1992). No research catchment data are available in South Africa to demonstrate the importance of this effect on catchment yields. The relatively dry conditions in South Africa force foresters to optimize the local water regime to promote economic yields and crop survival during droughts. Thus, employing different site management practices is likely to be an impractical means of modifying catchment discharge patterns.

Chapman and Scott (1999) have drawn attention to the hydrological effect of age class distribution of plantations in quaternary catchments. A modelling study revealed that the hydrople wlogical impact could be lessened and spread more evenly over time by striving for equal area age classes, and avoiding the single age-class situation seen in some quaternary catchments.

Integrated water resource management

By the 1990's, researchers had developed new empirical and process models, and acquired a better understanding of hydrological processes. These developments were parallelled by increasing awareness of the concept of IWRM, and the need to consider all relevant economic, social, environmental, and sustainability issues.

Bosch and Gadow (1990) suggested an approach that brings together physical and non-physical considerations in water use management related to forestry. They proposed tools that were suited for a participative management process. Two such proposed tools were the Analytic Hierarchy Process (AHP) for combining quantitative and qualitative elements of the impacts of forestry on water yield, and linear programming for optimising the planting patterns for best return on water used. This process was also suited to considering forestry in the wider context of water demand by other users. However, the supportive structures (institutional arrangements) and decision processes at that stage were not flexible enough for implementing such ideas. Bosch and Gadow also put the emphasis on formal decision tools supportive of the developed sector of the society, which limited the applicability of their ideas in the new political dispensation in South Africa.

Recognising the fact that sustainable management of water in the forestry sector cannot be accomplished in isolation of other land-use options and without considering the trade-offs in terms of economics, water quality impacts, environmental impacts and social impacts, the CSIR (Forestek) developed a Catchment Resource Assessment Model (CRAM) (Chapman *et el.*, 1995). Using water as a „currency“, CRAM simulates the impacts of land uses on hydrological, economic, environmental and social conditions in a catchment and allows for an objective evaluation of alternative land uses. CRAM was designed to answer to the requirements of objectivity, fairness, and the balancing of social, environmental and economic factors in the larger catchment. Although CRAM as a model was never adopted by decision makers, the CRAM concepts are now embedded in the new approach to water management in forestry and form the basis for water allocation.

The original APS remained in use through the 1990's, albeit with certain reforms, such as the introduction of afforestation advisory committees (to reflect the interests of non-forestry sectors), the introduction of environmental impact assessments of afforestation (under certain conditions), and the introduction of Strategic Environmental Assessments (SEAs) as the strategic framework within which afforestation could be considered. These were significant improvements, but there was nevertheless a slow acceptance of more sophisticated decision tools for managing the forestry water-use issue, and specifically for assessing the hydrological impacts of forest management. We suggest three major reasons:

- the lack of clear strategic direction and the overarching statutory support
- the „inadequacy" of the existing hydrological models in dealing accurately with the components of the water cycle which have most relevance to water demand management.
- Lack of coordination among Government departments.

The new statutory framework

The new constitution of 1996, which accompanied the establishment of the first democratic Government, paved the way for a totally new approach, new legislation and new policies for environmental management. This fresh start has been highly beneficial in permitting a complete reappraisal of water management, and allowing the incorporation of fresh concepts. A major change of emphasis brought about by the new National Water Act (NWA) was that sustainable management of forestry is now considered within the wider context of integrated water resource management (IWRM). The following are the main principles underpinning the new philosophy of IWRM in South Africa.

- The NWA has been designed and written specifically to support the national water policy, and forms the main instrument towards achieving the objectives of the policy, i.e. to „...ensure that water is allocated equitably and used beneficially in the public interest, while promoting environmental values" (Section 3(2) of the NWA).
- The NWA is based on the principle that water allocation should take more cognisance of the entire water cycle.
- It is recognized that water resources can only be effectively and efficiently managed within

a larger basin or catchment, to balance the interdependent roles of resource protection and resource utilization.

- A catchment is a dynamic and integrated social, economic and physical system, containing people who are involved in many types of activities. There is a generally accepted change from the formal physical modelling approach to management of water resources, to a more people oriented management approach. Values such as transparency and equity are a high priority in the decision process.
- The NWA, and others laws concentrate on enabling the people to participate in the decision process. This is a shift away from authoritarian decision making by a government body, towards one of participation and contribution to the decision process by interested and affected parties. The basic premise is that the water belongs to the people, but that the Department of Water Affairs and Forestry, as the custodian, will facilitate a process of water resource management by the people.

Forestry water resource management is now dealt with in this framework. The NWA has maintained the emphasis on regulation of afforestation at the catchment level, but has put forestry water use on an equal footing with other dryland activities that may use significant amounts of water before it enters the streams. Such activities are termed Stream Flow Reduction Activities (SFRAs). The Act defines a SFRA as „...any activity (including the cultivation of any particular crop or other vegetation) to be a stream flow reduction activity if that activity is likely to reduce the availability of water in a water course to the (ecological) Reserve, to meet international obligations, or to other water users significantly". Presently only forestry is listed in the Act as an SFRA, but studies are afoot to identify other potential SFRAs.

Much debate is currently taking place to plan how best to implement the new Act and major prerequisites have been summarized by Kruger and Bosch (1999):

- Catchment management strategies need to be established for each water management area, and these in turn depend on the establishment of the National Water Resource Strategy and the Classification System for Water Resources; the management strategies in turn provide for the specifics of the strategies and plans for the protection, use, development, conservation, management and control, of water resources.
- The institutions to implement the strategies must be established and properly empowered;

Catchment Management Agencies (CMAs) are the key institutions for this purpose. A Catchment management agency is a self regulatory body corporate with a Governing Board and an administrative structure that has the statutory responsibility, power and financial autonomy to perform a range of catchment management functions in a declared catchment area. Other statutory and non-statutory supporting institutional structures are described and discussed by Görgens *et al.* (1998). A number of these institutions have already been established.

- The Catchment Management Agencies will need, among other things to see to water allocation and licensing and this in turn requires that they should have available and apply appropriate procedures, supported by relevant information, criteria relevant to each of the considerations of the NWA, and methods for evaluating applications for licences; they will also need to raise water charges to finance their operations, which will require an appropriate pricing policy, and a consistent basis for charging each user of water. CMAs will operate on the basis of negotiation. They will require a suite of tools, methodologies and procedures suited for that purpose and for unique conditions of geography and scale.
- The Catchment Management Agencies will need to see to the protection of the water resources in their water management areas, which requires appropriate resource quality objectives from the national classification system, and ways for Agencies to assess and control the effects of uses on these objectives.
- Finally, the Agencies must be able to monitor and control the uses within their water management areas.

Recent changes in research focus

Implementation of the bold new vision requires a broader information base as well as improved decision support tools. Researchers are responding to these needs, and much progress is being made in providing the necessary information required by Government and CMAs. Noticeable trends have been the following:

- comparisons of water use and water use efficiency among different land-use options, permitting plantation forests to be compared to alternative crops.
- Improvements in models to support IWRM and improve predictions of water use and yields of forests

- greater networking by researchers across disciplines and interest groups, and a much wider awareness of the full range of stake-holder views through multi-disciplinary workshops

The current framework for IWRM is believed to ensure effective water-focussed catchment management. It nevertheless falls short of the internationally accepted concept of integrated catchment management, which is aimed at a sustainable balance between utilization and protection of all environmental resources in a catchment. For water-focussed catchment management to become more integrated and effective, other national and provincial government departments, local authorities and water-related stakeholders need to be brought into the catchment management process. It is envisaged that this will be a gradual process. In the mean time, prospects for managing catchments for sustainable water yield in an integrated and equitable manner have significantly improved.

Acknowledgements

We gratefully acknowledge Mr. Dirk Versfeld and anonymous reviewers for helpful comments on the manuscript.

Literature

Banks, C.H., 1961: The hydrological effects of riparian and adjoining vegetation. *Forestry in South Africa* 1: 31-46.

Boden, D.I., 1992: The relationship between soil water status and the growth of *Eucalyptus grandis.* ICFR Bulletin Series, No. 25.92, ICFR, Pietermaritzburg, South Africa.

Bosch, J.M., 1979: Treatment effects on annual and dry period streamflow at Cathedral Peak. *South African Forestry Journal* No. 108: 29-38.

Bosch, J.M. and Hewlett, J.D., 1982: A review of catchment experiments to determine the effect of vegetation changes on water yield and evapotranspiration. *Journal of Hydrology* 55: 3-23.

Bosch, J.M. and Smith, R.E., 1989: The effect of afforestation of indigenous scrub forest with *Eucalyptus* on streamflow from a small catchment in the Transvaal, South Africa. *South African Forestry Journal* No. 150: 7-17.

Bosch, J.M. and Gadow, K., v., 1990: Regulating afforestation for water conservation in South Africa. *South African Forestry Journal* 153: 41-54.

Bosch, J.M., Berliner, D, and le Maitre, D.C., 1995: Guidelines for the delineation and management of riparian zones. Unpublished contract report, CSIR Division of Water, Environment and Forestry Technology, Stellenbosch.

Burger, C., 1999: Comparative evaporation measurements above commercial forestry and sugarcane canopies in the KwaZulu-Natal midlands. PhD Thesis, Faculty of Science and Agriculture, University of Natal, Pietermaritzburg, South Africa.

Chapman, R.A. , Manders, P.T., Scholes, R.J. and Bosch, J. M., 1995: Who should get the water ? Decision support for water resource management. *Water Science Technology*. Vol. 32, no. 5-6, pp 37-43.

Chapman, R.A. and Scott, D.F., 1999: The implications of using average plantation age when assessing the hydrological impacts of afforestation. (In preparation).

Denison, N.P. and Kietzka, J.E., 1993: The development and utilization of vegetative propagation in Mondi for commercial afforestation programmes. *South African Forestry Journal*, No. 166: 53-60.

Department of Water Affairs, 1986: Management of the Water Resources of the Republic of South Africa. CTP Book Printers, Cape Town.

Dye, P.J., 1987: Estimating water use by *Eucalyptus grandis* with the Penman-Monteith equation. In: Forest Hydrology and Watershed Management. Proceedings of the Vancouver Symposium, August 1987, IAHS Publication no. 167:329-337.

Dye, P.J. and Versfeld, D.B., 1992a: Rainfall interception by *Eucalyptus grandis* canopies: a proposed modelling procedure for use in catchment water balance simulations. CSIR report FOR-DEA 369.

Dye, P.J. and Versfeld, D.B., 1992b: Rainfall interception by a ten-year-old *Pinus patula* plantation. CSIR report FOR-DEA 424.

Dye, P.J. and Mostert, G., 1994: Development of a rainfall interception model for forest litter under stands of *Eucalyptus grandis, Pinus patula* and *Pinus taeda*. CSIR report no. FOR-DEA 757.

Dye, P.J. and Poulter, A.G., 1995: A field demonstration of the effect on streamflow of clearing invasive pine and wattle trees from a riparian zone. South African Forestry Journal No 173: 27-30.

Dye, P.J., Poulter, A.G., Soko, S. and Maphanga, D., 1997a: The determination of the relationship between transpiration rate and declining available water for *Eucalyptus grandis*. Water Research Commission Report No. 441/1/97. 103 pp.

Dye, P.J., Soko, S. and Maphanga, D., 1997b: Inter-annual variation in water use efficiency of three clones in kwaMbonambi, Zululand. CSIR Report ENV/P/C 97048, 35 pp.

Dye, P.J., 1999: A preliminary verification of the 3-PG forest growth and water use model for *Pinus patula* plantations. Milestone report, Division of Water, Environment and Forestry Technology, CSIR, Pretoria, South Africa. 29 pp.

Everson, C.S., Molefe, G.L. and Everson, T.M., 1998: Monitoring and modelling components of the water balance in a grassland catchment in the summer rainfall area of South Africa. Water Research Commission Report No. 493/1/98. 138 pp.

Görgens, A. and Lee, J., 1992: Hydrological Impacts of Forestry: A Research Planning Study. Report to the Water Research Commission, KV 37/92.

Görgens, A, Pegram, G, Uys, M, Grobicki, A, Loots, L, Tanner, A and Bengu, R., 1998: Guidelines for catchment management to achieve integrated water resources management in South Africa. WRC Report No. KV 108/98, Pretoria, South Africa.

Hewlett, J.D. and Pienaar, L., 1973: Design and analysis of the catchment experiment. In: White, E.H. (Ed.) Proceedings of the Symposium on use of small watersheds in determining effects of forest land use on water quality. University of Kentucky, Lexington, Kentucky. Pp 88-106.

Hibbert, A.R., 1967: Forest treatment effects on water yield. In: Sopper, W.E. and Lull, H.W. (eds.) International Symposium on Forest Hydrology, Pergamon Press, Oxford.

Jewitt, G.P.W., 1992: Process studies for simulation modelling of forest hydrological impacts. Unpublished MSc dissertation, University of Natal, Pietermaritzburg, Department of Agricultural Engineering. Pp 145.

Kruger, F.J and Bosch, J.M., 1999: The development of a research programme to support the implementation of legislation and policy related to stream flow reduction activities. Contract report env-p-c 98159.Water Research Commission Report ISBN No. 1 86845 483 5, 41pp.

Le Maitre, D.C., Scott, D.F. and Fairbanks, D.H.K., 1997: Impacts of timber plantations on runoff in South Africa: a handy reference manual. Unpublished contract report, ENV/S-C 96068, CSIR Division of Water, Environment and Forestry Technology, Stellenbosch and Pretoria, South Africa.

Lesch, W. and Scott, D.F., 1997: The response in water yield to the thinning of *Pinus radiata, Pinus patula* and *Eucalyptus grandis* plantations. *Forest Ecology and Management* 99: 295-307.

Malherbe, H.L.W., 1968: Afforestation and Water Supplies in South Africa. Report of an Interdepartmental Committee, Department of Forestry, Pretoria, South Africa.

Nänni, U.W., 1956: Forest hydrological research at the Cathedral Peak Research Station. *Journal of South African Forestry Association* 27: 2-35.

Nänni, U.W., 1970: Trees, water and perspective. *South African Forestry Journal* 75: 9-17.

Nänni, U.W., 1972: Water use by riparian vegetation at Cathedral Peak. *South African Forestry Journal* 80: 1-10.

Olbrich, B.W., 1994: The application of the heat pulse velocity technique to the study of transpiration from *Eucalyptus grandis*. PhD thesis, University of the Witwatersrand, Johannesburg, South Africa.

Pitman, W.V. and Kakebeeke, J.P., 1991: The Pitman model - into the 1990s. Proceedings of the fifth South African national Hydrological Symposium, Stellenbosch. South African Committee for the International Association of Hydrological Sciences.

Prinsloo, F.W. and Scott, D.F., 1999: Streamflow responses to the clearing of alien invasive trees from riparian zones at three sites in the Western Cape province. Southern African Forestry Journal No 185: 1-7.

Rycroft, H.B., 1955: The effect of riparian vegetation on water loss from an irrigation furrow at Jonkershoek. *Journal of South African Forestry Association* 26: 80-88.

Schulze, R.E., Scott-Shaw, C.R. and Nänni, U.W., 1978: Interception by *Pinus patula* in relation to rainfall parameters. *Journal of Hydrology* 36: 393-396.

Schulze, R.E. and George, W.J., 1987: A dynamic, process-based, user-orientated model of forest effects on water yield. *Hydrological processes* 1: 293-307.

Schulze, R.E., 1989: ACRU: background, concepts and theory. WRC Report No. 154/1/89. Water Research Commission, Pretoria, South Africa.

Schulze, R.E., 1995: Hydrology and Agrohydrology. A text to accompany the ACRU 3.00 agrohydrological modelling system. WRC reportTT69/95.

Scott, D.F. and Lesch, W., 1995: The water yield gains obtained from clearfelling riparian zone vegetation. Proceedings of the Seventh South African National Hydrological Symposium, Grahamstown, South Africa. 4-6 September 1995.

Scott, D.F. and Lesch, W., 1997: Streamflow responses to afforestation with *Eucalyptus grandis* and *Pinus patula* and to felling in the Mokobulaan experimental catchments, Mpumalanga Province, South Africa. *Journal of Hydrology* 199: 360-377.

Scott, D.F. and Smith, R.E., 1997: Preliminary empirical models to predict reductions in total and low flows resulting from afforestation. *Water SA*, 23: 135-140.

Scott, D.F., Le Maitre, D.C. and Fairbanks, D.H.K., 1998: Forestry and streamflow reductions in South Africa: a reference system for assessing extent and distribution. *Water SA* 24: 187-199.

Scott, D.F., Le Maitre, D.C. and van Wyk, E., 1999: The effects of forestry on low flows: further studies on low flow in wet and dry cycles and in relation to catchments of increasing size. Contract report No. ENV/S-C 99059, CSIR, Pretoria, South Africa.

Smith, R.E. and Bosch J.M., 1989: A Description of the Westfalia Catchment Experiment to Determine the Effect on Water Yield of Clearing the Riparian Zone and Converting an Indigenous Forest to a Eucalypt Plantation. *South African Forestry Journal* 151:26-31.

Smith, R.E. and Scott, D.F., 1992: The effect of afforestation on low flows in various regions of South Africa. *Water SA* 18: 185-194.

Tarboton, K.C. and Schulze, R.E., 1991: The ACRU modelling system for large catchment water resource management. In: Van de Ven, F.H.M., Gutknecht, D., Loucks, D.P. and Salewicz, K.A. (eds.) Hydrology for the water management of large river basins. IAHS Publication No. 201: 219-232.

Van der Zel, D.W., 1970: The effect of thinning on flow in a Jonkershoek stream. *Forestry in South Africa* 11: 14-43.

Van der Zel, D. W., 1995: Accomplishments and dynamics of the South African afforestation permit system. *South African Forestry Journal* No. 172: 49-58.

Van Lill, W.S., Kruger, F.J. and Van Wyk, D.B., 1980: The effect of afforestation with *Eucalyptus grandis* Hill ex Maiden and *Pinus patula* Schlecht. Et Cham. on streamflow from experimental catchments at Mokobulaan, Transvaal. *Journal of Hydrology* 48: 107-118.

Van Lochem, M.D., 1986: The leaf area index of *Eucalyptus grandis* at different stages of the production cycle. Research Report, Saasveld School of Forestry, George, South Africa.

Van Wyk, D.B., 1977: Die invloed van bebossing met *Pinus radiata* op die totale jaarlikse afvoer van die Jonkershoek strome. Unpublished M.Sc. Thesis, University of Stellenbosch.

Van Wyk, D.B., 1987: Some effects of afforestation on streamflow in the Western Cape Province, South Africa. *Water SA* 13: 31-36.

Versfeld, D.B. and Bosch, J.M., 1982: The options for catchments - how to make the right decisions: another view. Foris. Faculty of Forestry, University of Stellenbosch. Communication No. 98: 427-446.

Versfeld, D.B., 1988: Rainfall interception in stands of *Pinus radiata* and *Protea neriifolia*: Measurement and prediction. In: van Laar, A. (ed.) Research Contributions to Plantation Forestry. University of Stellenbosch, Stellenbosch.

Versfeld, D.B., 1994: The forest industry and management for water conservation. In: Southern African Forestry Handbook 1994. South African Institute of Forestry. pp 657-674.

Wicht, C.L., 1967: Forest hydrology research in the South African republic. In: Sopper, W.E. and Lull, H.W. (ed.) International Symposium of Forest Hydrology, Pergamon Press, Oxford.

A Workers' View On Sustainable Forestry

Jill Bowling[1]

Abstract

Nearly 500 million people around the world rely on forests for their livelihoods, among them a high number of forest and wood workers. Unfortunately, not enough attention has been paid to the conditions of forest and wood workers and the contributions that they can make to sustainable forest management. Sustainable forest management must include safe, stable jobs with adequate wages and working conditions. In addition, workers, because of their close relation to the forest, can also play a key role in implementing sustainable practices. The goals of this paper are: to articulate why it is in the interests of workers to contribute to sustainable practices, to demonstrate how they can contribute, and to develop criteria and indicators for international forest policy efforts that reflect worker concerns.

Keywords: *Forest worker, wood worker, union, criteria, indicator.*

[1] Jill Bowling is Coordinator, Global Forestry Programme, International Federation of Building and Wood Workers, P.O. Box 1412Wood, 54 route des Acacias, CH-1227 Carouge GE, Switzerland, Tel: 41 22 827 3777, Fax: 41 22 827 3770, e-mail: jill@ifbww.org

K. von Gadow et al. (eds.), Sustainable Forest Management, 121–152.

Introduction

It is widely acknowledged that sustainable forestry must include ecological, economic and social elements. Despite this, in reality the social elements of forest policy are often neglected or poorly developed. In particular, forest and wood workers' issues have generally received little attention and they have been a blind spot in many international initiatives.

Because of the direct interest that workers have in sustainability, the management of forests and the long term supply of timber is a central issue of concern for forest and wood worker unions around the world. These unions have, as a result, over the last years broadened their traditional activities of industrial relations, trade union education and trade union solidarity to include active involvement in education and training on sustainable forestry issues, policy debates, and implementation of sustainable forestry. For example, unions are directly involved in sustainable forestry training for workers, national certification initiatives and reforestation activities at the community level.

Unions are, by their very nature, organisations that focus on people. For unions, it is apparent that sustainable forestry does not stop at the forest edge. Firstly, there is interchange in the workforce between forest jobs and wood (or other forest resource) processing jobs and many of the social issues these two groups of workers face are either the same or similar. Secondly, the way in which downstream processing of forest products is carried out can have a huge effect on recovery, and ultimately harvested volume. Finally, for unions, the linkage between poverty and over-use of resources is clear, and therefore for unions it is clear that sustainable forestry must be considered as an integral part of overall sustainable development activities where social welfare, economic growth and environmental protection are all inextricably linked.

This chapter examines the contribution of workers and their labour organisations to sustainable forest management and it develops criteria and indicators that could be used in international processes to more effectively incorporate workers' issues into sustainable forestry initiatives. The chapter is based on three fundamental premises:

1. The people who work in forests and who derive their incomes from forest products have a vested interest in the long-term sustainablility of the forests they work in; if there are no forests there are no forestry-based jobs.
2. Forests are only sustainably managed when workers are treated in a socially responsible

way, meaning that they have safe, stable jobs with adequate wages and working conditions.

3. Workers implement sustainable forestry practices; they have unique skills and knowledge to contribute to sustainable forest management and it is vital that this knowledge is both developed and used in forest management activities.

Who are the forest and wood workers and how are they organised?

Who are the forest and wood workers?

For the International Labour Organisation (ILO, 1998a: 246), the economically active population is divided into employed, unemployed and underemployed. Basically, the category "employed" comprises "paid employment", i.e. persons who are performing some work for wage or salary, in cash or in kind, and "self-employment", i.e. persons with an enterprise, which may be a business enterprise, a farm or a service, who are performing some work for profit or family gain, in cash or in kind. But, as pointed out by Poschen (1997), the notion of employment is not adequate for persons for whom forests are the main source of livelihood. In his paper about forests and employment, he introduces a broad definition of forest-based employment: "It encompasses timber production, fuelwood, wood and non-wood products produced for sale or subsistence in the formal as well as the informal sector of the economy, and also attempts to take a look at services, namely recreation and tourism".

Forest and wood workers are not a small marginal interest group. The World Bank (1999) has stated that there are nearly 500 million people around the world who are dependent on forests for their livelihoods. This includes various degrees of dependency with, for example, forest workers and many indigenous peoples being entirely dependent on forests for their livelihoods and others, such as local community members or farmers, being partially dependent. Just how many of the 500 million people can be termed forest and wood workers is a little more difficult. Poschen (1997) notes that recorded forest-based employment in the world is about 15 million and he estimates that including subsistence and informal employment, forests provide an estimated 45 million full time jobs around the world.

The term "forest work" may cover many activities from timber harvesting, tree planting and tree tending to fire fighting and the gathering of fuel-wood and non-timber forest products. The distinction between forest work and timber processing may be blurred for example with the use of portable mills in a forest setting. In addition to these problems work may be highly seasonal, and it may be coupled to work in other sectors, non-paid work, or unemployment. Forest workers may have no long-term employment stability. The increasing use of contracts and subcontracts, the informalisation of work and the use of migrant workers who may be outside the law vastly complicate data gathering. Workers in remote locations may be invisible for the purposes of data gathering.

Even where forestry employment figures are available they may not reflect the true economic dependence of people on their forests. Non-monetary returns such as those from firewood or food gathering and indirect employment for example in tourism and the services sectors are also important. Wealth generated from these activities has flow-on effects particularly at the level of the local community.

Forest-based work is important at two levels. In a number of countries such as Finland, Sweden, Canada and countries in the tropics such as Brazil, Indonesia and Malaysia it is one of the main branches of economic activities. In other countries the forest sector may be less important nationally but very important to local timber-dependent communities. Even a small number of jobs in an area without other economic options may be very important to the local economy, especially where communities are isolated and their ability to generate other forms of income is limited. In addition, the indirect effects of forest-based employment on other sectors can be substantial.

How are forest and wood workers organized?

At the national level, many forest and wood workers are organised into unions. Unions are grass roots organisations, which represent the interests of workers. National unions may join together at the regional or international level, for example forest and wood unions have formed the Nordic Federation of Building and Wood Workers, the European Federation of Building and Wood Workers and the International Federation of Building and Wood Workers (IFBWW). The latter organisation is headquartered in Geneva and has a membership of 300 unions with 11 million workers, the bulk of which is in construction.

The union movement across the world is facing tremendous change and is in a process of adapting priorities and developing strategies to deal with changes in the workplace (Olney, 1996).

Within the union movement itself there has been a decline in membership in many countries. Labour markets across the world face globalisation, deregulation and increased informalisation of work. The twin problems of unemployment and unsatisfactory employment (where workers do not find themselves in safe, stable jobs with adequate compensation) are increasing. Even when workers are in principle protected by legislation, employers are able to circumvent the law to undermine basic worker rights. For example, contract work is increasingly used to pressure workers, reduce their rights and avoid payments of basic benefits and social security. Within the labour force, migrant workers, women and children are particularly vulnerable and open to exploitation.

How have unions adapted to changing conditions to increase their effectiveness in protecting workers rights? While globalisation and deregulation have enabled companies to exploit workers more effectively by bringing working conditions down to the lowest common denominator, the labour movement has countered with efforts aimed at a globalisation of social responsibilities thereby uplifting working conditions around the world to the highest standards. Unions have also systematically tried to get full and effective implementation of the core ILO standards which offer protection of basic fundamental human rights.

Strategically, unions have been merging to reflect the increased inter-sectoral connections at the enterprise level and they have broadened their traditional portfolios for example by linking labour issues to other sectoral issues such as those of the environment. They have actively sought to engage women in leadership positions and to target programmes to workers with special needs such as youth. They have also begun to actively explore partnerships not only within the union movement but also with partners such as civil rights, community, church and environment groups. For example, in June 1999, the General Secretary of the IFBWW met with the Director General of the World Wide Fund for Nature (WWF), which is the largest environment group in the world, to discuss issues of mutual concern and set a framework for collaborative efforts in the future (IFBWW and WWF, 1999).

Why are forest and wood workers interested in sustainable forestry and what do they have to offer?

Why are forest and wood workers interested in sustainable forestry?

Over the last years forest and wood workers around the world have been showing increasing concern about the loss of the world's forests. This concern is both for the loss of forest area and for the loss of forest quality. In 1995 the world's forests covered 3,454 million hectares, slightly more than half of which were in developing countries. Between 1990 and 1995, there was an estimated net loss of 56.3 million hectares of forest world wide which represented a decrease of 65.1 million hectares in developing countries and an increase of 8.8 million in developed countries (FAO, 1998).

Defining and measuring forest quality is difficult. While noting that quality may mean different things to different people and that perceptions of quality may change over time, Dudley and Stolten (1999: 21) have defined forest quality as "the significance and value of all ecological, social and economic components of the forest landscape" and have developed criteria for measuring these components which include indicators on direct and indirect employment and subsistence.

The issues surrounding forest loss and degradation around the world are controversial, providing a highly charged and emotional debate with many disparate groups increasingly weighing in as stakeholders. Policy makers, forest owners and forest managers have come under increasing scrutiny. Professional foresters have found themselves being challenged by non-foresters, and in many places the public has demanded a place at the decision-making table. Stakeholders with different objectives and visions of forest quality have clashed. Forest workers have often found themselves on the front line in these controversies as their jobs and communities have been threatened.

In the heat of highly charged, emotional debates over forest management where loggers are often portrayed as the "bad guys" with a chainsaw in hand it is easy to portray forest workers as the enemies of sustainability. High profile clashes between loggers and environmentalists such as those of the Pacific Northwest of the USA and Canada have perpetuated this myth. Particularly in areas of high forest dependency and rapid change, job loss and the associated pain of dislocated families and communities have caused profound social problems. These problems cannot and should not be ignored. If society wishes to protect forests it is fair that damage to the life-time investments of forest based workers should be compensated.

What such conflict hides is the fact that forest and wood workers have a fundamental interest in sustainable management. After all if there are no forests then there are no forestry jobs. Furthermore it is in the interests of forest and wood workers to actively engage in debates on sustainability in order to ensure that sustainability truly includes social components as well as ecological and economic components.

Leaders of forest and wood unions began to actively engage in the debates on sustainable forestry in the late 1980's when they realised that workers had a clear, vested interest in sustainable forest management. They had seen first hand that unsustainable harvesting in many countries had already led to reductions in the allowable harvest, and people and communities who had invested in the forest industry had had their expectations of forest based income crushed. Job loss also meant loss of union members.

For example in 1934, the Philippines had about 17 million hectares of forest covering 57% of its land base. In the 1960's, the forest area dropped from 12.9 to 10.4 million hectares, the logging industry was at its peak and 100,000 workers were employed. These workers were highly unionised. By 1990 the remaining forests covered only 6.1 million hectares and union records showed negligible unionised forest workers. Unions today in the Philippines are now helping former employees to claim ownership over forest lands and they have a policy of community based forestry (Philippines Department of Environment and Natural Resources. 1990; Malonzo, 1998: pers comm.).

What did the unions learn from this experience? For union leaders the story of what happened in the Philippines demonstrates many of the classic features associated with poverty, loss of land rights and land tenure, uncontrolled exploitation, corruption, inability and unwillingness to enforce regulations, or no regulations. The consequences are a forest resource which is depleted in both area and quality. Furthermore, the reduction of the forests has led to the reduction in the number of workers and a reduction in union membership. One of the main reasons for this is that the forests were managed for the short term and there was no plan for sustainable production. The clear conclusion for unions in the Philippines has been that union interests are closely related to sustainable management (Malonzo, 1998: pers comm.).

Contribution of knowledge and skills

At the landscape level, forests that are managed for timber production and harvest can play an important role in complementing protected areas. Careful planning and skillful execution of work can add significant value to all components of a managed forest thereby increasing its contribution to the goal of overall sustainability.

Because of their close relationship with the forest, forest workers have unique knowledge that they can use to make valuable contributions to sustainable management practices. Their knowledge may include: better ways of doing work to reduce negative impacts on the environment, knowledge of public use patterns, including illegal activities such as timber theft, and knowledge of the local ecology including the location and behaviour of rare species. Forest workers are also in a good position to monitor forest activities and they can and should be used as part of an early warning system to detect problems in the forest such as forest fires and respond quickly to these problems. This knowledge is often not used to advantage by forest managers. Similarly, workers who transform and transport forest products are in a position to make suggestions on how work can be done efficiently, effectively and in a safe and environmentally responsible way.

It is important not to overstate the contribution that workers can make - after all they are not the ultimate decision-makers for many forest management decisions. However, because they implement the work and because they are physically on the work-site they can make valuable contributions. Because many workers are poorly educated or outside traditional power bases, it is easy to overlook their contributions and this is a mistake that needs to be corrected.

In order to make the most effective contributions to sustainable forest practices workers need to be well-equipped, well trained and included as part of the decision-making process. They need to be provided with equipment which is well-maintained and technically appropriate to the job they undertake. They need to be trained on how to use their equipment efficiently and safely, and educated on what sustainable forestry is and why sustainability is important. Training and retraining should be part of a process of continuous lifelong learning which is provided within the context of stable long term jobs where workers have the opportunity to build on their skill base over time (Odgaard, 1999). Finally, participation of stakeholders is a recognised way to manage and workers need to be included at all levels of forest management decision-making including in the boardrooms of corporations. Worker participation offers an important contribution to define what should be sustained, which is, according to Schlaepfer (1997), a fundamental question to be addressed.

Let us illustrate these points using some examples. The use of technically sophisticated logging equipment can lead to high recovery rates with little damage to the remaining forest. Low impact logging can considerably reduce damage to the stand and improve recovery rates through better felling and bucking practices. The timing of operations can be used to avoid or minimise compaction of soils and erosion. Effective and careful use of chemicals and fertilisers can also make significant contributions to the overall health of the forest. Particularly in diverse tropical forests many "non-commercial" species are "wasted". High-grading commercial species can change overall species composition in a forest and put great pressure on the most valuable species. Research on new uses for these currently non-commercial species is urgently needed to reduce pressure on valuable species and reduce waste of non-valuable species. Further along the production process, keeping value-added production as close as possible to where timber is grown can help to strengthen local economies and create jobs and in so doing it can take pressure off the forest by over-exploitation. This is particularly so when there are few economic options available to people and short-term overuse of the forest outweighs long term benefits.

Effective co-ordination between the entire forest management team including those people engaged in planning, nursery production, planting operations, silvicultural activities, final harvest, and marketing can make significant contributions to sustainable practices. This coordination can for example, include ensuring that adequate sources of the required species are available for reforestation soon after harvest, planning silviculture treatments to minimise the use of harmful chemicals and minimise damage to the remaining stand, and timing of harvest operations to minimise the negative impacts of bad weather such as compaction and erosion.

In timber harvest operations delegation of authority coupled with appropriate training can lead to more appropriate cutting decisions rather than a "cookie cutter approach" formulated at a distance with less direct knowledge of local conditions. Where decision making can be delegated and this is backed up by appropriate training, jobs become more interesting and fulfilling, and workers are able to make better decisions, which more adequately fit the local conditions. For example, logger select thinning based on guidelines rather than single tree marking has been used effectively by the Oregon State Department of Forestry in the USA and it is cost effective. Forest workers must understand that sustainability involves obligations of the present generation to preserve natural resources for future generations. And it involves training not only on the importance of sustainable forestry but also on the practical implementation of sustainable resource management.

Involvement of trade unions

In the late 1980's, forestry unions in many countries saw similar patterns of forest loss and as a result worked together to develop a vision for union involvement in sustainable forestry. This vision included engagement of their memberships in proactive steps to prevent depletion of forest resources. Wood processing and construction unions, many of which were closely associated with forestry unions or included forest workers in their memberships, also acknowledged a stake in the process. They recognised that not only were jobs in the forest at risk but also downstream jobs that relied on the raw materials such as processing and associated jobs in transport and construction were also at risk. For these reasons unions saw forestry issues in the context of wider concerns.

As they developed their policy, unions recognised the very important role that they could play in protecting forests from over-exploitation. They recognised that in many places over-use of resources was inextricably tied to poverty and part of the traditional union agenda was the elimination of poverty through direct action to improve working conditions and wages.

At the international level, the IFBWW recognised the importance of sustainable forest management to its affiliates in the wood and forestry sectors. In the late 1980's it held a series of regional conferences in Africa, Asia and Latin America to explore key problems and develop solutions. This led to the publication of a global union Tropical Forest Action Programme (IFBWW, 1992). As discussions continued within the union movement to include more affiliates the programme was revised a year later to include temperate forests. Horst Morich, Chairman of the IFBWW Wood and Forestry Committee summarised the position: "In recent years the trade unions have become aware of this new challenge and we too are under an obligation to examine the ecological repercussions of our own actions. This necessary reappraisal has already begun in the trade union movement. Protecting the natural foundations of our existence is a task for humanity as a whole. We as trade unionists want to make our contribution to this effort." (IFBWW, 1993a: 7-8).

In 1993 the IFBWW established a Forest Programme to guide and co-ordinate global union activities (IFBWW, 1993b). This programme outlined the role of trade unions in protecting forests and identified the key areas for trade union intervention as improving training, occupational health and safety, promoting secure and permanent jobs, safeguarding workers rights, trade union representation, building up independent trade unions, and improving market access of timber and wood from sustainably managed forests. Subsequent work with unions in several countries found that while the forest and union situation varied between countries, there were a number of common factors: low awareness of forestry issues, limited technical expertise, prioritisation of traditional trade unions concerns (such as low wages and poor working conditions), and poor organisation. These factors helped to explain the low level of involvement in the national and international policy initiatives.

In 1997 the IFBWW secured funds from the Dutch government and established a Global Forestry Programme. The goals of this programme were to ensure that union concerns on global forestry issues are incorporated into international initiatives, to initiate and develop national union programmes, and to improve working conditions (IFBWW, 1997).

The following examples show that at a very practical level unions have been engaged in reforestation activities and in the promotion of good forest practices. In Ghana, a country with high dependence on forest resources and rapid deforestation the Timber and Wood Workers Union has created a nursery and demonstration reforestation project. Training is carried out to raise environmental awareness and sensitise workers and the general public to initiate actions for the protection, conservation and restoration of the environment, with particular reference to forestry. The nursery was commissioned by the President of Ghana in July 1998. In Peru, the Peruvian Amazon Wood Workers Federation has established a medicinal plant nursery. The Federation is working with the local university to develop a marketing plan for the medicinal plants and is working with local farmers who will grow the plants to supplement their traditional incomes.

How far are International Processes taking account of Workers' Issues?

Forestry Initiatives at International Level

The Earth Summit held in Rio de Janeiro in 1992 provided the foundation for many subsequent international forestry processes. The Summit produced several important documents. The Declaration on Environment and Development includes 27 principles defining the rights and responsibilities of nations as they pursue human development and well-being. Agenda 21 reflects a global commitment and consensus at the highest political level on how to make development socially, economically and environmentally sustainable. The non-legally-binding, authoritative Statement of Forest Principles provides guidance for the management and sustainable development of all types of forests. Principle 2 of the Statement stipulates among other things that "Forest resources and forest lands should be sustainably managed to meet the social, economic, ecological, cultural and spiritual human needs of present and future generations." It includes employment as one of these needs. It also states that governments should promote and provide opportunities for the participation of interested parties including labour and it recognises the rights of indigenous peoples and forest communities (United Nations Conference on Environment and Development, 1992).

The UN Commission on Sustainable Development (CSD) has continued its work on sustainable forestry through the Intergovernmental Panel on Forests (IPF) which met between 1995 and 1997. The IPF focused on the following programme elements: national forest and land-use plans, underlying causes of deforestation, traditional forest-related knowledge, ecosystems affected by desertification and pollution, needs and requirements of countries with low forest cover, financial assistance and technology transfer, forest assessment, scientific knowledge and valuation of forest benefits, criteria and indicators, trade and environment, international institutions and instruments, and legal mechanisms (Secretariat to the Intergovernmental Panel on Forests, 1996).

The IPF report and recommendations were forwarded to the UN General Assembly Special Session in 1997 and based on this the UN General Assembly decided to continue the intergovernmental dialogue on forests and establish an ad hoc open-ended Intergovernmental

Forum on Forests (IFF) under the aegis of the CSD. The IFF has met three times and is scheduled to report to the CSD in 2000. Work has been divided into three categories. Category I includes two items; promoting and facilitating the implementation of the IPF's proposals for action and monitoring progress. Category II consists of matters left pending on financial resources, trade and environment and transfer of environmentally sound technologies to support sustainable forest management, other issues from the IPF process needing further clarification, and the forest related work of international and regional organisations. Category III relates to international arrangement and mechanisms and includes discussion on a possible forest convention or similar legal instrument (United Nations, Commission on Sustainable Development, 1997). It is obvious that workers' issues have not been important for either the IPF or the IFF and union involvement in these processes has been minimal.

Major initiatives in global sustainability have also come from the International Tropical Timber Organisation (ITTO). The ITTO has set for its members a Year 2000 Objective, which is the goal of having all tropical timber entering international trade to come from sustainably managed sources by 2000. Forest and wood worker unions have Observer status within the ITTO, and union representatives started to attend meetings and actively participate in the process in early 1998.

At the regional level, there have been a number of important forest processes including the Helsinki Process (now Pan European Process), the Montreal Process (for temperate and boreal forests), the Tarapoto Proposal for the Amazon and regional initiatives for Dry-Zone Africa, the Near East, Central America and the African Timber Organisation. It is beyond the scope of this chapter to deal with these processes in any detail. However, worker issues have generally had a low profile in these initiatives, and union or worker participation has been minimal.

Outside the government sector the World Bank, which has a mission of poverty alleviation, has long been involved in forestry projects with the financing of forestry projects at nearly $2.5 billion from its inception to 1991 (World Bank, 1991; Johnson, 1999). With this scale of investment bank policy has a tremendous effect on forest practices around the world. The Bank's Forest Policy adopted in 1991 is used to guide its investments. This policy is currently being revised. Despite the fact that the World Bank has a mission of poverty alleviation, its forest policy has very little to say about forest workers and the input of forest workers and their representative organisations has not been actively sought by the Bank.

Forest certification is another approach aimed at improving forest management practices. Certification is based on market incentives and has been defined by Elliott and Hackman (1996: 9) "as a voluntary process which results in a written certificate being produced by an independent third party attesting to the location and management status of the forest from where the forest product originated. Certification involves independent assessment of a forest management operation on the ground using specified social, ecological and economic criteria or standards". Although the trade in certified products is still small and market trends are not clear (Hansen, 1998) there is considerable interest in certification from a number of different directions including unions.

To date, several different forest certification schemes have been developed including those of the International Standards Organisation (ISO), the Canadian Standards Association (CSA) and the Pan European Forest Certification (PEFC) process. Currently most certified forests around the world are certified by the Forest Stewardship Council (FSC). The FSC has developed ten Principles and 56 Criteria covering economic, ecological and social aspects to guide forest management activities and provide a consistent framework for the development of locally determined forest management standards. Principle 4 addresses community relations and workers rights and states that "Forest management operations shall maintain or enhance the long-term social and economic well-being of forest workers and local communities" (FSC, 1999). The recent election of the Chairman of the IFBWW Wood and Forestry Committee to the FSC Board at the General Assembly held in June 1999 is an indication that this organisation is prepared to give workers' issues serious attention.

Unions, initially cautious of forest certification have especially over the last two to three years become more involved in certification efforts. They are now actively participating in a number of national initiatives including those in Sweden, Germany and Ghana, where they have been involved in the negotiation of national standards. At the global level, the IFBWW adopted a policy on forest certification in 1997. This policy has been subsequently used as a basis for developing training programmes on sustainable forestry and forest certification in several countries and for working to include unions in national certification processes.

The examples cited above show that ecological, economic and social elements are widely acknowledged as necessary to sustainable forest management. Despite this, in international forums, the social elements, which include the issues of communities, indigenous peoples and workers are generally under-represented or not articulated well. In particular, workers issues have generally not been well considered in major policy initiatives. This has to change. To do this the labour movement has to actively participate in the deliberations either as observers, or preferably, as working members to put workers' needs on the agenda.

How are workers issues considered in international criteria and indicators for sustainable forestry?

International processes have recognised that it is necessary to measure progress towards sustainable forestry and this has led to the development of principles, criteria and indicators. Criteria and indicators can be at the national level or the level of the forest management unit. They have been developed in the context of international processes and by the research community for example the Center for International Forestry Research (Colfer et al., 1995).

It is beyond the scope of this chapter to analyse the criteria and indicators developed by all the international processes. However, to illustrate the position of workers issues in the international processes this chapter uses as case studies the criteria and indicators developed by the ITTO, the Montreal Process, and the Pan European Process. Appendix 1 presents the social criteria that relate to worker protection.

The ITTO has developed seven criteria and sixty-six indicators. Economic, Social and Cultural Aspects are grouped together under Criterion 7 which has 18 indicators grouped into the headings of socio-economic aspects (13 indicators), cultural aspects (1 indicator) and community participation (4 indicators). The original criteria and indicators developed by the ITTO (ITTO, 1992) contained a criterion on Socio-Economic Effects at the National level with indicators of employment patterns and trends, income generation and distribution patterns, national revenue and expenditure budgets for forest management and availability of environmental assessment procedures. Also in this document was a criterion at the forest management unit level on socio-economic benefits with indicators on the number of people employed and the nature and extent of benefits from forestry activities.

The updated criteria and their indicators, published in July 1998, made several important moves forward. Firstly the indicators cover wage rates. They also include worker health and safety - injury rates and the existence and implementation of procedures to ensure the health and safety of forest workers. Finally, they point to the importance of downstream processing by including the ratio of domestic log production to processing capacity of wood-based industries and the efficiency of utilisation in terms of the percentage of felled volume processed (ITTO, 1998).

The Montreal Process has developed seven criteria and 67 indicators. Worker issues are covered in Criteria 6: Maintenance and enhancement of long-term multiple socio-economic benefits to meet the needs of societies. The indicators are grouped under subheadings of production and consumption, recreation and tourism, investment in the forest sector, cultural, social and spiritual needs and values, and employment and community needs. The latter covers direct and indirect employment, average wage rates and injury rates.

The Pan European criteria and indicators were approved at the Third Ministerial Conference for the Protection of Forests in Europe which met in Lisbon in June 1998. They contain criteria and quantitative and descriptive indicators for sustainable forest. Criterion 6 which is the maintenance of other socio-economic functions and conditions contains indicators grouped under the concept areas of: significance of the forest sector, recreational services, provision of employment, research and professional education, public awareness, public participation and cultural values. The indicators deal with income levels and training but do not explicitly mention health and safety.

We note that the ECE/FAO/ILO Team of Specialists on Social Aspects of Sustainable Forest Management in Europe provided a comprehensive overview and analysis for input to the Third Ministerial Conference. One of the major recommendations coming from this group was to "ensure that the workforce is well-trained and receives proper remuneration and health care, and is available, in terms of numbers, composition and skills, to put SFM into practice" (Joint FAO/ECE/ILO Committee on Forest Technology, Management and Training 1997: 29). The main topics of work agreed to in Lisbon for ongoing work are rural development and socio-economic aspects of sustainable forest management, biodiversity in forest ecosystems, and criteria and indicators for sustainable forest management. This is an opportunity to develop these issues further. Unions are now included as Observers at the Expert Level Meetings on the Follow-up of the Lisbon Conference.

In the three case studies presented above we can identify the following workers' issues:

. employment data (all three processes)

. health and safety (most fully developed in the ITTO)

. wages (all three processes)

The question is: is this enough? The next section seeks to answer this question.

How should workers' needs be better integrated into international processes?

Working conditions for forest and wood workers

"Working hours were unregulated and long, both in factories and in the countryside, especially in summer when daylight lasts longer. Industries that were operating 24 hours a day, applied 2 twelve hour shifts with no rest. The wages were set for bare survival; it was said that higher wages would only lead to idleness and immoral living.... Those who had no property and no employment were considered vagrants and legally defenceless. They could therefore be sent to forced labour or put in prison." (Wallin, 1994).

"Working in the woods has given Hamonagan deep lines in his face, making him look older than his 30 years. Between 6:30 a.m. and 5 p.m. each day, he fells trees, earning 700 Rupees per cubic metre, good money in Indonesia. Pay is on piece-rates, without allowances for the quality of the logs and other forest or safety work. So, the more he cuts, the more he earns for himself and his family. At 165 cm. and 55 kg., he weighs slightly more than three times his chainsaw, and is less than twice as tall as its 90 cm. guide bar. The chain saw is dangerous; without essential safety devices like a chain break and catcher, and poor vibration dampening, it would be illegal in most industrialized countries. The company issues him a hard hat; steel-toed boots, ear plugs, visor and cut-protecting trousers he would have to buy himself, but he prefers to save the money and take the risk. The trees he cuts are huge: balancing precariously on a scaffold two meters up the trunk, he and his helper must jump 1.5 meters to the ground as 50-meters and 25 tons of wood crash to the earth with a shudder. It's no use running away; as the tree falls, branches break and rain from the sky. Better to stay put and look out for them. With ten years experience Hamonagan is an "old-timer". He is also very lucky, claiming never to have had a serious accident. Some of his co-workers have not been so fortunate; the company had ten logging fatalities last year among 700 employees. If Hamonagan is injured, though, few would know. Chainsaw operators like him work for contractors, so their accidents are not even recorded."(ILO, 1997).

The first quote above describes working conditions in Sweden late last century. The second quote describes current working conditions in Indonesia.

Getting information on current working conditions is difficult. Workers are often afraid to report problems because they fear reprisal or dismissal. Anecdotal information from unions however provide rich insights into the way that forest-based work is carried out around the world. The conditions that they report in many places, especially in developing countries and in Central and Eastern Europe, are far from satisfactory. Each day forest workers die while on the job. They do not get adequate training, they are injured, they are forced to retire early because of their injuries, perhaps with no social protection, and they work in very poor conditions for very low wages. They live in forest camps with inadequate facilities and may move from one logging site to another far away from their families for long periods of time. They may work for subcontractors with no security of employment and no right to collective bargaining. In many places, they have inferior legal protection and the protection given to them on paper may not be translated to reality. These conditions are not a legitimate part of sustainable forestry.

Forest work is some of the most dangerous work in the world with accident frequencies and death rates two to three times higher than in other sectors. Accident frequency and death rates vary considerably from country to country, for example the number of deaths per one million cubic meters of wood produced is 0.11 in Finland, 1.87 in Switzerland, and 4.4 in Sarawak, Malaysia (ILO, 1997).

Women in forest and wood processing jobs are particularly vulnerable and they face special problems that are different and additional to those faced by men. In France, Faugere (1998:1) found that although women performed a great diversity of activities ranging from heavy forestry work to other forest-related tasks, they were absent from the local network of associations and trade unions and their work "seems to be regarded rather as a pastime or assistance than as something that has and contributes value". Subsequent work in Brazil, Ghana, and Zimbabwe (Faugere and Bowling, 1999) with women in forest and wood processing jobs found active discrimination against women and illegal working conditions with women trapped into low status, low paying jobs and lacking the resources to move elsewhere.

Working conditions for socially responsible management

Over the last years through a series of workshops, unions have identified the working conditions that would contribute to sustainability. This list has now been presented as an evolving document in a number of forums (Bowling and Fitzpatrick, 1998):

1. Right to organise

Employers should allow workers to have the right to organise and form democratic trade unions, and to engage in collective bargaining.

2. Remuneration and living and working conditions

Employers should provide fair compensation and living and working conditions, and governments should pursue an active labour market policy designed to promote full, productive and freely chosen employment.

3. Health and safety

Employers should work to eliminate preventable accidents and diseases, and ensure safe working conditions.

4. Equality

Employers should respect and pursue policies designed to promote equality of opportunity and treatment in employment, eliminating discrimination based on race, colour, sex, religion, age, political opinion, national extraction or social origin. Employers should ensure that work sites are safe environments for women, free from sexual harassment and other forms of discrimination and abuse, and they should promote women's access to jobs in forestry and promote affirmative action in skills training, recruitment and career development.

5. Child labour

Child labour should be abolished.

6. Forced labour

Forced or compulsory labour should be abolished.

7. Participation

Employers should ensure full worker participation in all decision-making that affects workers' working conditions and conditions of employment.

8. Training

Employers should ensure workers have the required competence in the area of work they are to undertake so that work is carried out safely and effectively and with attention to environmental protection. Employers should regularly assess skill needs and, where necessary, they should provide formal training and skills tests.

9. Job security

Workers should be offered long-term job security and, unless otherwise agreed, they should be employed on a permanent basis.

10. Contract workers

Whenever possible, conditions for contract workers and regular employees should be standardised so that they have equality of terms and conditions. Contractors should not be used to avoid or deny legal rights or obligations.

11. Migrant workers

Locally trained workers, who are living locally, should be used where possible. Where migrant workers are employed, employers shall protect these migrant workers and their families.

12. Indigenous people

The rights of indigenous peoples should be respected. Employers should promote access to jobs in forestry and affirmative action in skills training, recruitment and career development.

13. Community involvement

Employers should respect the social fabric of local communities, and consult with local communities on decisions, which affect the well-being of these communities. Employers should where possible use locally trained people who are living locally and they should not undercut local wages structures by using cheap imported labour. Local communities should get fair share of the economic returns generated by their local forests.

Many of these items are inter-linked, and improvements in one area will necessarily lead to improvements in other areas. For example, without long-term job security it is difficult to make investments in worker training, and worker skills do not have a chance to develop fully. This has implications for health and safety. In a recent study of wood processing operations in Brazil (Alves,1999) 43 percent of accidents occurred within 6 months of starting to work with a specified piece of machinery and this dropped to 8 percent when workers had been working with the machinery between 18-24 months. Piece-rate remuneration encourages workers to attain high outputs as quickly as possible without regard to the ecological consequences and basic safety standards. This form of remuneration creates stress on workers and stress on the forest. Providing workers with long-term employment and a stable basic wage improves not only the health and safety of the workers but also the health of the forest.

There may be a strong link between worker issues and the issues of communities and indigenous peoples. For example, a worker may be indigenous and live in the local community. Providing community members with good jobs and a fair share of the economic returns from a forest will necessarily give value to the forest and hence an incentive for sustainable use of resources.

International Labour Organisation Conventions: international standards for the protection of workers

The ILO Conventions and their supporting Recommendations provide essential international guidance and protection on many workforce issues. The ILO Conventions are adopted by the International Labour Conference after consultation with member states (about 170 member states). The Conference is a tripartite body composed of government, employer and worker representatives. Once a member state ratifies a Convention it becomes subject to legally binding international obligations.

While the ILO's 180 Conventions cover a wide range of labour issues, the most fundamental human rights are covered in the seven core ILO Conventions which aim to prevent the very worst forms of repression, exploitation and discrimination. These core Conventions, which are amongst the most highly ratified of all ILO Conventions, are Conventions 87 and 98 on the rights to freedom of association and to bargain collectively, Conventions 29 and 105 on the abolition of forced labour, Conventions 111 and 100 on the prevention of discrimination in employment and equal pay for work of equal value, and Convention 138 on child labour. Taken together, these core Conventions form a basic minimum level of protection for workers (ILO,1999).

The relevance and importance of the seven core ILO Conventions was stressed in June 1998 when the International Labour Conference adopted the ILO Declaration of Fundamental Principles and Rights at Work. The Declaration is important because it reinforces the principle that Membership of the ILO represents a basic commitment to the Constitution of the ILO, and particularly to the fundamental rights embodied in the core conventions. The Declaration also reaffirms that the ILO has a reciprocal obligation to member states to assist them in attaining these objectives (ILO, 1999).

The ILO Conventions and their supporting Recommendations form a strong basis for protecting workers. In addition, the ILO Code of Practice on Safety and Health in Forestry Work (ILO, 1998b) provides comprehensive guidance and is now forming the basis for national initiatives in several countries. Unions who along with governments and employers were involved in the development of the code are supporting implementation of the code by providing training sessions in South Africa, Brazil, Chile, Indonesia and Malaysia.

Proposed criteria and indicators for workers' issues in sustainable forest management

We see that there is a big gap between the criteria and indicators developed at the international level and the needs of forest and wood workers. It is obvious that more attention has to be given to workers' issues to ensure that sustainable forestry truly covers the workforce.

We propose in this section principles, criteria and indicators that could be used for this purpose. To do so we have taken the input of hundreds of workers from around the world who have direct experience with the problems that they face. These workers have articulated to us very clearly and directly through seminars and meetings, and discussions at the workplace, the key issues that they face and the conditions that they want. We also rely heavily on the ILO Conventions to provide guidance as they provide widely accepted standards that have been through extensive negotiations between governments, employers and workers.

Finally we use the theoretical framework developed by Lammerts van Bueren and Blom

(1997). They have defined principles, criteria and indicators as follows:

" A principle is a fundamental law or rule, serving as a basis for reasoning and action. Principles have the character of an objective or attitude concerning the function of the forest ecosystem or concerning a relevant aspect of the social system that interacts with the ecosystem. Principles are explicit elements of a goal, e.g. sustainable forest management or well managed forests."

"A criteria is a state or aspect of the dynamic process of the forest ecosystem, or a state of the interacting social system, which should be in place as a result of adherence to a principle. The way criteria are formulated should give rise to a verdict on the degree of compliance in an actual situation.

"An indicator is a quantitative or qualitative parameter which can be assessed in relation to a criterion. It describes in an objectively verifiable and unambiguous way features of the ecosystem or the related social system, or it describes elements of prevailing policy and management conditions and human driven processes indicative of the state of the eco-and social system." (Lammerts van Bueren and Blom, 1997: 18-22)

Using this approach we propose the following principle, criteria and indicators for forest and wood workers for use at the national level. We list here examples of both qualitative and quantitative indicators acknowledging that the list is not comprehensive and needs further refinement and field testing.

Principle

Within the context of sustainable forestry, the fundamental rights of forest and wood workers are respected and their contributions are encouraged.

Criteria and indicators

Criterion 1: Forest and wood workers are important actors in sustainable forest management and their employment and income opportunities are optimised.

Indicator: 1. 1 Number of jobs by job category, age and sex.

Criterion 2: Forest and wood workers have the right to organise and form democratic trade unions, and to engage in collective bargaining.

Indicators: 2.1 Number and percentage of forest and wood workers who are members of a trade union.

2.2 Number and percentage of forest and wood workers who are covered by a collective agreement.

2.3 Existence of a legal basis giving forest and wood workers the right to organise and form democratic trade unions.

Criterion 3: The wages of forest and wood workers are fair and equitable, and provide an adequate standard of living.

Indicators: 3.1 Average wage (by time or by piece), stratified by job undertaken, age and sex.

3.2 Existence of legally-binding minimum wage regulations.

3.3 Fulfillment of the minimum wage regulations.

3.4 Minimum wage for an acceptable quality of life.

Criterion 4: Forest and wood workers are provided with social benefits which are consistent with national standards.

Indicators: 4.1 Existence of national standards covering benefits in case of sickness, accident, invalidity, unemployment, maternity, old age and death.

4.2 Compliance with national standards covering benefits in case of sickness, accident, invalidity, unemployment, maternity, old age and death.

4.3 Compliance with relevant ILO Conventions (Convention 102 on social security, Convention 118 on equality of treatment and Convention 157 on maintenance of social security rights).

Criterion 5: Forest and wood workers are provided with safe working conditions and employers aim to eliminate preventable accidents and diseases.

Indicators: 5.1 Number of work related accidents per year per person, categorised by type of accident, job type, output, age, number of years in the job and sex.

5.2 Number of days taken off per year and per person due to accidents and disease by job type, output, age, number of years in the job and sex.

5.3 Existence of a legal basis for protecting health and safety.

5.4 Compliance with ILO Code of Practice of Safety and Health in Forestry Work.

5.5 Existence of training programmes for health and safety.

5.6 Provision of adequate and well-maintained equipment and protective clothing and effective use of this equipment and clothing.

Criterion 6: Forest and wood workers are provided with training and retraining so that their work is carried out competently and conforms to best practice.

Indicators: 6.6 Number of days per year and per person given by employers for training.

6.7 Existence of training programmes for forest and wood workers.

Criterion 7: Where possible, forest and wood workers are provided with job stability.

Indicators: 7.1 Median length of time that workers are employed by a forest or timber producing company.

7.2 Percentage of the workforce by job category on contracts that are less than one year.

Criterion 8: Forest and wood workers are full participants in all aspects of the management, conservation and sustainable development of forests.

Indicators: 8.1 Existence of regulations giving forest and wood workers the right to participate in forest management.

8.2 Type of participation of forest and wood workers in management (for example simple information to involvement in decision-making).

Criterion 9: Forest and wood workers have basic social rights.

Indicators: 9.1 ILO Convention 138 on child labour is respected.

9.2 ILO Conventions 29 and 105 on forced labour are respected.

9.3 ILO Conventions 100 and 111 on equality of opportunity and treatment are respected.

We note that the issues surrounding long-term job security are very complex and need further work. The ILO is in the process of developing a Convention on Contract Labour.

Discussion and conclusions

Forestry around the world is facing profound changes. Increasing population pressures, deforestation and loss of forest quality, technological developments, and increased interest of the public in forestry issues are inducing a dynamic situation in forestry which should not be missed by forest and wood workers.

Decisions, resolutions and policy on sustainability mean nothing unless they are implemented. And it is the people who are working in the forests who must ultimately implement these concepts. Forest planning will fail unless the people responsible for implementation have understood the relevance of sustainable forestry, are included in the decision-making process, and are willing to implement sustainable practices.

Working conditions for forest and wood workers in many places around the world are less than satisfactory and they stay behind the conditions found in many other sectors. The fundamental needs and rights of workers have been identified and translated into international conventions and regulations such as those of the ILO. Unfortunately the ILO Conventions are not ratified by all countries and among those countries which have ratified the Conventions, implementation at the management unit level may be less than satisfactory. In the different international processes dealing with sustainable forest management, despite clear progress, workers' issues are still underdeveloped.

Global initiatives are complemented at national, state and local levels through a variety of processes. These include legislative efforts, codes of practices, industry standards for self-regulation, and national forest certification processes. The linkage between international and national efforts may not always be obvious: often local managers may be unaware of the commitments their national governments have made on their behalf and, either consciously or not, the implementation of international agreements may not occur.

In addition, when the industry perspective is sought at local levels, all too often it is only the owners of businesses, and the corporations which are represented, not the workers or their representative groups. This is an important oversight; on many issues the interests of employers and workers are not aligned.

According to our analysis, sustainable forest management cannot be reached without giving forest and wood workers good working conditions, wages that assure a reasonable life quality and the possibility to participate in forest management decision-making. The way these conditions are realised are very different in different places in the world. In many cases improvements are necessary. This implies constant effort at both international and national levels.

Workers' issues need to be clearly articulated in all efforts to define sustainable forest management and mechanisms should be developed to assure that these issues are implemented at the forest management unit level. For this purpose it is essential that workers are actively participating in the different forestry processes, including definition of criteria and indicators for sustainable forest management and forest certification.

In order to be successful workers need a good information basis about the conditions of forest and wood workers in the world. Official statistics are very important but often lack data about issues like participation, social security or health and safety. A good data basis is essential for political argumentation. It would be useful for the research community and administration to invest more in this direction and to ensure that the research results are readily available to workers.

Evolution during these last years shows that different stakeholders with different objectives and visions of the forest like environmental NGO's and forest and wood workers have similar interests in sustainable forest management. They all want forests which are able to fulfill their ecological, economic and social functions for present and future generations. It is therefore clear that forest and wood workers could become stronger if they build alliances with groups that share their interests.

We have seen that workers' issues are often overlooked in forest policy. One way to improve the situation is to more actively disseminate information about their needs among the public including policy makers. This could increase the awareness about the importance of forest and wood workers issues in forest management.

Forest and wood workers are essential actors in sustainable forest management. They are working under difficult conditions, they know the forest and they have a stake in sustainable forest management. We believe that they deserve more support in their work and also that their skills are better used in forest management by including them in decision-making.

References

Alves, C. G., 1999: Survey of Occupational Accidents in Sawmills and Plywood Plants of Manaus and Itacoatiara. Report prepared for IFBWW, Panama, 17pp.

Bowling, J. and Fitzpatrick, S., 1998: Workers contribute to sustainable forest management. Proceedings of International Conference on Indicators for Sustainable Forest Management 24-28 August 1998, Melbourne, Australia p74-75.

Colfer, C.J.P., Prabhu, R. and Wollenberg, E., 1995: Principles, Criteria and Indicators: Applying Ockham's Razor to the People-Forestry Link. CIFOR Working Paper No.8 October 1995, 16pp.

Dudley, N. and Stolten, S., 1999: Evaluation of Forest Quality. Towards a Landscape Scale Assessment. Equilibrium Consultants, Bristol, 155 pp.

Elliott, C. and Hackman, A., 1996: Current issues in forest certification in Canada. A WWF Canada discussion paper. WWF, Canada, 63pp.

Faugere, I., 1998: The Role of Women on Forest Properties in Haure-Savoie: Initial Research. Report prepared for the United Nations Economic Commission for Europe Timber Section, Geneva, Switzerland, 11pp.

Faugere, I and Bowling, J., 1999: What is Work Like for Women in Forestry? Results of an IFBWW questionnaire on women in forestry. Report prepared for IFBWW, Geneva, 8pp.

Food and Agriculture Organisation, 1998: *State of the World's Forests 1997*. FAO, Rome, Italy, 200pp.

Forest Stewardship Council, 1999: Principles and Criteria for Forest Stewardship Council Revised January 1999. Forest Stewardship Council, Oaxaca, Mexico, 6pp.

Hansen E., 1998: Certified Forest Products Marketplace. *Forest Products Annual Review, 1997-1998* United Nations, Geneva, 17-28.

International Federation of Building and Wood Workers, 1992: IFBWW Tropical Forest Action Programme. IFBWW, Geneva, 10pp.

International Federation of Building and Wood Workers, 1993a: Forest Action Programme. IFBWW, Geneva, 28pp

International Federation of Building and Wood Workers, 1993b: IFBWW Forest Programme: Protecting forests - a trade union task. IFBWW, Geneva, 16pp.

International Federation of Building and Wood Workers, 1997: Global Forestry Programme, IFBWW, Geneva, 2pp.

International Federation of Building and Wood Workers and World Wide Fund for Nature, 1999: Press Release: Unionists and Conservationists Agree to set New Eco-Label Guidelines: IFBWW, Geneva, 2pp.

International Labour Organisation, 1997: New directives for protecting loggers and preserving the environment. *World of Work* 21: 8-11.

International Labour Organisation, 1998a: *World Employment Report 1998-99: Employability in the global economy - How training matters*. ILO, Geneva, 258pp.

International Labour Organisation, 1998b: *Code of Practice on Safety and Health in Forestry Work*. International Labour Office, Geneva, 78pp.

International Labour Organisation, 1999: Website: http://www.ilo.org/public/english/index.htm

International Tropical Timber Organisation, 1992: Criteria for the Measurement of Sustainable Tropical Forest Management. ITTO Policy Development Series No.3, March 25, 1992 Yokohama, Japan, 5pp.

International Tropical Timber Organisation, 1998: Criteria and Indicators for Sustainable Management of Natural Tropical Forests. ITTO Policy Development Series No. 7, ITTO, Yokohama, 22pp.

Johnson, I., 1999: Statement by Ian Johnson, Vice President, The World Bank Intergovernmental Forum on Forests Third Session Commission on Sustainable Development. May 11, 1999.

Joint FAO/ECE/ILO Committee on Forest Technology, Management and Training, 1997: *People, Forests and Sustainability. Social Elements of Sustainable Forest Management in Europe.* ILO, Geneva, 216pp.

Lammerts van Bueren, E. M. and Blom, E. M., 1997: *Hierarchical Framework for the Formulation of Sustainable Forest Management Standards.* The Tropenbos Foundation, Leiden, 82pp.

Malonzo, L.,1998: Past President of the National Federation of Labour, Philippines. Personal communication.

Montreal Process Working Group, 1998: The Montreal Process. Website: http://www.mpci.org/whatis/criteria_e.html

Odgaard, G., 1999 Lifelong Learning and Sustainable Managed Forests. Paper for the ECE/FAO/ILO Seminar Forest Operations of Tomorrow, 20-24 September 1999.

Olney, S. L., 1996: *Unions in a Changing World: Problems and Prospects in Selected Industrialised Countries.* ILO, Geneva, 99pp.

Philippines Department of Environment and Natural Resources, 1990: Master Plan for Forestry Development Main Report. Philippines Department of Environment and Natural Resources, Quezon City, Philippines, 465pp.

Poschen, P., 1997: Forests and employment - much more than meets the eye. Proceedings of the XI World Forestry Congress 13-22 October 1997 Antalya, Turkey. Volume 4: Economic contribution of forestry to sustainable development: 61-78.

Schlaepfer, R., 1997: Ecosystem-Based Management of Natural Resources: A Step Towards Sustainable Development IUFRO Occasional Paper No. 6, Vienna, 32pp.

Secretariat to the Intergovernmental Panel on Forests, 1996: Forests in the Global Debate. The Intergovernmental Panel on Forests: its mandate and how it works. Commission on Sustainable Development, Intergovernmental Panel on Forests, New York, 5pp.

Third Ministerial Conference on the Protection of Forests in Europe, 1998: General declaration and resolutions adopted. Ministry of agriculture, Rural Development and Fisheries of Portugal, Ministerial Conference on the Protection of Forests in Europe, Lisbon, 64pp.

United Nations Commission Sustainable Development, 1997: Proposed Programme of Work of the Intergovernmental Forum on Forests. E/CN.17/IFF/1997/2.

United Nations Conference on Environment and Development, 1992: Agenda 21:Programme of Action for Sustainable Development, Rio Declaration on Environment and Development, Statement of Forest Principles. The final text of agreements negotiated by Governments at the United Nations Conference on Environment and Development (UNCED), 3-14 June 1992, Rio de Janeiro, Brazil. United Nations Department of Public Information, New York, 294 pp.

Wallin, K., 1994: *The Origins of Trade Unions in Sweden.* LO/TCO, Bistandsnamnd, 64pp.

World Bank, 1991: A World Bank Policy Paper The Forestry Sector. World Bank, Washington DC, 57 pp.

World Bank, 1999: Website: http://www.worldbank.org/html/extdr/fspr.htm

Appendix 1. Social criteria and indicators of immediate concern to workers developed by ITTO, Montreal Process and Helsinki Process (Montreal Process Working Group, 1998; Third Ministerial Conference on the Protection of Forests in Europe, 1998; ITTO, 1998).

Process	Criteria	National Level Indicators	FMU Level Indicators
ITTO	7: Economic, Social and Cultural Aspects	7.1 Value and percentage contribution of the forestry sector to the Gross Domestic Product.	-
		7.2 Quantity (volume) and value of wood and non-wood forest products traded in (a) the domestic market, and (b) the international market	+
		7.3 Quantity (volume) and value of wood and non-wood forest products for subsistence use, including fuelwood	+
		7.4 Ratio of domestic log production to the processing capacity of wood-based industries	-
		7.5 Efficiency of utilisation in terms of the percentage of felled volume processed	+
		7.6 Existence and implementation of mechanisms for the effective distribution of incentives and the fair and equitable sharing of costs and benefits among the parties	+
		7.7 Existence and implementation of procedures to ensure the health and safety of forest workers	+
		7.8 Employment in the forestry sector: (a) number employed, (b) percentage of total workforce, (c) average wage rate, and (d) injury rate	(a) + (b) - (c) - (d) +
		7.9 Number and extent of forest sites available primarily for: (a) research, (b) education, (c) the direct use and benefit of local communities, and (d) recreation	(a) - (b) - (c) + (d) +
		7.10 Number of people dependent on the forest for subsistence uses and traditional and customary lifestyles	+
		7.11 Area of forest upon which people are dependent for subsistence uses and traditional and customary lifestyles	+
Montreal Process	6: Maintenance and enhancement of long-term multiple socio-economic benefits to meet the needs of	Production and consumption a. Value and volume of wood and wood products production including value added through downstream processing; b. Value and quantities of production of	

	societies	non-wood forest products; c. Supply and consumption of wood and wood products, including consumption per capita; d. Value of wood and non-wood products production as percentage of GDP; e. Degree of recycling of forest products; f. Supply and consumption/use of non-wood products	
		Investment in the forest sector a. Value of investment, including investment in forest growing, forest health and management, planted forests, wood processing, recreation and tourism; b. Level of expenditure on research and development, and education; c. Extension and use of new and improved technologies; d. Rates of return on investment	
		Employment and community needs a. Direct and indirect employment in the forest sector and forest sector employment as a proportion of total employment; b. Average wage rates and injury rates in major employment categories within the forest sector; c. Viability and adaptability to changing economic conditions, of forest dependent communities, including indigenous communities; d. Area and percent of forest land used for subsistence	
Pan European Process	Criteria 6: Maintenance of other socio-economic functions and conditions	6.1 Share of the forest sector from the gross national product	
		6.3 Changes in the rate of employment in forestry, especially in rural areas (number of people employed in forestry, logging, forest industry)	
		6.3.1 Existence of a legal/regulatory framework, and the extent to which it provides for legal instruments for securing income levels in forest sector	
		6.3.2 Existence and capacity of an institutional framework to: develop and maintain human resource skills in all relevant tasks	
		6.3.3 Existence of economic policy framework and financial instruments, and the extent to which it supports programmes to ensure employment in rural areas in relation	

		to forestry	
		6.3.4 Existence of informational means to implement the policy framework, and the capacity to: . secure a fair share of income form non-wood products coming from rural sources of income	

Making Rio Work

The Vision of Sustainable Development and its Implementation through Forest Certification

D. Burger [1]

Abstract

The aim of this paper is to establish whether and how forest certification is an appropriate instrument for making sustainable development a reality in forest management. The history and role of the vision of sustainable development are first presented. The vision is designed to perform functions of guidance, focus, legitimisation and identification. The guidance for action it provides is identified with reference to 5 principles of sustainable development: precaution, efficiency, social justice, partnership, consistency. These principles can orient the search for paths to sustainable development. FSC forest certification is strongly influenced by the Rio Conference on Environment and Development, and makes a declared attempt to turn the vision of sustainable development into a reality. The key elements of the certification system are identified, and the issue is examined as to which principle of sustainable development each element complies with. Forest certification can help make sustainable development a reality; whether it succeeds in doing so will depend on five factors: consistency with sustainable development, credibility, legality, financial viability and acknowledgement by business partners.

Keywords: *sustainable development, sustainable forest management, forest certification, Forest Stewardship Council*

[1] Gesellschaft für Technische Zusammenarbeit (GTZ), 65760 Eschborn, Germany.

K. von Gadow et al. (eds.), Sustainable Forest Management, 153–192.

Introduction

At the UN Conference on Environment and Development held in Rio de Janeiro in 1992, 178 countries agreed on a joint vision of sustainable development. Rio was not only the largest conference of all time, but was also probably one of those whose outcome has been most seriously misunderstood and underestimated.

The recommendations and undertakings produced at Rio have often been dismissed as noncommittal clichés by people believing the Conference to have been of little use "except as a free holiday for the participants". One reason for that might be that the results were recorded in a large number of very long documents, written in "diplomatese" that was very difficult to read.

Forestry professionals tend to underestimate the significance of the Rio Conference. This is because they believe that in Rio, other fields of professional specialisation finally began to adopt the concept of sustainability that foresters had already been developing for over two hundred years. In doing so, however, they overlook the fact that the vision of sustainable development produced in Rio is far more wide-ranging than mere strategies for sustainable forestry. When understood as an integral component of implementing the vision of sustainable development, sustainable forest management means far more than just managing forests in compliance with the rules of sustainable forestry.

Forest certification was developed as an instrument to push forward and implement sustainable forest management. Yet although the number of certified forests is increasing at an astonishing rate, there remains a broad lack of clear understanding as to precisely what forest certification involves. In particular, forest certification is often seen as involving a set of forest management standards laid down on a more or less arbitrary basis by environmental groups.

This paper will attempt to present the holistic vision of sustainable development with reference to five guiding principles of action for sustainable development. These principles are designed as a framework to orient the search for paths to sustainable development. The issue will then be addressed of whether and how forest certification reflects these principles of sustainable development, i.e. whether forest certification translates the vision of sustainable development into the forest management context, thus helping turn it into reality. Finally, the issue will be raised of the significance of forest certification for the further development of sustainable forest management as a pillar of sustainable development.

The Vision of Sustainable Development

From catch-up to sustainable development

For a long time, development was understood as an exclusively economic process that could be adequately described in terms of growth in Gross National Product (GNP). Thus the United Nations defined the target for the First Development Decade from 1961 to 1970 as being 5% and that of the Second Development Decade from 1971 to 1980 as being 6% growth in GNP.

In his 1949 address to the nation, President Truman spoke of “underdeveloped” areas that had not yet reached the state of economic development of the Northern countries, particularly the USA, and whose development needed to “catch up” (Sachs,W., 1997). This idea of “catch-up development”, which was prevalent for a long period, was based on the notion that all peoples move on the same track and that development thus means catching up on the same form of economic growth already achieved by the pioneers.

After the 1960s, the interrelationships between economic development and the environment gradually became a key issue on the development agenda. In particular, the report to the Club of Rome entitled “The Limits to Growth”, submitted by Meadows in 1972, aroused keen interest. This report illustrated very clearly that exploitation of the environment poses a threat to economic growth. The relationship between economic growth and environment was discussed in detail at the 1972 United Nations Conference on the Human Environment held in Stockholm, albeit then still under the heading: environment *or* economic development?

The one-dimensional idea of catch-up development was gradually superseded by a more holistic, multi-dimensional concept of development. The term “sustainable development” was first used in 1980 in the World Conservation Strategy of IUCN, WWF and UNEP: “For development to be sustainable, it must take account of social and ecological factors, as well as economic ones; of the living and non-living resource base; and of the long-term as well as the short-term advantages of alternative action“.

In its 1987 report entitled “Our Common Future”, the World Commission on Environment and Development (WCED) set up by the United Nations in 1983 and led by Brundtland, gave the following definition: “Sustainable development is development that meets the needs of the present without compromising the ability of future generations to meet their own needs.”

In 1992, the United Nations held the World “Conference on Environment and

Development" (UNCED) in Rio de Janeiro. At this largest conference ever held, the delegations of 178 countries agreed on a shared vision of sustainable development.

Sustainable development is described as a
"global partnership for

- economically viable,
- socially just and
- environmentally sound

development, not only for the present, but also into the future" (Agenda 21, Preamble).

1949	Address to the nation Truman	concept of catch-up development
1961	1st UN Development Decade (1961 – 70)	goal of development: economic growth (+5% GNP)
1962	Silent Spring by R.Carson	life on earth is threatened
1971	2nd UN Development Decade (1971 – 80)	goal of development: economic growth (+6% GNP)
1972	Limits to Growth by D&D.Meadows Report to the Club of Rome	limited resources and capacity of earth to withstand burdens set limits to growth
1972	UN Conference on the Human Environment (Stockholm)	environment <u>or</u> development
1980	IUCN: World Conservation Strategy	the term "sustainable development" used for the first time
1983	World Commission for Environment <u>and</u> Development (Brundtland Commission.)	Final Report (1987): Our Common Future definition of sustainable development
1992	UN Conference on <u>Environment and Development</u> (Rio de Janeiro)	vision of sustainable development (embracing the economy, society and the environment)

Table 1. From "catch-up" to "sustainable" development

Sustainable development is understood as a

- normative or ethical principle: The development interests and opportunities of future generations are placed on an equitable footing with those of the present generation. This entails an obligation to manage natural resources as well as economic and social resources, such that future generations encounter a stock of resources that offers them at least equal opportunities to satisfy their needs and undergo development;
- all-embracing process:

- The nature of the process is such that it embraces all spheres of life, i.e. not only the economic, but also the ecological and social dimensions. These dimensions are very closely interlinked (see Fig. 1), and interact accordingly. Viewing only one dimension in isolation cannot but result in a highly fragmentary picture of the development potentials.
- In terms of location, sustainable development embraces the entire globe. Global sustainability in turn embraces and presupposes regional, national, sub-national and local sustainability.
- In human terms, all people, social groups and governments participate in sustainable development, both as actors and as affected parties.

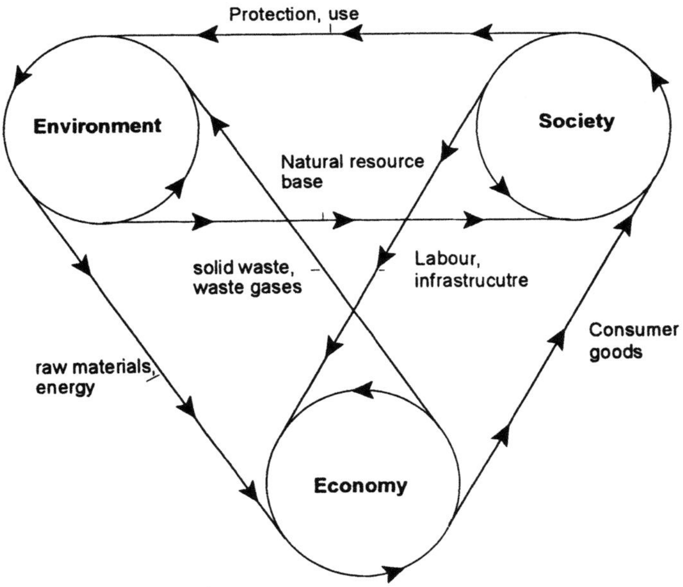

Figure 1. Interactions between the economy, the environment and society

The complex and comprehensive vision of sustainable development, embracing all spheres of life, is set out and elaborated, in some cases in very great detail, in the 5 Documents adopted in Rio:

- the Rio Declaration, with its joint commitment to the principles and dimensions of sustainable development;

- Agenda 21, the strategy for action which "aims to prepare the world for the challenges of the next century";

- the Biodiversity Convention;

- the Framework Convention on Climate Change;

- the Forest Principles, precursor of a Convention.

For an overview of the scope and content of the vision of sustainable development as described in the Rio documents see Fig.2, Table 2 and Table 3.

In Conferences and Conventions after Rio, for instance in the Desertification Convention or at the World Conference on Women, repeated attempts have been made to broaden the scope of life domains geared to the vision. The Commission on Sustainable Development (CSD) was instituted to facilitate and support implementation of the resolutions adopted at Rio and the vision of sustainable development. This body in turn instituted the Intergovernmental Panel on Forests (IPF), which was designed to prepare and support implementation of the resolutions relevant to the forestry sector. At the UNGASS Conference 5 years after Rio it put forward a wealth of proposals for implementation. The realisation thereof is to be supported by the Intergovernmental Forum on Forests (IFF).

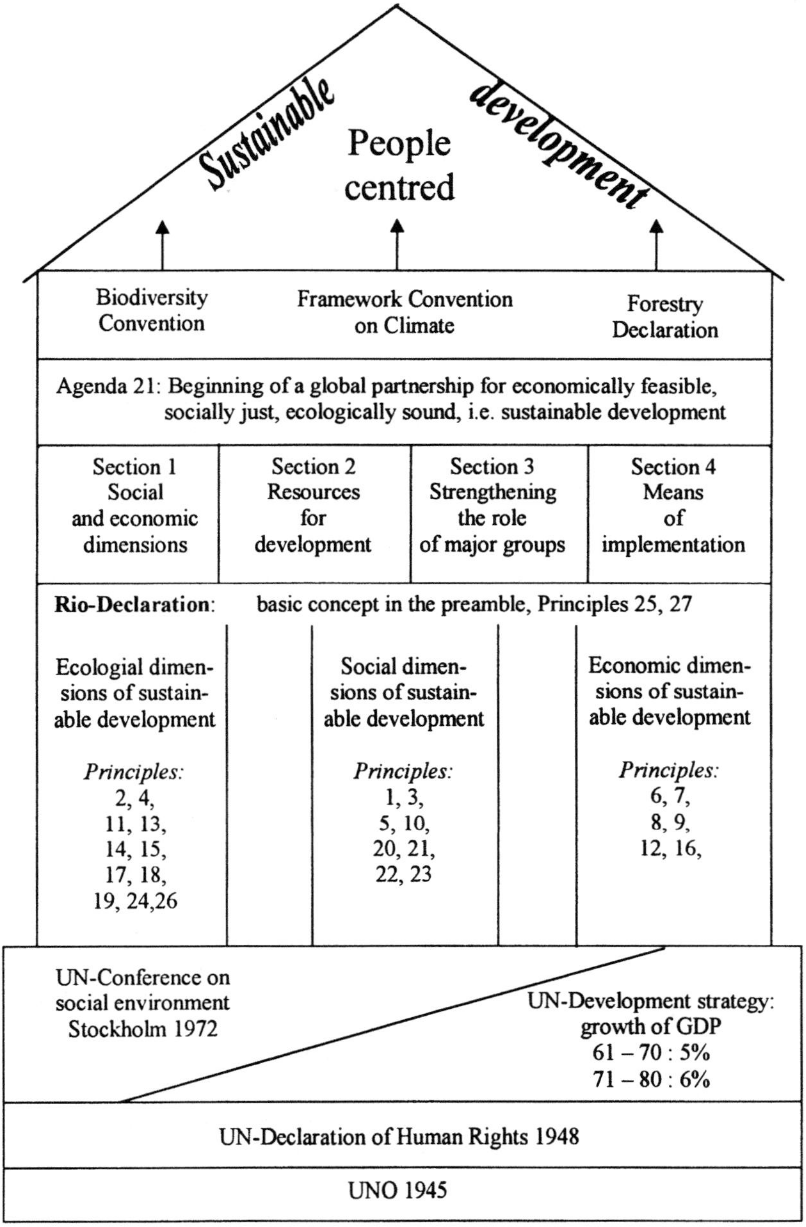

Figure 2. The Rio agenda for Sustainable Development

Aim: to protect the global environmental and development systems through new, equitable, global partnerships among states, key sectors of societies and people	
Principles:	
1.	Human being at the centre of concerns for sustainable development
2.	Sovereignty and responsibility of states to exploit their own resources
3.	Right to development for present and future generations
4.	Environmental protection an integral part of development processes
5.	Eradication of poverty an indispensable requirement for sustainable development
6.	Priority for developing countries; international actions address the interest of all countries
7.	Global partnership, common but differentiated responsibilities, responsibility of developed countries
8.	Elimination of unsuitable production and consumption patterns; demographic policies
9.	Strengthening of endogenous capacities through exchange and transfer of technologies
10.	Full sharing of information and participation of citizens in environmental decision-making
11.	Environmental legislation applied to specific environmental and development contexts
12.	Open international economic system; no unilateral, discriminatory trade policies
13.	National and international law on liability and compensation for environmental damage
14.	Prevention of the transfer of activities and substances damaging to the environment or to health
15.	Environmental protection according to the precautionary approach
16.	Internationalisation of environmental costs; polluter pays principle
17.	Environmental impact assessments as a national instrument
18.	Obligation to provide notification of natural and environmental disasters; international help
19.	Transboundary environmental effects: timely information and consultations with affected states
20.	Participation of women in environmental management and development
21.	Mobilisation of youth in a global partnership for sustainable development
22.	Recognition of the identity, culture, interests of indigenous peoples; participation in sustainable development
23	Protection of the environment and natural resources of oppressed, dominated, occupied peoples
24	Respect of international environmental protection laws in times of armed conflict
25	Peace, development, environmental protection are interdependent and indivisible
26	States resolve environmental conflicts peacefully and in accordance with the Charter of the UN
27	Implementation of the Declaration in a spirit of good faith and partnership

Table 2. The Rio Declaration in catchwords.

Preamble: Agenda 21 addresses poverty, hunger, disease, illiteracy and deterioration of the ecosystems, at preparing the world for the challenges of the next century; represents the "beginning of a global partnership for an economically viable, socially just, ecologically sound development, not only for the present, but also for the future; in other words, sustainable development".	
Section 1: Social and economic dimensions	**Section 3: Strengthening the role of major groups**
2. International cooperation	23. Preamble
3. Combating poverty	24. The importance of women for sustainable development
4. Production and consumption	25. Children and young people
5. Demographic dynamics	26. Indigenous people
6. Human health	27. Non-governmental organisations
7. Human settlements	28. Local authorities
8. Environment and development	29. Workers and their trade unions
	30. Business and industry
Section 2: Resources for development	31. The scientific and technological community
9. The atmosphere	32. Farmers and government policies
10. Land resources	**Section 4: Means of implementation:**
11. Deforestation	33. Financial resources and mechanisms
12. Desertification and drought	34. Technology transfer
13. Mountains	35. Science for sustainable development
14. Agriculture and rural development	36. Education and training
15. Biological diversity	37. National mechanisms and international cooperation
16. Biotechnology	
17. Oceans, seas, coastlines	38. International institutional arrangements
18. Freshwater resources	39. International legal instruments and mechanisms
19. Toxic chemicals	40. Information for decision-making
20. Hazardous wastes	
21. Solid waste and sewage	
22. Radioactive wastes	

Table 3. Agenda 21 in catchwords

The vision: magic formula, empty formula or framework for orientation?

The holistic, i.e. all-embracing understanding implicit in the vision of sustainable development gives rise to serious misunderstandings and controversies. Given that the vision claims to embrace all people, life domains and biotopes, it is then concluded that the vision claims to be able to resolve all tensions and conflicts between and within societies and sectors, and to establish clear and consistent directives for action by all countries and social groups. In other words, it is believed to be a magic formula for solving all problems.

However, the vision cannot, nor is it designed to, eliminate or "make disappear by magic" conflicting interests, such as those between industrialised and developing countries, or between the urban and rural populations, or conflicting goals, such as those between short- and long-term profit, between private and macroeconomic profit, or between agricultural and forestry production goals. Nor can it or is it designed to relieve countries and social groups of the responsibility for identifying and negotiating their paths to development by prescribing for each actor precisely who should do what, when, where and how. The vision is designed neither to resolve conflicts, nor to prescribe developments on a deterministic basis. Having said that, it can serve as an orientation aid to help identify solutions to conflicts and paths to development.

Given that the vision of sustainable development contains no prescriptions to determine the actions of individual actors, the conclusion tends to be drawn that it is an empty formula, i.e. that any action whatsoever would be compatible with the vision. This conclusion overlooks the fact that a vision can indeed serve as a guide to action, even though it does not prescribe any specific actions. The vision of sustainable development is neither a magic formula nor an empty formula. It should be understood not as a deterministic, but as a heuristic principle, i.e. as a framework for orientation to help seek and identify paths to development in negotiations between partners in development.

Guide rail function: By helping decision-makers decide whether or not an action will be sustainable, the vision acts to a certain extent as a guide rail to help them find their way along a corridor of permissible actions. Section 1 of the German Road Traffic Regulations also functions in the same way: The mere prescription that each road user must act such as not to endanger any other, is not in itself sufficient to determine the correct action in each specific situation. Having said that, the rule does not permit any action whatsoever, but sets a framework within which to judge whether certain actions are contrary to traffic regulations.

Focusing function: Whilst the vision of sustainable development presented here cannot be achieved in full, it helps focus attention on key elements, and helps those seeking solutions form a clear understanding of what they are looking for. The vision's principle that sustainable development is based on partnerships, for instance, can be helpful in identifying paths to development for a specific region, in that it focuses attention on alliances between key actors in the regional development process that might still need to be strengthened.

Legitimisation function: Sustainable development cannot be prescribed or planned by technocrats, but must be negotiated between key actors. Here, the vision of sustainable development can serve as an aid to negotiation by recalling to the actors' minds the consensus achieved in Rio. For instance, negotiations on development cooperation measures can be facilitated by making explicit reference to the basic consensus already agreed on by the partner countries at the Rio Conference and at subsequent events, as for instance those to promote participation by minority groups, to improve the position of women, or to promote sustainable forest management.

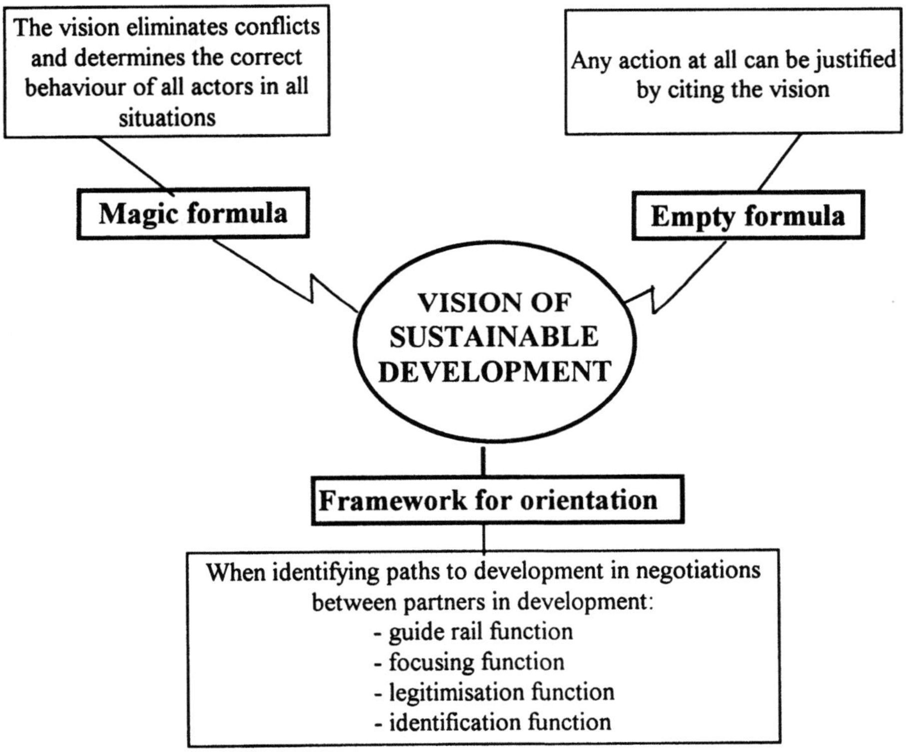

Figure 3. The role of the vision of sustainable development

Identification function: Human action is geared to different types of orientation, which can be hierarchically organised: From general, ethically sound visions, more specific principles or rules of conduct can be derived. These are geared to objectives, which can be described by criteria. Indicators specify the extent to which each criterion should be fulfilled. Thus sustainable forest management, for instance, can serve as a vision for actions in the forestry sector; maintenance of the ecological functions of the forest would be a principle derived from that, the maintenance of representative natural plots a criterion, and the proportion of total forest area accounted for by reference plots the indicator.

As Fig.4 shows, the further down the hierarchy one moves and the greater the degree of operationalisation, the more the orientation loses its validity but gains in specificity: The vision of sustainable development is designed to serve as an orientation everywhere, locally and globally, for all life domains and for all time. Yet it remains unspecific and vague. By contrast, an indicator describes in precise detail the target status of clearly defined parameters within a circumscribed spatio-temporal locus.

The different levels of orientations are not mutually contradictory, but complementary and interdependent. The more specific the objectives, the more swiftly they need to be adapted to changing ambient conditions. In such situations, however, the overall orientation can easily become lost, and an opportunistic arbitrariness become the norm, unless the specific objectives are integrated into more general visions that are binding in the long term.

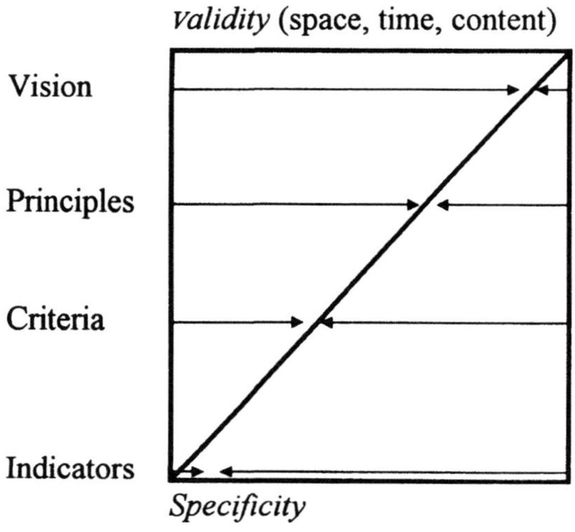

Figure 4. Validity and specificity of various orientations

This results in the very important identification function of the vision (cf. Bleicher,K., 1990, p.

98 ff.): Companies which make a binding commitment to a particular vision remain reliable for their partners, even when specific objectives are changed, because the overall orientation remains intact. Staff too can continue to identify with the long-term corporate vision, despite changes in the short-term objectives. A convincing vision has a strong impact on staff motivation, and increases their creative learning capability (Senge, P.M., 1990, p. 206). The motivating function of a vision, which motivates by generating a meaningful framework for action, is expressed in the following proverb: "If you want to build a ship, don't summon people to hand out tools and chop wood. Awaken their desire to sail the open seas."

Guiding principles of action for sustainable development

The guiding principles of action for sustainable development were derived from the vision of sustainable development, and constitute a first step towards the formulation of operational goals. They describe rules of behaviour for people to manage their resources and deal with each other. These principles do not relate only to natural resources, but also to all other material and intangible resources that people need in order to live. They describe rules of behaviour not only for governments and public development agencies but for all actors in development.

Resources for sustainable development are:

Environmental resources:

> biosphere, landscapes and biotopes comprising air, water, soil;
> vegetation, fauna, raw materials;

Economic resources:

> anthropogenic real capital such as machinery, roads, buildings;

Social resources:

> human capital: people with their education, training and health;
> social capital: the established body of rules, institutions and collective knowledge, which determines "the way in which the economic actors interact and organise themselves to generate growth and development" (World Bank, 1997 : 3, 77ff.). The Enquete Commission on the "Protection of Humanity and the Environment" speaks of "institutional capital", and the "shaping of conditions for the life of the individual and society as a whole" (Deutscher Bundestag, 1998: 43)

Actors in development:

> The principles of sustainable development ought ideally to be complied with by all actors in the development process. A large number of interrelationships and dependencies exist between those actors. Action by any actor will inevitably affect

others. If the actors' actions are to be harmonised more closely with the principles of sustainable development, be it through legal provisions, training, extension work or through more conducive general frameworks, then the entire field of actors needs to be taken into account, together with the mutual interrelationships between them.

Fig.5 is an example of a complex field of actors influencing directly or indirectly the sustainability of forest management. Obviously, these actors will never comply completely with the principles of sustainable development. In order to promote sustainable development through forestry, key actors must be identified, their commitment to must be supported. This is precisely the strategy of forest certification.

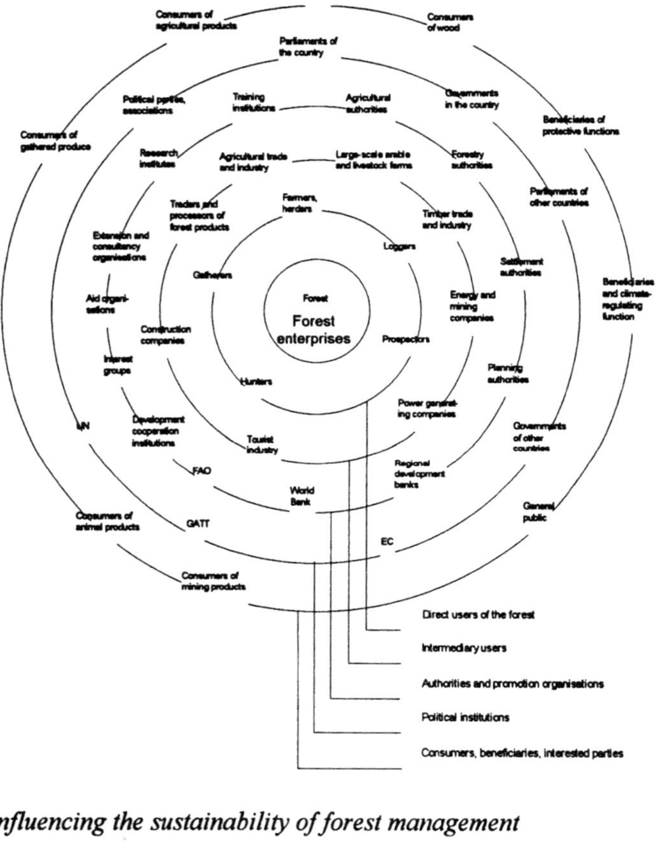

Figure 5. Actors influencing the sustainability of forest management

The precautionary principle

This principle is based on the idea that measures should be taken to protect the resources of future generations, such that at least the same life opportunities are preserved for those generations, that are now available to the present generation. Formulated initially with respect to natural resource management, the principle is understood in many societies in a traditional context, i.e. the context of hunting and fishery, as well as the management of soils and trees, and was the point of departure for the development of sustained-yield forest management. Having said that, the principle is not confined to environmental resources, but must also be applied with respect to economic and social resources (Deutscher Bundestag, 1998: 43).

On the basis of the rules formulated by the Enquete Commission "The Protection of Humanity and the Environment" for the ecological dimension, the following more general rules can be drawn up for the precautionary use of all types of resource:

Regeneration rule: The rate of extraction of renewable resources should not exceed their rate of regeneration. This is a necessary but not yet a sufficient condition for the maintenance of regenerative capability. To maintain the regenerative capability of forest, it is also necessary to preserve species diversity, as well as natural structures, and growth and regeneration dynamics. This rule applies not only to natural resources. Social capital, too, such as is embodied in traditional social security systems or in conflict management mechanisms, must be reproduced to at least the same extent at which it is destroyed by changed framework conditions. Otherwise, riches are lost.

Substitution rule: When non-renewable resources are used, a maximum rate of recycling is to be ensured. Where these resources are consumed without recycling, an at least equivalent substitution by other – preferably renewable – resources needs to be ensured. "The society of the 21st century will be more dependent than ever on the development of social, intellectual, creative and cooperative skills and capabilities. Material resources must be substituted by an expanded body of knowledge." (German Bundestag, 1998: 42)

Burden intensity rule: Just as natural resources can only be burdened by substance discharges within certain limits without losing the capacity to perform their essential functions, so also can social resources not be burdened unlimitedly. For instance, social cohesion, which is a part of social capital, cannot be exposed unlimitedly to the burden of unequal opportunities without suffering damage, i.e. a loss of social capital. Burden intensity is a function of the level and the rate of repetition of a burden. It must correspond to the capacity of the resource to react.

The rule of precautionary hazard containment: Principle 15 of the Rio Declaration on Environment and Development states: "where there are threats of serious or irreversible damage, lack of full scientific certainty shall not be used as a reason for postponing cost-effective measures to prevent environmental degradation". (Earth Council, 1994: 19). Social resources should also be handled with a similar awareness of risks.

These rules are necessary, but certainly not sufficient conditions for the protection of resources for future generations. They can provide pointers and a structure for the process of seeking and negotiating paths to sustainable natural resource management.

The efficiency principle

According to this principle, resources should be managed such that a certain benefit is achieved with minimum input, or that a certain input delivers maximum benefit. This can be termed the efficiency principle.

There are indications that the efficiency of resource management can still be considerably improved, which would allow economic growth without increasing resource degradation. In their work "Factor Four", Weizsäcker and his co-authors (Weizsäcker et al., 1997) describe the possibility of "doubling wealth whilst halving resource use".

Efficiency is not synonymous with saving. Under certain circumstances, efficiency can be increased by saving. Having said that, when savings are made on investment in the future, this can lead to inefficient resource use.

According to economists, the principal mechanism of control for efficient resource use is the market. "The allocation of resources for various purposes at the appropriate locations using the most efficient methods is considered the real task of market control" (Deutscher Bundestag, 1998: 38). Having said that, the control function cannot be left to markets on an unrestricted basis, because competition is often limited and prices distorted. Consequently, in order for the efficiency principle to be fulfilled, at least the following two additional rules need to be observed:

> "Prices must perform the basic function of steering on markets, on a permanent basis. As far as possible they should reflect the scarcity of resources, sinks, production factors, goods and services."
>
> "The frameworks for competition should be designed such that functioning markets emerge and are maintained, innovations are stimulated, a long-term approach pays off, and the social change needed to adapt to future requirements is promoted." (Deutscher Bundestag, 1998: 48)

Despite the eminent significance of the efficiency principle for the future viability of all societies, two points need to be borne in mind:

- Firstly, even if the efficiency of resource use is maximised, the earth's resources remain limited, which is why the efficiency principle cannot override the precautionary principle.

- Secondly, efficiency has no direction. Even if efficiency is maximised, the inappropriate use of resources leads to negative developments, according to the motto: "having lost sight of their goal, they doubled their speed".

The efficiency principle can only contribute to sustainable development when applied in conjunction with the other principles, and in particular only when it is regulated by consensus-based, i.e. negotiated, systems of objectives.

The social justice principle

This principle requires that the benefits and drawbacks of resource use, in other words profits and costs, as well as opportunities and risks, be distributed fairly among the concerned social groups. Three aspects are of major importance here:

- The distribution of financial or ecological income: These benefits should be allocated such as to enable all actors to cover their basic needs. The significance of ecological income, e.g. from gathering, hunting or fishing, is often underestimated in comparison to wage income. Thus new forms of land use in which traditional forms are displaced or prevented entirely, e.g. the introduction of plantation agriculture, can result in major social injustice. The share of income considered appropriate is dependent on the level of entitlement of the actors, which must be negotiated. In order that a fair distribution of income can be achieved through negotiation, it is necessary both that transparency be created concerning costs and benefits, and that policy frameworks permit that transparency.

- Protection against risks: The risks to be protected against include not only jobs that are unsafe and hazardous, but also the indirect risks to which affected persons are exposed as a result of resource use, such as the risk of flooding to which villages situated below a forest clearance are exposed. The term protection embraces both the prevention of damage, and compensation and the assumption of risk by mutual-aid groups and risk protection organisations, whether that involve traditional forms of organisation such as family, clan or village community, or modern social security systems such as insurance companies. Protection against risks also needs to be negotiated; this requires transparency, especially with respect to the nature and scope of the risks, and the performance capability of existing and alternative security systems.

- Opportunities for self-development: The right to training and equal opportunities is also dependent on negotiation and policy frameworks. "The creation of justice or equality of opportunity to serve primarily social policy interests is not just a social objective, but is also conducive to good long-term economic performance, and is thus also an economic objective" (Deutscher Bundestag, 1998, p. 33)

The three aforementioned aspects of social justice all need to be negotiated, because it is only possible for all actors to receive enough if some actors are prevented from receiving too much. To achieve this, however, policy frameworks conducive to development must be in place. Several development cooperation organisations have compiled lists of corresponding conditions, shown below in Table 4.

World Bank	OECD/DAC	BMZ*
Good Governance (1989):	*Participatory Development and Good Governance (1995)*	*Criteria for Frameworks Conducive to Development (1991)*
Accountability concerning the use of public funds Transparency of decision-making Predictability of the behaviour of public sector representatives Openness of information for all actors in a national economy Rule of law for all	Rule of law Efficiency of public administration Stemming of corruption Limitation of excessive military expenditure	Respect for human rights Participation of the population in political decision-making Rule of law and guaranteed legal security Introduction of a social market economy Governance for development

*Table 4. Socio-political prerequisites for development Source, after Hammel (1997). *BMZ:* Bundesministerium für wirtschaftliche Zusammenarbeit *(Federal Ministry for Economic Cooperation and Development).*

The partnership principle

Partnerships between states, key sectors of society and persons are called for in many places in the Rio documents, e.g. in the Preamble to and Principle 27 of the Rio Declaration. As understood at Rio, partnerships are an essential characteristic of sustainable development. Without partnerships, the changes necessary for sustainable development can neither be brought about nor maintained.

Partnership means sharing, not in the sense of breaking up and thus destroying a whole, but in the sense of participating in a whole that should in principle be preserved and further developed. Partnership goes beyond common experience, requiring that the visions that give orientation be shared to at least some minimum extent. The visions of the partners can certainly be different, but must contain a certain overlap of common values. If, for instance, a German municipality enters into a partnership with an indigenous village in Amazonia, their visions can be very different, but they must at least share some notions such as that the forest is worthy of conservation. Partners need to have a minimum of knowledge about each other and communication with each other. Partnerships also require a certain, though not necessarily formal, level of agreement concerning mutual expectations and definition of roles.

Partnerships are intensified by increasing the common ground between the partners with respect to: visions, values, goals, expectations, information on the other partner(s) and the setting, by potentials for communication, and by formal and informal agreements.

Typical features of viable and resilient partnerships are respect for the expertise and culture of the other partner(s), transparent and reliable definition of roles, an ability to engage in dialogue and openness between the partners.

Sound partnerships rarely arise spontaneously. Generally they grow gradually, typically passing through the following phases:

- In the information phase the actors merely share information, initially concerning questions of common interest, then increasingly also concerning themselves.
- In the communication phase additional feedback is given, including feedback on mutual perceptions.
- In the cooperation phase tasks and objectives are shared on a case-by-case basis.
- The coordination phase is characterised by case-by-case agreements concerning mutual expectations.
- The partnership/alliance phase is reached when longer-term, formal or informal agreements concerning mutual expectations have been achieved that extend beyond individual cases.

The importance of formal agreements for a partnership is often overestimated. They only say something about the quality of a partnership insofar as they correspond to common ground that has grown up through time. Partnerships, be they in the form of associations, cooperatives or Forest Partnership Agreements, should therefore be given enough time to grow.

The process of globalisation, coupled with the growing recognition of both the "failure of the state" and the "failure of the market", are leading to a global restructuring of the roles of development actors. In the field of tension between the state and the market, neither of which is now able to perform the tasks of control on its own, the formation of new, viable alliances guided by the vision of sustainable development has taken on special significance.

The consistency principle

This principle calls for consistency or compatibility among sub-systems. In a certain sense it represents the hub of sustainable development, in that the ultimate aim of the vision is to design the sub-system of human economic activity such that it becomes compatible in the long term with the overall system of our planet. The motto is: "Act locally, think globally".

Sustainable development is a holistic concept that embraces all spheres of human life. However, this holistic aspiration conflicts with the limited human capacity for coping with complexity: people are only able to perceive and tackle sub-systems, which are sections of the whole. They are then at risk of mistaking the part for the whole (reductionism), and of overlooking the interrelationships between the sub-system and its surrounding system, including the inconsistencies and threats harboured there.

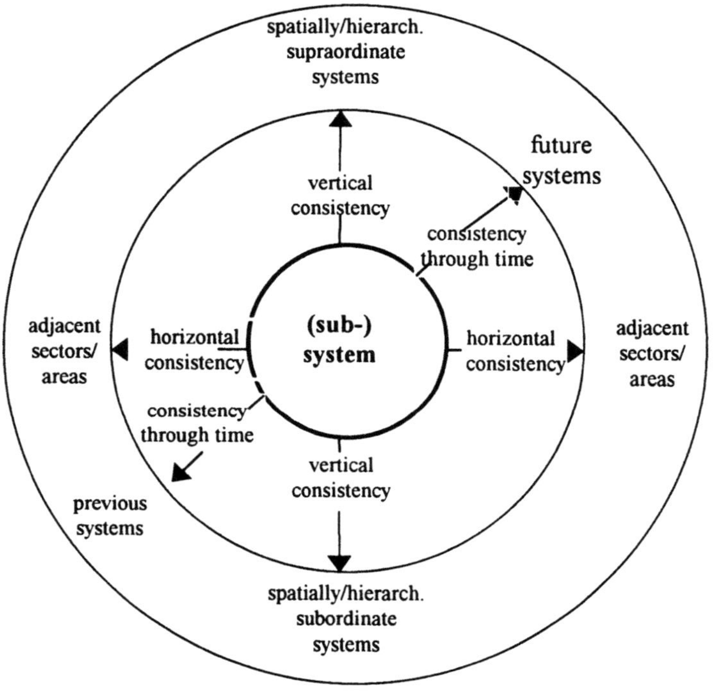

Figure 6. Compatibility of a (sub-)system with its neighbouring and surrounding systems

One way out of the dilemma created by the fact that holistic perception is impossible, and reductionism dangerous, is offered by a constant switch of perspectives. This means that, at intervals, the focus is taken from the sub-system under consideration and directed outwards, so as to perceive both the framework conditions and extraneous factors affecting the sub-system in question, and the impacts of that sub-system upon the outside. In this repeated switch of focuses from the inside to the outside, the outward perspective must take various directions: to the right and left to the adjacent sectors, in order to check horizontal consistency, upwards and downwards to the supra- and subordinate planning action levels, in order to ensure vertical consistency, and also backwards and forwards to the past and the future, in order to check the compatibility of planning with developments to date, i.e. temporal consistency.

Great circumspection is thus necessary in order to perceive inconsistencies. Perception must extend far beyond the boundaries of the sub-system within which the subject is operating. The approach of switching perspectives may appear trivial, and indeed, it means nothing else than "looking beyond the end of one's nose". However, the implementation of this approach is by no means trivial, because for someone to raise his sight beyond the sub-system in which he is working generally means he needs to acquire an additional perceptive faculty: the production engineer, for instance, usually has not learnt which parameters need to be observed when analysing a social system; he will need to enter into an exchange with the specialists in the neighbouring disciplines.

If, while moving between perspectives, contradictions or ruptures between a sub-system and its neighbouring systems or the surrounding system are perceived, then the significance of these inconsistencies must be evaluated according to whether they pose a threat to sustainable development, in order, if this is found to be the case, to design the sub-systems differently. Such modifications must be negotiated between the affected actors and sub-systems.

Horizontal inconsistencies between various sub-systems or sectors, e.g. between agriculture and forestry, arise very frequently in land-use planning. Efforts are often then made to solve the problem by means of better, in many cases very expensive data collection and mapping, but this regularly fails. Conflicts between different perspectives and interests can be resolved in three different ways:

- Possibly, improved information and communication can lead to improved understanding, and reveal that the conflict is in fact illusory; in this case, win-win solutions are possible from which both sides profit.

- Possibly, one view can be enforced against the other; this, however, harbours the risk that the conflict may erupt anew at any time when the relations of power change.

- Resolving conflicts by compromise presupposes common systems of objectives or visions that are supraordinate to the conflicting goals. In planning processes, the investment of time and money in forums where common visions are elaborated is often more worthwhile than the effort put into additional data collection. However, planning experts are usually much less well acquainted with the techniques of conflict management than with those of data collection.

In the event of vertical inconsistencies between planning and implementational bodies at different levels, an agreement to observe the subsidiarity principle can be helpful. In the Constitution of the Federal Republic of Germany, this principle regulates the relationship between the federation and the individual federal states. In the context of decentralisation processes, this can also be helpful in resolving potential conflicts between central and decentral authorities. The principle states that the higher community or institution may only assume such tasks that cannot be performed by the lower level. In addition to this "prohibition of withdrawal", the subsidiarity principle also contains a "requirement to provide assistance" (Stockmayer, 1996): to assist the decentral unit in realising its potentials, appropriate frameworks (i.e. an enabling environment) and in particular security must be guaranteed. Instead of intervention, the subsidiarity principle demands "the courage to let grow".

Temporal inconsistencies can be understood as contradictions or ruptures between previous and future developments. Their importance for the development process is particularly difficult to assess. On the one hand, development necessarily implies change, and thus certain breaks with the past; on the other hand, there is a danger that the breaks brought about by change will cause damage, e.g. if elements are substituted without an awareness of their true value, or if too many elements are substituted in too short a period.

Particularly in the social sphere, temporal inconsistencies can be easily overlooked and can have serious consequences for sustainable development. For instance, a new social security system may in theory be far more efficient than a traditional system; but if the users cannot perceive the relationship of the new system to the previous one and do not have enough time to learn how to handle the new system, the innovation can be ineffective or damaging.

Possibly the most serious errors in development planning have been committed through attempting to introduce innovations without giving enough attention to temporal inconsistencies, to dangerous ruptures between the past and the future, particularly in the social sphere. Dangerous ruptures that might occur during change processes can be avoided or mitigated by observing the following three rules:

- Innovations should be based on a careful analysis of existing practice. This seemingly trivial rule is rarely observed in reality. Thus very often new forms of land use are introduced because they appear to be more efficient with regard to marketable products, without the ecological and social functions of the forms of land use practised hitherto having been examined and understood.

- Innovations should be designed and put across as a further development of existing practice, such that it is made clear which technological, cultural, organisational or economic elements are being retained, and which substituted.

- The speed of innovation and the speed of learning should be harmonised. This rule is equally important with respect to both the development and the implementation of innovations:

- The development of innovations always commences with hypotheses that need to be subjected to theoretical or practical verification, i.e. tested. It is dangerous, although not unusual, for innovations to be introduced before the test or learning experience has been processed, and thus for unverified hypotheses to be viewed as innovations.
- The implementation of an innovation is not complete until the innovation is put into actual practice, as opposed to merely being laid down in a plan or in a legal regulation. Experience shows that innovations are learnt faster and better by those who are to practice them, the more intensively these individuals actively participated in the development of the innovations. The participation of target groups in the development of innovations is undoubtedly a time-consuming process. Nonetheless, it has very often transpired to be an error not to develop innovations on a participatory basis, in order to save time. Such time gains are frequently illusory, because implementation then takes place all the more slowly, if at all.

The consistency principle contains two components. The first, which might be termed the searchlight component, consists of illuminating the development process in all directions in search of inconsistencies. The second, the roundtable component, serves to address and analyse the contradictions detected through the first, and can lead to changes of course being made.

Understood thus, the consistency principle performs the function within the development process analogous to that performed by the central nervous system in the body: It detects disturbances in the form of inconsistencies, processes this information, and notifies any necessary changes in course to the implementing organs. The institutionalisation and design of the consistency principle is of paramount importance to the development process. It requires competence at several levels: technical competence in information collection and processing, because otherwise the contradictions will not be illuminated properly; communicative competence, because otherwise the negotiation process will not take place smoothly; and political competence, because otherwise it will not be possible to take decisions concerning changes of course. Institutionalisation of the consistency principle provides a substantial addition to the stock of social capital.

The principles as a guide to action when seeking paths to sustainable development

The development of a region, or of an individual enterprise, is influenced directly and indirectly by a very large number of actors, as a result of which framework conditions are constantly changing. Development as complex as this cannot be designed sustainably by planning a programme on a one-off basis. Gearing efforts to the vision of sustainable development rather calls for an ongoing process of adjustment and learning, because new and different threats to sustainability will constantly arise. At the same time, new opportunities to shape the process in the interests of sustainability will also keep emerging.

The five principles are designed to make the general vision of sustainable development more readily manageable as a guide to action. The principles are designed to help planners perceive and identify key aspects of development as a whole, and especially risks and opportunities for sustainability, without focusing exclusively on more specific, individual aspects. For instance, the principle of efficiency is designed to help focus attention on aspects of efficient resource use, whilst at the same time providing a reminder that this perspective is incomplete, and that e.g. the principles of precaution and social justice should also be considered. The principles should focus perception on the key aspects of sustainable development, whilst at the same time keeping sustainable development as a whole in view.

Similarly, the principles are designed to help evaluate key aspects of development, by establishing a link to the norms and values on which consensus was achieved in the vision of sustainable development. The principles should help create a situation in which for instance a development is not judged solely according to criteria of efficiency, but appropriate value is also attached to considerations of precaution and social justice.

Finally, the principles are designed to provide guidance in the search for the appropriate design of measures, by identifying the direction in which key aspects of development can be steered in order to move closer towards the vision, e.g. by intensifying partnerships.

Thus the principles should not be understood as operational directives, but as guiding principles to foster holistic perception, evaluation and design of complex processes geared to the vision of sustainable development. The principles thus lend orientation and structure to the ongoing learning process of development at the enterprise, regional and global development levels. The process of sustainable development is a constant balancing act amongst the different principles of sustainable development.

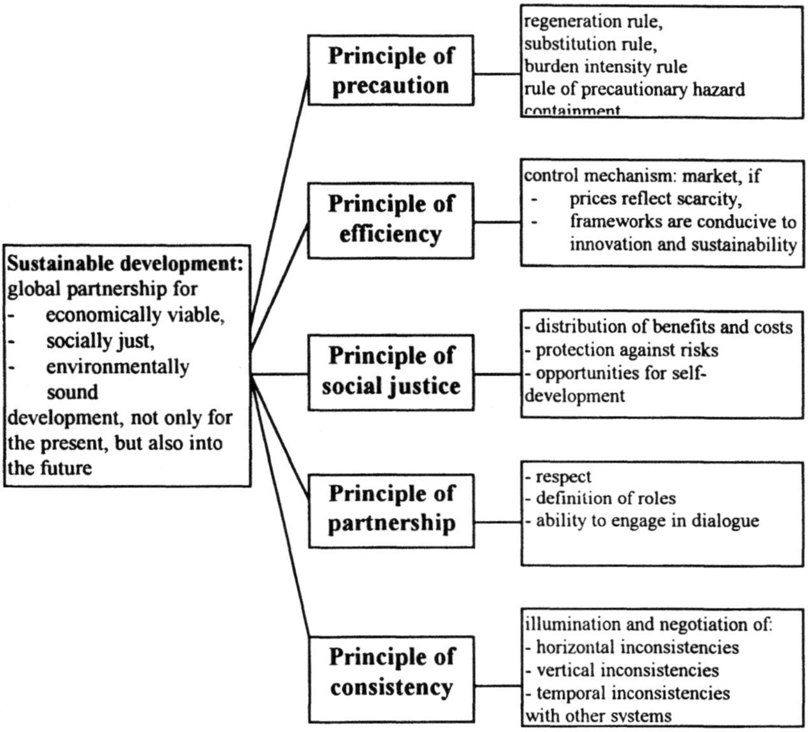

Figure 7. The principles of sustainable development – an overview

Forest Certification

The roots: tropical timber boycott and Rio Conference

In the 80s, the rate of destruction of tropical forests was continuing to increase on a dramatic scale: whereas 11.3 million ha p.a. were cleared in the tropics in the years prior to 1980, in the decade to 1990 the annual figure was 15.4 million ha. During that decade, 8.1% of all tropical forests were lost (FAO,1993).

In 1983 the "Committee on Forest Development in the Tropics" of the FAO (Food and Agriculture Organization of the United Nations), which as early as 1967 had already drawn attention to the ecological and economic consequences of forest destruction, proposed an international strategy to save the tropical forests. In 1985 the FAO put forward the Tropical Forestry Action Plan (TFAP) elaborated by the World Bank, the United Nations Development Programme (UNDP) and the World Resources Institute, which was adopted in 1986. This triggered a veritable avalanche of studies and conferences. Yet after only a few years it became clear that the commitment and willingness to change, both on the part of the tropical forest countries and on the part of donors, was not serious enough to have a perceptible impact on the development of tropical forest destruction. The rate at which this destruction was taking place was increasing year by year.

Much more than by the statistics on tropical forest destruction, the public and politicians in Europe were disturbed by the growing awareness of the fact that tropical forest destruction is also a major contributory factor to the greenhouse effect, and thus can also become a direct threat to the industrialised countries. The Enquete Commission of the German Bundestag "Preventive Measures to Protect the Earth's Atmosphere" therefore addressed the issue of protecting tropical forests in three hearings and a very comprehensive report (Deutscher Bundestag, 1990). Its recommendations led to the German Government's tropical forest programme.

Influenced by the dramatic destruction of tropical forests, environmental NGOs propagated a tropical timber boycott, in which many local authorities in Germany, as well as the German ministries for transport and construction, participated. Tropical timber importers defended themselves on the one hand with the sound argument that only a small proportion of tropical forest destruction is directly attributable to trade in tropical timber. However, they also put forward the ill-considered argument that the concession management practised in tropical forests is sustainable.

As early as 1989 the German Agency for Technical Cooperation (GTZ[2]), which in numerous projects supports the sustainable creation of value added from the tropical forest in the interests of regional development, distanced itself both from an indiscriminate tropical timber boycott, and from an unconsidered defence of the concession management then practised. It argued that "the boycott should be lifted as quickly as possible for tropical timber proven (i.e. labelled) to have been produced under conditions of sustainable forest management". "A tropical timber boycott directed only against timber from forests not managed sustainably would increase considerably the likelihood of sustainable forest management being introduced" (Burger and Faller, 1989). The GTZ also took this position in the report on the protection of tropical forests for the Enquete Commission (Deutscher Bundestag, 1990, p. 785).

[2] *Deutsche Gesellschaft für Technische Zusammenarbeit.*

With a view to acknowledging good forest management, as opposed to punishing the bad, the Smartwood Program of the Rainforest Alliance in the USA introduced independent forest certification in 1989. At the same time, countless homegrown certificates began to appear on the market based on nothing whatsoever. In a random sample of 80 environmental declarations on timber and paper products, WWF/UK found only three that could be at least partially verified (Vallejo and Hauselmann, 1999).

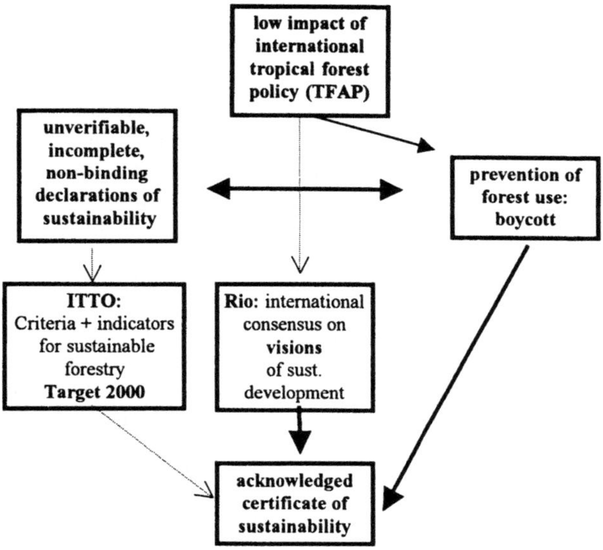

Figure 8. Roots of forest certification.

The conceptual and institutional foundations for forest certification were laid in the early 90s. The definition of sustainable forest management became largely standardised with the criteria and indicators of the International Tropical Timber Trade Organisation (ITTO), (ITTO, 1992, 1998), even if that definition was not yet operational and was still incomplete with regard to social criteria. This definition is consistent with the more comprehensive vision of sustainable development. The institutional foundations for internationally acknowledged, independent forest certification were laid in 1993, when 130 representatives of mainly environmental, economic and social organisations from all over the world founded the Forest Stewardship Council (FSC). This body accredits certifiers. On the basis of the internationally valid principles and criteria of sustainable forest management, declared to be derived from the Rio vision, and on the basis of the procedural regulations for the establishment of national standards and for certification, the FSC works to help bring about realisation of sustainability in the holistic sense, i.e. in accordance with the 5 aforementioned principles. The direction taken by the FSC on forest certification continues to be consolidated:

- By April 1999, more than 16 million ha in a total of 30 countries were already certified in accordance with FSC.

- The rejection of independent certification is decreasing; even organisations of European forest owners who one year ago refused to accept certification, now accept it and are currently developing a pan-European certificate (PEFC).

- Whilst ITTO has never come out in favour of independent certification, it does support it indirectly, whether intentionally or unintentionally, in two ways: The criteria and indicators developed for national reporting provide an initial basis for a national certification system, especially since their revision in 1998. Malaysia and Viet Nam for instance are currently developing national certification standards on the basis of the ITTO criteria. Secondly, the ITTO member countries have, by committing to Target 2000, i.e. the voluntary obligation to trade only in sustainably produced tropical timber from the year 2000 onwards, committed in principle to the same goal as forest certification. Even though ITTO has to date made no attempt whatsoever to support the introduction of proof that objectives have been achieved, it has made a significant contribution towards strengthening the legitimacy of the goal of forest certification.

- The national certification systems initially independent of the FSC, e.g. Indonesia and Malaysia, are making increasing efforts to achieve mutual recognition with FSC. It is to be hoped that mutual recognition will also be achieved between PEFC and FSC, in view of the fact that there are a large number of certificates which are confusing consumers, and jeopardising certification as a result.

In the coming years it is to be assumed that other organisations alongside FSC will also shape the development of forest certification. This development will be a delicate balancing act, in which there will be a constant risk of leaning too far in the direction of unverifiable (i.e. lacking clear criteria), incomplete or non-binding declarations of sustainability, or too far in the direction of a blocking of forest utilisation, i.e. by imposing conditions which cannot be met, which would de facto amount to a boycott.

The certification system – an instrument for implementation of the principles of sustainable development ?

The fact that forest certification has its roots in the Rio process does not necessarily mean that it is consistent with the vision of sustainable development. A forest certification system, however, in order to be accepted by consumers and to be supported by development organisations, must comply with the principles of sustainable development. When comparing different certification systems or when negotiating mutual recognition between different certification systems, one should thoroughly analyse how seriously each certification system is committed to the principles of sustainable development.

Since the FSC system is currently the only fully elaborated and internationally acknowledged forest certification system, only that system will be analysed in the following chapters. This analysis might serve as a reference for the evaluation of other systems once they are readily developed, e.g. PEFC in Europe or LEI in Indonesia.

Whether the certification system is consistent with the vision of sustainable development can be measured on the basis of the five principles of sustainable development. Each of the key elements and procedures of the FSC certification system will be examined to establish which principle of sustainable development (PSD) it complies with (see Fig. 9).

FSC principles and criteria for sustainable forest management

The principles and criteria (P&C) apply to the management of tropical, temperate and boreal forests. All certifiers accredited by FSC are required to apply them. In the introduction to the Principles and Criteria it is explained that the FSC and FSC-accredited certification organisations will not insist on perfection in satisfying the P&C, but that major failures will normally disqualify a candidate from certification, or will lead to de-certification. These decisions will be taken by individual certifiers. “Some flexibility will be allowed to cope with local circumstances”.

FSC Principle 1 requires compliance with national and international laws, treaties and agreements, as well as the FSC P&C. In Criterion 1.5 it also incorporates protection against illegal harvesting and settlement, and Criterion 1.6 requires forest managers to demonstrate their long-term commitment to the FSC P&C; in other words, forest managers cannot comply with the principles in one forest to be certified, and contravene them elsewhere. The principle is clearly designed to serve the purpose of consistent incorporation of forest management into the local and supra-local legal context (PSD 5).

FSC Principle 2 prescribes that long-term rights of land tenure and use be defined, documented and legally established. Criterion 2.2 refers specifically to the protection of traditional rights of the local population. Criterion 2.3 calls for the employment of appropriate mechanisms to resolve disputes over land tenure claims and use rights. At issue here are both PSD 3 (social justice) and PSD 5, (consistency through prevention or resolution of conflicts with neighbours).

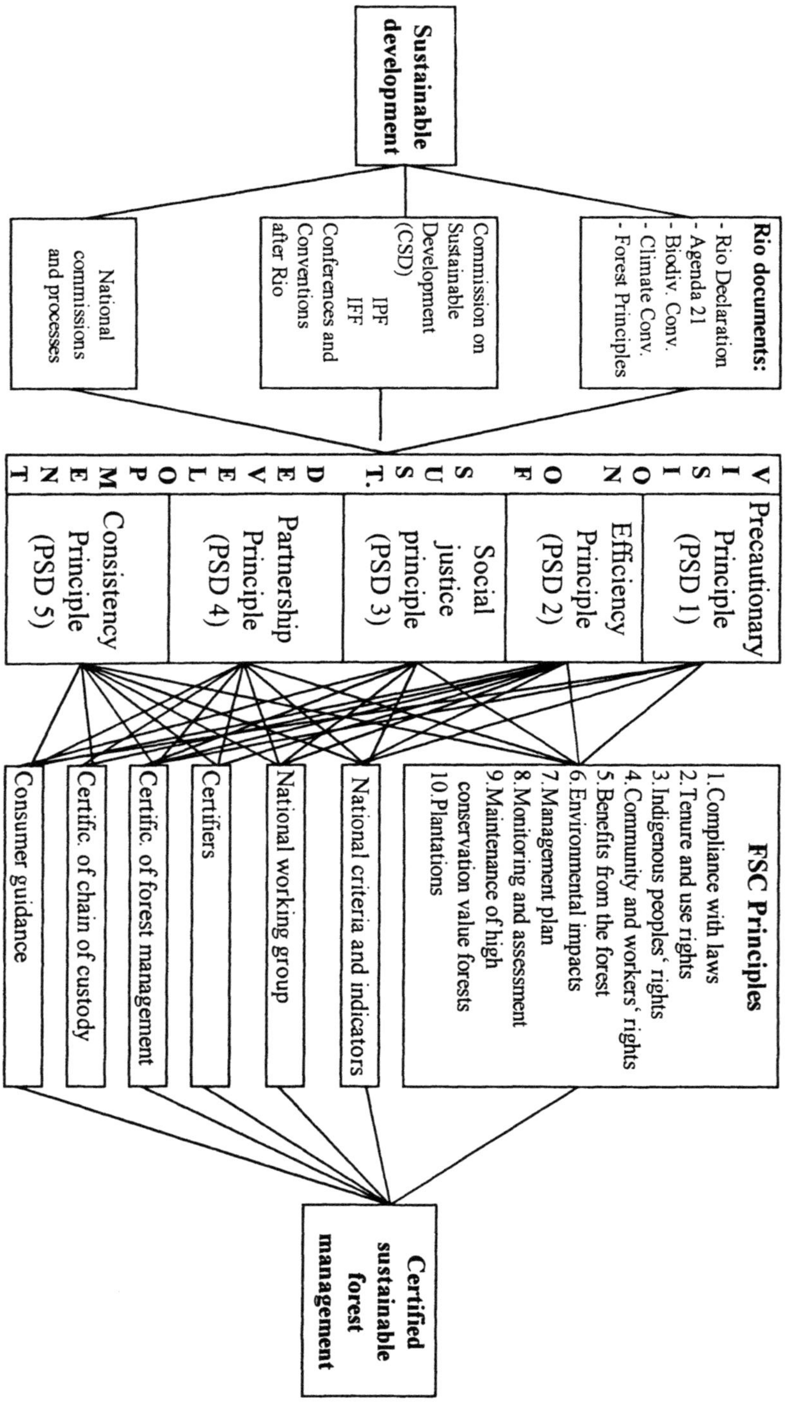

Figure 9. Consistency of the FSC forest certification system with the vision of sustainable development.

FSC Principle 3 is designed specifically to protect the land and resource use rights of indigenous peoples. According to Criterion 3.2, forest management shall not threaten or diminish, either directly or indirectly, the resources or tenure rights of indigenous peoples. Criterion 3.3 states that sites of special cultural, ecological, economic or religious significance to indigenous peoples shall be clearly identified in cooperation with such peoples, and recognised and protected by forest managers. Pursuant to Criterion 3.4, indigenous peoples must be compensated if their traditional knowledge is utilised in management of the forest. This principle serves the purpose of social justice (PSD 3), and the contractual incorporation of forest management into the cultural and social setting (PSD 5).

FSC Principle 4 relates to the long-term social and economic well-being of forest workers and local communities. It concerns employment opportunities, health and safety at work, freedom of organisation, ILO Conventions, social impact studies, consultation, participation and compensation. These issues address detailed aspects of social justice (PSD 3) and efficiency (PSD 2), and in particular efficient human resource management. They also address detailed aspects of the definition of roles and participation (PSD 4), and compatibility with international Conventions (PSD 5).

FSC Principle 5 requires that forest management operations encourage the efficient use of the forest's multiple products and services to ensure economic viability, and a wide range of environmental and social benefits. Aspects are addressed such as economic viability, production costs including environmental and social costs, reinvestment, establishment of local processing capacities for forest products, prevention of damage and waste, strengthening and diversification of the local economy by avoiding dependence on a single product. The issues addressed chiefly involve efficiency (PSD 2), but also the precautionary principle (PSD 1).

FSC Principle 6 aims to conserve biological diversity, water, soil, unique and fragile ecosystems and landscapes, and thus maintain the ecological functions and integrity of the forest. Criterion 6.4 calls for representative samples of existing ecosystems to be left in their natural state, one of the most controversial points in the negotiation of the German standards. This FSC principle corresponds to the principle of precaution (PSD 1).

FSC Principle 7 calls for a management plan to be written, implemented and kept up to date, clearly stating the long-term objectives of management and means of achieving them. As well as explaining the rationale for the sustainable rate of annual harvest, it should also contain plans for the identification and protection of rare, threatened and endangered species. A summary of the primary elements of the plan is also required to be published. The management plan is designed primarily to ensure efficient use of the operation's resources (PSD 2), whilst its publication is designed to ensure the participation of interest groups, and can be a first step towards the formation of new alliances (PSD 4). The rationale for the sustainable rate of harvesting, and the incorporation of nature conservation measures into the management plan, correspond to the precautionary principle (PSD 1).

FSC Principle 8 prescribes monitoring to assess the condition of the forest, yields of forest products, chain of custody, management activities and their social and environmental impacts. Criterion 8.5 requires publication of a summary of these reports. These surveys form the basis for continuation or adjustment of the measures with respect to precaution (PSD 1), efficiency (PSD 2), social justice (PSD 3), and participation and possible partnerships (PSD 4).

FSC Principle 9 used to cover the protection of primary forests, and was modified in January 1999. It now requires that high conservation value forests be managed such that the attributes defining such forests be maintained. All relevant decisions are to be considered in a precautionary context. Criterion 9.2 requires that the consultative portion of the certification process must place emphasis on the identified conservation attributes, and options for the maintenance thereof. Pursuant to Criterion 9.3, the maintenance and/or enhancement measures contained in the management plan shall be specifically included in the published summary of the management plan. This principle serves the purpose primarily of precautionary maintenance of resources (PSD 1). Special emphasis is placed on participation (PSD 4).

FSC Principle 10 relates specifically to plantations which provide an array of social and economic benefits, and can contribute to satisfying the world's needs for forest products. At the same time, they should complement the management of, reduce pressures on, and promote the restoration and conservation of natural forests. The criteria contain a number of prescriptions, e.g. concerning the design of natural forest conservation objectives, wildlife corridors, preference for native tree species, and a proportion of the overall forest management area is to be managed such as to restore the site to a natural forest cover. Pursuant to Criterion 10.9, plantations established in areas converted from natural forest since 1994 cannot qualify for FSC certification. This principle seeks primarily to ensure the efficient use of resources (PSD 2), and the precautionary conservation of natural resources (PSD 1).

National working groups, national criteria and indicators

National working groups render the FSC Principles and Criteria (P&C) more specific, and draw up national criteria and indicators (C&I) that correspond more closely to the economic, ecological and social conditions prevailing in the respective country. In some countries, separate standards are developed for different forest formations and products. In Brazil, for example, two national standards are currently being completed, one for the Amazonian terra firma forest, and one for plantations. In Bolivia, one national standard for timber production has been completed, and one standard for the sustainable production of Brazil nut is about to be so.

As is the case with the FSC at the international level, the national working groups must also include representatives of the private sector, and social and environmental groups. These three constituent divisions vote on an equal footing, regardless of the number of members. National standards are usually drawn up in a process of protracted and difficult negotiation between the various interest groups, interspersed repeatedly with hearings, workshops and field tests. It is becoming increasingly apparent that this process of negotiation, quite regardless of its outcome, is per se of major significance for forestry policy. These national working groups are emerging as key forestry policy fora, at which both new alliances are formed, and better harmonisation of the forestry sector and its policy frameworks (e.g. fiscal regulations) can be negotiated.

The establishment of a certification system, and in particular the national FSC working groups, are having a catalytic effect with regard to the reallocation of roles in the forestry sector. The state is usually present as an observer, although it has very broad scope to define that observer role, either choosing to step back and merely observe, or electing to play a more active, facilitating role, as for instance in Bolivia. There, certification standards and forestry law are so closely harmonised that, with the exception of social criteria, virtually all certification criteria are prescribed by law. In Bolivia the state is also making use of the potential offered by the certification system to actively alleviate its own burden, by forgoing certain monitoring and reporting functions. The FSC working group fora are also becoming increasingly important to trade unions in various tropical countries, as the fora take on more active and negotiated roles in the forestry sector.

In elaborating national criteria and indicators, the national working groups help make more specific all 5 principles of sustainable development, on a basis commensurate with their relevance to the individual FSC Principles. Having said that, the working groups play a particularly significant role with respect to the implementation and further elaboration of the principles of partnership (PSD 4) and consistency (PSD 5).

Certifiers

Certifiers are accredited by FSC after careful examination of their independence, reliability and competence. This means that accredited certifiers are then allowed to use and award FSC certificates. They issue the certificates in their own name and on their own responsibility, but in so doing are allowed to make known their accreditation by FSC, and use its name. The certified enterprise may use a certificate bearing the name of the certifier and FSC. The certification of forest management and that of the chain of custody are two entirely different procedures. They may be carried out by the same certifier, but need not be. For both types of certification, the certifier must be accredited by FSC.

Certifiers provide a guarantee to customers that the information provided by the producer concerning the production process is correct. They thus create a basis for communication between producers and consumers that is based on trust. They enable the consumer to gain significant insight into the forestry sector, and to intervene (through purchasing decisions), even though s/he may never have set foot on the forest land in question. The certifier establishes a link between the system of forest management and the system of the consumer economy, creating a basis for the two systems to be harmonised and made mutually compatible. The certifier thus plays a key role in implementing the principle of consistency (PSD 5).

Certification of forest management

The certification of forest management always has several components:

- In the examination component the compliance of procedures within the operation with the FSC P&C is examined in an objective and verifiable manner. If national FSC C&I are recognised for the country in question, compliance with these also needs to be examined. The examination is carried out not only on a categorical basis, i.e. to establish whether the principles, criteria and indicators are "satisfied" or "not satisfied", but also on a gradual basis, to establish the extent to which they are satisfied.

- In a consultative component, recommendations and directives are issued as to when and to what extent procedures within the operation should be adjusted to fit certain criteria. Having said that, directives on the method of adjustment are usually not part of the certification process.

- In a compulsory participatory component the certifier must, within the scope of consultations, give all stakeholders an opportunity to comment on compliance with the FSC C&I (and national C&I) by the forest management being certified. The form of consultation must be appropriate to the culture of the stakeholders. In future, the national working groups will certainly need to provide the certifiers with more detailed methodological directives for these consultation processes. In Bolivia, for instance, the certification process in a concession has currently become deadlocked, because the certifier has obtained a written declaration of consent by leaders of the indigenous community, but the indigenous groups do not recognise this as a consultation, because in their culture such decisions have to be taken with all those affected participating.

- In a systems linking component, relations between the forest management to be certified and neighbours (land conflicts), other sectors (e.g. competition for labour, access to markets, funds), regional development, national policy and legislation are examined, and shortcomings in the harmonisation of systems identified. Directives and deadlines for improvements may also be issued.

The certification of forest management thus embraces not only an examination of the ecological dimension (PSD 1), the economic dimension (PSD 2) and the social dimension (PSD 3) of forest management, but also makes a significant contribution towards implementation of the partnership principle (PSD 4) and the consistency principle (PSD 5).

Certification of the chain of custody

Certification of the chain of custody only involves an examination of the reliability of information on the provenance of raw materials, semi- and finished products from certified forest management. It does not examine compliance with ecological, social or economic standards by operations along the chain of custody. The latter would be examined for the issue of an eco-label or in the context of a life cycle analysis (LCA), but not for an FSC certificate, which is thus limited in terms of its significance, to information on the sustainability of the mode of production of raw materials.

Certification of the chain of custody makes all producer, processing and trading enterprises along the chain of custody mutually dependent on one another. Each enterprise undertakes to behave in a certain manner. A breach of that obligation jeopardises the entire chain. A chain of partnerships arises from the forest management operation to the consumer. Certification of the chain of custody thus contributes to implementation of the partnership principle (PSD 4).

Consumer guidance

Consumer guidance is a key element of a certification system. The actions of an individual are determined by that individual's perceptions and value system. Perceptions are determined not only by the objective circumstances, however, but also to a very large extent by the way in which those circumstances are communicated, the aspects on which attention is focused, and how the objective circumstances are related to ethical values. Communication with consumers on the ethical and political significance of the vision of sustainable development, on the role of sustainable forest management as a pillar of sustainable development, and on the role of responsible consumer behaviour is thus of crucial significance for a certification system. The consumer's opinion is formed as a result of what is communicated to him, how, and by whom. One major strength of the FSC certification system lies in the fact that information on it is conveyed to consumers by organisations considered credible and competent opinion-formers in the public domain. Consumer guidance on FSC forest certification helps implement all 5 principles of sustainable development, but especially to the assumption by consumers of an active role in implementation of the partnership principle (PSD 4), and in building a bridge between producer and consumer systems (PSD 5).

Conclusions

The vision of sustainable development can be described more specifically through the five guiding principles explained in the previous chapters. Any program or instrument claiming to contribute to sustainable development should strive for greater compliance with these principles.

Forest certification aims to strengthen sustainable forest management, defined as forest management according to the demands of sustainable development. Although the area of certified forests is increasing rapidly, having reached more than 16 million hectares worldwide in mid 1999, this is still only about 0,5 percent of the world's forest area. Whether forest certification will contribute significantly to strengthening sustainable forest management is dependent on five factors:

1. The consistency of forest certification systems with the vision of sustainable development: Certification must satisfy the principles of sustainable development as far as possible, both with respect to the criteria and the procedures applied. In particular this calls for participation in forest management, in the establishment of the certification system, and especially in the definition of national standards as well as in the implementation of certifications.

   As demonstrated in chapter 3.2 and in table 5 below, the FSC certification system is highly consistent with the principles of sustainable development. Any alternative system should produce at least the same degree of consistency.

Elements of the certification system	Principles of sustainable development				
	PSD 1	PSD 2	PSD 3	PSD 4	PSD 5
FSC P&C : Principle 1					X
Principle 2			X		X
Principle 3			X		X
Principle 4		X	X	X	X
Principle 5	X	X			
Principle 6	X				
Principle 7	X	X		X	
Principle 8	X	X	X	X	
Principle 9	X			X	
Principle 10	X	X			
National working groups				X	X
National criteria and indicators	X	X	X	X	X
Independent certifiers					X
Certification of forest management	X	X	X	X	X
Certification of the chain of custody				X	
Consumer guidance	X	X	X	X	X

Table 5. Consistency of the FSC certification system with the Principles of Sustainable Development (PSD).

2. The credibility of certification: This determines acceptance by purchasers, and calls not only for objective and publicly known criteria, but also independence on the part of certifiers. Credibility

also requires that the operation from which the certified timber originates can be identified. The certification of all operations in a region will certainly not be seen by consumers as credible. Certification must therefore take place at the level of the individual enterprise, and not on a regional basis. The credibility of certificates is reduced as the number of competing certificates increases. It is therefore important that certificates be made comparable through mutual recognition.

3. The legality of the certification system: Pursuant to the rules of WTO (World Trade Organisation) certification must not discriminate against certain countries or types of forest. Were it to be applied only for tropical timbers, it would probably contravene the WTO provisions. It is therefore important that certification embrace compliance with the criteria of sustainable forest management for all forests. Certification must also be voluntary, as it might otherwise constitute an "unnecessary trade barrier" according to WTO.

4. The financial viability of certification: As a market instrument, certification must not impose untenable costs on economic actors. Having said that, a distinction needs to be drawn between the costs of conversion to sustainable forest management, to which all countries represented at the Rio conference and also most national forestry laws are committed – in other words these costs are not attributable to certification – and the costs of certification per se, i.e. the certifiers' costs and the costs of providing special documentation for certification. At present the costs fluctuate widely, particularly as a function of the size of operation. In Latin America, costs of between 0.55 and 21.33 US$ per ha have been calculated; in operations of 25,000 ha and more the figure is under 1 US$ per ha (De Camino, Alfaro, 1998). The costs of certification are entirely warranted for larger operations, but for smaller operations the instrument of group certification will need to be further developed, through which costs can be significantly reduced.

5. The acknowledgement by business partners: Forest certification is based on the idea that consumers of forest products committed to the vision of sustainable development should act as active business partners of forest managers, honouring certified sustainable management through preferential purchasing or even through higher prices. This rationale is starting to work. At the moment, demand for certified timber is higher than supply and many producers of certified timber are achieving a price premium. However, as seen in figure 5, there are many more business partners of forest management who could demonstrate their commitment to the vision

of sustainable development by acknowledging and honouring sustainable forest management: investors, e.g. banks and shareholders, could provide credits under preferential conditions recognising that certification provides a clear indication of compliance with general performance standards required for investments; insurance companies could offer lower tariffs to certified companies taking into account that risks of fires, grievances, accidents and disturbances through conflicts with neighbours could be lower than in uncertified companies; supervising government agencies could reduce the number of reports and audits because compliance with legislation is regularly examined by the certifier.

Compliance with the principles of sustainable development is not just a question of philosophical or ethical attitude but rather a question of sound far-sighted management. Many business partners receive benefits from sound forest management and it seems to be fair that well managed forest enterprises request partners to honour such benefits. Thus the alliance between forest management and consumers could be extended to other partners, resulting in a broader and stronger alliance for sustainable forest management.

Notwithstanding the high potential of forest certification, it can only be one instrument among many to put sustainable development into practice in the forest sector. Appropriate legal and economic frameworks and appropriate expertise among those employed in forest management are also essential.

Strengthening sustainable forest management is an important contribution to sustainable development because the forest sector, due to its ecological, economic and social functions, must be considered a pillar of sustainable development. It should be borne in mind, however, that sustainability in the forest sector is a necessary but not the sole condition for sustainable development. Other sectors, too, need to comply with the principles of sustainable development.

Literature

Bass,S., 1998: Forest Certification – The Debate about Standards. Rural Development Forestry Network. Network paper 23b, ODI, London

Bleicher,K., 1996 Das Konzept Integriertes Management. 4.revid. Aufl.,Campus,Frankfurt,NewYork

Burger,D., 1997: The vision of sustainable development. entwicklung + ländlicher raum, Frankfurt, 31.Jg., H.4,: 7-10

Burger, D., 1997: The political and social dimension of sustainable development. Presentation at the workshop

"Sustainable Development – Yeguare Programme" EAP/GTZ, Zamorano, Honduras, 24.-28.11.1997, mimeo, GTZ, Eschborn

Burger,D. and Faller,M. 1989: Tropenholzimportbyokott und Technische Zusammenarbeit. Forstprojekte und Bykott-ein Widerspruch?. gtz-info.Zeitschrift für Technische Zusammenarbeit,Eschborn,4/98:19-23

Burger, D. and Happel, J., 1997: Das Leitbild nachhaltiger Entwicklung - handlungsleitende Orientierung der GTZ? Diskussionspapier, GTZ, Eschborn

De Camino,R. and Alfaro, M., 1998:Certification in Latin America: Experience To Date. Rural Development Forestry Network, Network Paper 23c, ODI,London

Deutscher Bundestag, 1990: Schutz der tropischen Wälder: eine internationale Schwerpunktaufgabe. Bericht der Enquete-Kommission des Deutschen Bundestages "Vorsorge zum Schutz der Erdatmosphäre", Bonn

Deutscher Bundestag, 1998: Konzept Nachhaltigkeit. Vom Leitbild zur Umsetzung. Abschlußbericht der Enquete-Kommssion "Schutz des Menschen und der Umwelt – Ziele und Rahmenbedingungen einer nachhaltig zukunftsverträglichen Entwicklung" des 13. Bundestages. Bonn

Earth Council, 1994: The Earth-Summit - Eco 92: different visions. Inter-American Institute for Cooperation on Agriculture IICA, San José, Costa Rica

FAO, 1993: Forest Resources Assessment 1990, Rome.

Hammel,W., 1997: Entwicklungszusammenarbeit ist politischer geworden. E+Z.,Berlin, Jg,38,1:12-14

ITTO, 1989: Criteria and Indicators for Sustainable Management of Natural Tropical Forests. International Tropical Timber Organisation, Yokohama

ITTO, 1992: ITTO Guidelines for the Sustainable Management of Natural Tropical Forests.International Tropical Timber Organisation, Yokohama

Kruedener, B. v. and Burger,D., 1998: Certifying forest management. A consumer-driven market instrument with potential to promote sustainable development. gate, Eschborn, 2/98: 30 -35

Sachs,W., 1997:Sustainable Development - Zur politischen Anatomie eines internationalen Leitbildes. epd Entwicklungspolitik : d1 - d10

Senge,P.M., 1990: The Fifth Discipline. The Art & Practice of The Learning Organization. Currenca Doubleday, New York

Vallejo,N., and Hauselmann,P., 1999: La Certificacion, un Seguro de Credibilidad. Pnencia preparada para el Seminario Biotrade, Villa de Leyva-Colombia,23.-25.3.1999

Weizsäcker, E.U. v., Lovins, A.B. and Lovins, L.H., 1995: Faktor Vier – doppelter Wohlstand – halbierter Naturverbrauch; der neue Bericht an den Club of RomeDroemer, Knaur, Munich.

World Bank, 1997: Expanding the Measure of Wealth. Indicators of Environmentally Sustainable Development. Environmentally Sustainable Development Studies and Mongraphs Series No 17. Washington

Perceived images of various actors engaged in sustainability discussions

G. Hütte [1]

Abstract

Beginning in 1980, a new surge of interest seizes the more than two hundred year old sustainability discussion, initiated by a powerful coalition in the nature conservation sector. This contribution shows how important actors in the forestry and nature conservation sectors were able to influence images about sustainable resource management during this new surge of discussions. Based on the patterns of argumentation, the typical roles may be distinguished: the culprit causing the problem, the victim affected by the problem and the solver of the problem. The frequency of the image of an actor in these different roles is quantified and ranked for three countries: Germany, France and Spain. Despite high losses in positive image, the public forestry authority remains the most powerful actor, followed by the nature conservation NGOs. NGOs in forestry are found to be the weakest actor in the sustainability discussion. Proposals to overcome obstacles for discussion are made and a greater involvement by forestry scientists is recommended.

Keywords: *sustainability, discussion, content analysis, patterns of argumentation, policy field analysis.*

[1] Dr. Gero Hütte, Institute of Forest Policy, Forest History and Nature Conservation, Büsgenweg 3, D-37077 Göttingen, Germany

K. von Gadow et al. (eds.), Sustainable Forest Management, 193–216.

Introduction

Sustainability (*Nachhaltigkeit*) is a traditional as well as a modern term. Born as an idea in the age of idealism around the time of the French Revolution in Central Europe (1st lexical mention in Onomatologiae forestalis-pisdatorio-venatoriae Supplementum; Stahl, 1780:582) the idea of *Nachhaltigkeit* crossed the Atlantic ocean as „sustained yield“ (Duerr/Duerr, 1975; Alston, 1992, p. 20). Like in Central Europe the term had a career closely connected to the foundation and development of a centralized forestry authority, with the term being used as an integrating symbol or as "focal point of faith" (Duerr/Duerr, 1975, p. 36-41). Throughout its history the term has been discussed in waves with higher intensity discussion taking place in years of social or economic crisis or in years of change, and lower intensity discussion taking place in stable years (Duerr/Duerr, 1975, p. 40). Recently a new surge of discussions has been developing (Fig. 1).

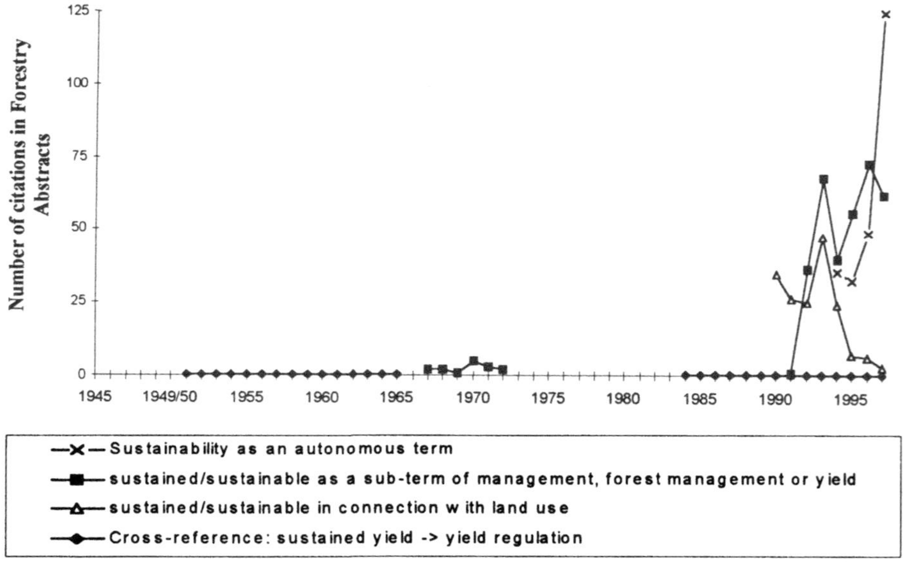

Figure 1. Development of the mentioning of sustained/sustainable in the „Forestry Abstracts"

New about this discussion is the fact that this time the "surge of discussions" was not initiated by the forestry sector itself, but by a coalition of organizations from the nature conservation sector, (IUCN, UNEP and WWF), who in 1980 published the „World Conservation Strategy: Living Resource Conservation for Sustainable Development" in English, French and Spanish. If the term is defined in a generally accepted way by groups outside the forestry sector, then perhaps the forestry organizations might loose their ability to define it themselves. In reality the agenda-setting process has been initiated by groups outside the forestry sector in three consecutive and gradually increasing waves, the first one developing around the time of the „World Conservation Strategy“ (1980), the second one around the time of the Brundtland-Report „Our common future“ (1987) and the third one around the time of the „II. United Nations Conference on Environment and Development“ (in 1992; see Fig. 2).

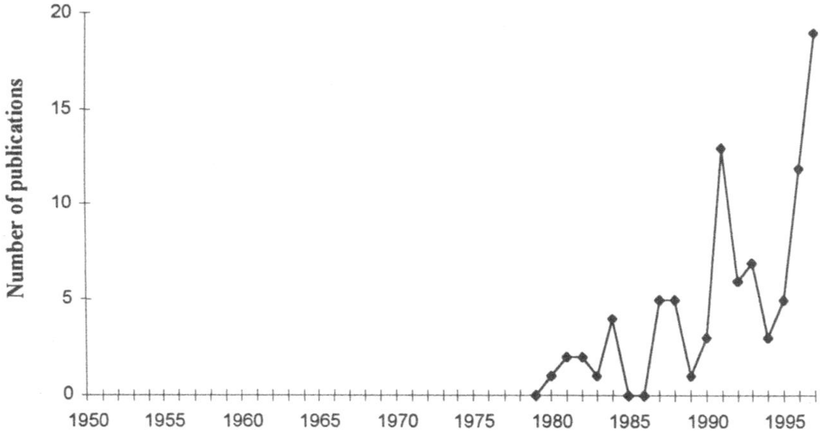

Figure 2. Development of the number of publications, provided by the IUCN-library in the web, concerning the term „sustain“, N=89, February 1998.

The enormous influence of the nature conservation sector on the forestry discussions is shown by the changes in expressions. The traditional word „sustained“, which had survived over a long period of time by encompassing the emotional sense of being a valuable concept, as well as having a concrete rational and technical sense, is more and more replaced by the word „sustainable“, which expresses more the emotional part and makes room for a redetermination of the technical meaning. The most obvious change can be seen in the Spanish language: the Spanish version of the World Conservation Strategy used the traditional word „sostenido“ (IUCN/UNEP/WWF, 1980), which has now been changed almost completely by the nature conservation discussion, as well as in the forestry discussion and even in legal documents, to the new word „sostenible“ (Hütte, 1999, p. 36-37).

However, the changes that are taking place do not only concern words and the move from a more rational view to a more emotional one. A change in image is also taking place. Yet the debate does not concentrate on the differences in image, but normally only on the thematic differences, for example sustainability of species conservation versus sustainability of wood production. The image of a problem entails more than just the image of a specific theme; it entails the image of a whole pattern of argumentation. The theme of course is a central issue in the pattern of argumentation, but the perceived role of the actors is also quite interesting. In the emotional dispute around the sustainability concept we may distinguish three kinds of roles (Prittwitz, 1990, 116-121, adapted):

1. the role of the culprit,
2. the role of the victim
3. the role of the problem solver.

Often these elements of patterns of argumentation are not taken into account during discussion. Even if the forestry sector is seen as a culprit of the problems of sustainability in the actual discussion, the actors in this field normally try to change the themes discussed, despite having only changed the image of the culprit under the same theme.

In terms of these three elements of patterns of argumentation, i.e. culprits, victims and problem solvers, the following text shows the actual images of sustainability in Germany, France and Spain and the changes since 1990.

Theoretical considerations

The discussion covering a specific theme can be seen as a *policy field*[2]. Within a given policy field, comparable perhaps to a historical roman arena, a variety of actors attempt to improve their position or to hold the current position which was attained in the past. The actors may be seen as being engaged in a subtle or open combat within an environment of constantly changing coalitions. The weapons are arguments and the battle takes place under the scrutiny of an interested public (Hütte, 1999, p. 16-20).

The best position for an actor to be in, as perceived by the interested public, is the position of a problem solver. To be accepted by the public as a problem solver one has to have the legitimacy to act which will result in a certain solidarity on the part of the observers. This solidarity may be expressed by the acceptance of a specific action. To achieve solidarity may be a major objective of an actor engaging in a public dispute.

The worst position for an actor to be in, is to be perceived as a culprit. The culprit looses legitimacy to act and frequently has to transfer resources to the victim and/or to the problem solver. A culprit is in danger of being isolated and of loosing power. So the actors, being at risk when they are perceived as a culprit causing a problem, try to play down the

[2] terminology used by Krott (1988)

importance of the problem or their role as a culprit. This is a defensive strategy, typical of the forestry profession in the current sustainability discussion.

Alternatively the actors could try to change their role into that of a victim or of a problem solver, which would represent a pro-active alternative. The strategy of creating the image of being a victim is ambiguous. On the one hand the victim cannot be the culprit and therefore needs support. So the role of being a victim is a very helpful one when there is a need for finding partners for coalitions. On the other hand, the image of being a victim may imply that the actor is weak and unable to act as a problem solver. On the one hand, to be perceived as a victim may provoke support and therefore the transfer of resources to the victim, while alternatively there may be the anticipation that the victim will transfer resources to an unknown problem solver in return for an acceptable solution of the problem. The role of a victim may be a desirable one for those perceived as culprits but undesirable for those perceived as problem solvers.

A practical strategy for an actor perceived as a culprit may be to first change his role into that of a victim and then into the role of a problem solver. The following sections explore the different roles of the main actors in the current sustainability dispute from 1990.

Definition

Sustainability discussion in this context is understood as “all linguistic expressions using the term “sustained/sustainable“ or one of its synonyms“. In the French language this definition applies to the term “durable“ and its synonyms, in the German language to the term “nachhaltig“ and in the Spanish language to the term “sostenible“. The sustainability discussion includes national and sectoral discussions.

Materials and Methods

Materials: Sectors, countries and texts analysed

The sustainability discussion was analysed covering the two sectors *forestry* and *nature conservation*. Within the forestry sector the sustainability concept fulfills an integrating function as symbolic „flag-word“ (Bachem/Battke, 1991, p. 298), its use dating back more than 250 years (v. Carlowitz, 1713, p. 105). The emotional content of the concept has gradually been replaced by operational planning and yield control techniques since the end of the 18th century (Zürcher, 1965, p. 104). Today, within the forestry sector, the sustainability concept is associated with both technical and emotional aspects (Eblinghaus/Stickler, 1996, p.

42f). The second important actor is represented by the nature conservation sector (IUCN/UNEP/WWF, 1980). Although the term sustainability has recently been incorporated in the terminology of several other industrial sectors, in sociology and psychology, the nature conservation sector is a main actor in the current sustainability discussion.

The European countries which were analysed are France, Germany and Spain. These three countries were very important for forestry and nature conservation before the expansion of the EU to Scandinavia and Austria. They had the greatest forested areas: Spain 15.65 million hectares, France 14.78 million hectares and Germany 10.38 million hectares (Statistisches Bundesamt, 1992, p. 50). Also at this time they had the highest mean annual increment: France 65.87 million m^3, Germany 57.44 million m^3 and Spain 27.75 million m^3 in 1990 (Kuusela, 1994, p. 74). Accordingly these three countries in 1989 had the highest wood production of the former EU: Germany 46.23 million m^3, France 43.73 million m^3 and Spain 17,17 million m^3 (Statistisches Bundesamt, 1992, p. 62).

The three countries had also been leaders in the field of nature conservation in the former EU. According to the IUCN-classification they had the greatest amount of protected areas, besides the United Kingdom: France 4.5 million hectares, Germany 2.8 million hectares and Spain 2.6 million hectares (Statistisches Bundesamt, 1992, p. 140).

Among the various possibilities of evaluating the sustainability discussion, articles from what are generally perceived to be leading technical journals were analysed. The advantages of this approch are:

1. Statements in the discussion can be analysed over a period of time which enables the scientist to develop a time series.
2. The statements in the discussion were made completely independently of the analysing scientist whose interests are aimed only at the observers during a certain period of time.
3. The analysis of the published material presents a good chance of obtaining a standard opinion of a given sector.

The following periodicals were examined:

Germany

a) nature conservation: "Natur und Landschaft“, edited monthly by the Bundesamt für Naturschutz (BFN), a governmental organisation, with 6400 editions (1998).

b) forestry: "AFZ/Der Wald – Allgemeine Forst Zeitschrift für Waldwirtschaft und Umweltvorsorge“, edited twice weekly by the Bayerischer Landwirtschaftsverlag (BLV) with 8500 editions (4/1997).

France

a) nature conservation: "Le Courrier de la Nature", edited twice monthly by the Société Nationale de Protection de la Nature (SNPN) with 4800 editions (1998) and "Aménagement et Nature", edited quarterly by the Association pour les Espaces Naturelles with 2000 editions (1998).

b) forestry: "Revue Forestière Française – Forêts – Bois - Nature" (science), edited every two months by the École Nationale du Génie Rural, des Eaux et des Forêts (ENGREF) with 2400 editions, "Arborescences" (national forests), edited twice monthly by the Office National des Forêts (ONF) with 17,000 editions (1998) and "Forêts de France" (private forests), edited ten times per year by the Fédération Nationale des Syndicats de Propriétaires Forestiers Sylviculteurs with 7300 editions (1998).

Spain

a) nature conservation: "Ecología", edited anually during the time of analysis by the Instituto Nacional para la Conservación de la Naturaleza (ICONA) (governmental) with approximately 1000 editions (1999) and "Quercus", edited monthly by the Servicios Informativos Ambientales S.L. with 16,000 editions (1999).

b) forestry: "Montes", edited quarterly by the Asociación y Colegios de Ingenieros de Montes e Ingenieros Técnicos Forestales with approximately 5000 editions (1999) and Revista Forestal Española (RFE), edited two to four times per year discontinuously by the GAESA with approximately 5000 editions (1999).

Methods

From all the articles in the above listed periodicals, mentioning sustained/sustainable or one of the synonyms, stratified random samples were taken in the following way: 90 articles per sector and country from the period 1960-1989 and 90 articles per sector and country from the period 1990-1995. Altogether 1050 articles were analysed. In the Spanish nature conservation sector the first journal, Ecología, was founded in 1972, so for this sub-discussion, a shorter period was analysed (1970-1989) and accordingly a reduced number of articles (60) were analysed.

Due to the question concerning the perceived role of important actors in the sustainability discussion and the changes in position, for each actor, the frequency of articles which were written by himself, his image as a culprit of problems, as a victim of problems and as a problem solver were counted. The difference between the frequency of being mentioned as a victim and as a culprit is defined as moral authority in discussion. In this way a preponderant image as a victim shows a positive moral authority and a preponderant image as a culprit a negative one, in a ustainability discussion .

The public nature conservation administration, nature conservation science, nature conservation NGOs, public forestry administration, forestry science, private forestry and other

NGOs were analysed as actors. By comparing the positions of the actors in the sustainability discussion a ranking was carried out. Furthermore the comparison of the position of an actor before 1990 with the position after 1990 shows whether a given actor had been able to improve his position in the ustainability discussion or whether position had deteriorated.

Results

Public nature conservation administration

Since the public nature conservation administration consists of defined institutions financed mainly from public budgets and represented by officials, they are mainly public executive institutions. Research institutes are part of this category if they are mainly dependent upon public financing without possessing the right to award the title of Dr./PhD.

The public nature conservation administration possesses a certain influence on all six sustainability discussions (Tab. 1). In Germany and Spain the leading technical journals of the nature conservation sector “Natur und Landschaft” and “Ecología” are both published by public institutions.

	Germany		France		Spain	
	nature conservation	forestry	nature conservation	forestry	nature conservation	forestry
Share of all articles	34% n=90 s=±0.051	1% n=90 s=±0.013	19% n=90 s=±0.043	1% n=90 s=±0.008	10% n=90 s=±0.035	8% n=90 s=±0.030
Image as culprit	2% n=94 s=±0.014	0% n=66 s=±0	0% n=137 s=±0	0% n=76 s=±0	0% n=147 s=±0	0% n=127 s=±0
Image as victim	4% n=63 s=±0.025	2% n=82 s=±0.013	0% n=132 s=±0	0% n=96 s=±0	0% n=97 s=±0	1% n=106 s=±0.011
Image as problem solver	19% n=121 s=±0.036	1% n=95 s=±0.008	10% n=164 s=±0.025	0% n=95 s=±0	11% n=139 s=±0.026	5% n=153 s=±0.017

Table 1. Position of public nature conservation administration 1990-95, shaded: deficits in self-projection ().

Therefore, it is not surprising that in the German sustainability discussion of the nature conservation sector 34% of all articles were written by authors from the public nature conservation administration which makes them dominant in this discussion. In France and Spain the public nature conservation administration is not dominant, but with 19% and 10% respectively this actor does have a certain influence on the sustainability discussion .

In the forestry sustainability discussions however, public nature conservation administration does not have a clear influence, with the exception of the Spanish sustainability discussion (8%). Nevertheless, the analysis shows that this actor does have the potential of expressing itself within the forestry sustainability discussions.

Within nature conservation sustainability discussions, public nature conservation administration is frequently perceived as being in the most desirable role, the role of the problem solver of problems. This indicates some legitimacy for action in all three sustainability discussions. However in the cases of the German and French sustainability discussions the percentage of time they are mentioned as a problem solver of problems is significantly lower than the percentage of articles written by this actor. This shows a certain deficit in self presentation. If every author from public nature conservation administration showed his own institution as a problem solver, then this institution would have more legitimacy for action.

In sustainability discussions conducted within the forestry sector, the public nature conservation administration is not accepted as a solver, with the exception of the forestry sector in Spain. Here lies a challenge for the nature conservation sector.

The moral authority of the public nature conservation administration is positive or neutral in all cases. It is hardly ever mentioned as a culprit of problems and when this does occur, the frequency of its image as a victim is higher than the frequency of it being perceived as a culprit. For this category then, the analysis shows no deficit – no challenge – for public nature conservation administration. The changes in position since the beginning of the current high surge in the sustainability discussion are shown in Tab. 2.

The public nature conservation administration is able to improve or maintain its position in virtually all of the sustainability discussions. Only in the Spanish sustainability discussion of nature conservation did this actor have to accept substantial losses. This can be correlated with the change in the political system after the Franco era, when there was a loss of confidence in traditional institutions and a growing influence of the NGOs on dicourse. The greatest improvement could be observed in the sustainability discussion within the French nature conservation sector. The influence of public nature conservation administration on the sustainability discussion of this sector has grown from a very low level (2%) to a respectable share of all articles (19%). With this growth, the image as a problem solver has grown from 6% to 10% of all problem solvers mentioned. In the German sustainability discussion of nature conservation a peculiar development was discovered. While a higher percentage of all articles are published by authors from public nature conservation administration, the image of this actor as a problem solver is decreasing. This means a considerable loss in legitimacy in spite of simultaneously greater efforts to influence the sustainability discussion.

	Germany				France				Spain			
	nature conservation		forestry		nature conservation		forestry		nature conservation		forestry	
	1960-89	1990-95	1960-89	1990-95	1960-89	1990-95	1960-89	1990-95	1970-89	1990-95	1960-89	1990-95
Share of all articles	**+ 6%**		**+ 1%**		**+ 17%**		**0%**		**- 10%**		**+ 6%**	
	28% n=90 s=±0.049	34% n=90 s=±0.051	0% n=90 s=±0	1% n=90 s=±0.013	2% n=90 s=±0.012	19% n=90 s=±0.043	1% n=90 s=±0.010	1% n=90 s=±0.008	20% n=60 s=±0.067	10% n=90 s=±0.035	2% n=90 s=±0.011	8% n=90 s=±0.030
Image as culprit	**- 2%**		**0%**		**- 1%**		**0%**		**0%**		**0%**	
	4% n=81 s=±0.027	2% n=94 s=±0.014	0% n=54 s=±0	0% n=66 s=±0	1% n=101 s=±0.005	0% n=137 s=±0	0% n=81 s=±0	0% n=76 s=±0	0% n=65 s=±0	0% n=147 s=±0	0% n=87 s=±0.003	0% n=127 s=±0
Image as victim	**- 3%**		**+ 2%**		**0%**		**0%**		**0%**		**+ 1%**	
	7% n=77 s=±0.036	4% n=63 s=±0.025	0% n=63 s=±0	2% n=82 s=±0.013	0% n=121 s=±0	0% n=132 s=±0	0% n=103 s=±0	0% n=96 s=±0	0% n=57 s=±0	0% n=97 s=±0	0% n=77 s=±0	1% n=106 s=±0.011
Image as solver	**- 5%**		**+ 1%**		**+ 4%**		**0%**		**- 5%**		**+ 5%**	
	24% n=83 s=±0.054	19% n=121 s=±0.036	0% n=54 s=±0	1% n=95 s=±0.008	6% n=114 s=±0.023	10% n=164 s=±0.025	0% n=97 s=±0.005	0% n=95 s=±0	16% n=80 s=±0.043	11% n=139 s=±0.026	0% n=119 s=±0.003	5% n=153 s=±0.017
Changes in position	0		▲		▲		0		▼		▲	

Table 2. Changes in position of public nature conservation administration 1960-89 to 1990-95, ▲: improvement, ▼: loss, 0: little change

In forestry sustainability discussions, public nature conservation administration gained territory in Germany and in Spain, but this actor achieved a certain influence and notable legitimacy as a problem solver only in Spain.

Public forestry administration

Public forestry administration consists of all forestry institutions which are financed mainly from public budgets or manage public property and are therefore mainly public executive institutions. Research institutes are a part of this category too, if they are mainly dependent upon public financing without possessing the right to award the title of Dr./PhD.

The public forestry administration is the most important actor in all analysed forestry sustainability discussions. In all three countries it has a much greater influence on the sectoral sustainability discussions than public nature conservation administration. In Germany 48% of all articles are written by authors who are members of public forestry administration (Tab. 3). This shows a very clear dominance of this actor in the sectoral sustainability discussion.

In the sustainability discussions conducted by the nature conservation sector in

Germany and particularly in Spain the public forestry administration has an influence as well. Nevertheless its share of the sustainability discussion within the nature conservation sector is very low when taking into account that sustainability is traditionally inherent to public forestry administration and part of its own specific image. The reason is either that public forestry administration is not accepted as, or asked as the expert concerning this term, or that the authors from forestry only concentrate on their own sectoral sustainability discussion. Whichever the case public forestry administration is obviously not able to take advantage of the sectoral tradition which gives the term its technical and rational content.

The public forestry administration is the main solver of problems in all forestry sustainability discussions, but nevertheless the big difference between its involvement in discussion and its image as a problem solver shows a considerable lack of self-projection. It would be very easy to increase the legitimacy for action for public forestry administration by mentioning this actor more frequently as a solver of problems. In the sustainability discussions conducted by the nature conservation sector, public forestry administration is mentioned as a solver of problems too. In France and Spain the confidence in public forestry administration is high enough so that even the other actors portray it in the desired role of a solver of problems.

	Germany		France		Spain	
	nature conservation	forestry	nature conservation	forestry	nature conservation	forestry
Share of all articles	2% n=90 s=±0.015	48% n=90 s=±0.050	0% n=90 s=±0	36% n=90 s=±0.051	6% n=90 s=±0.017	28% n=90 s=±0.050
Image as culprit	4% n=94 s=±0.019	7% n=66 s=±0.032	3% n=137 s=±0.011	7% n=76 s=±0.033	2% n=147 s=±0.011	6% n=127 s=±0.021
Image as victim	0% n=63 s=±0	20% n=82 s=±0.043	1% n=132 s=±0.006	18% n=96 s=±0.051	1% n=97 s=±0.006	7% n=106 s=±0.026
Image as problem solver	1% n=121 s=±0.007	35% n=95 s=±0.049	3% n=164 s=±0.016	32% n=95 s=±0.051	8% n=139 s=±0.023	18% n=153 s=±0.033

Table 3. Position of public forestry administration 1990-95, shaded: projection deficits ().

The moral authority of the public forestry administration in forestry sustainability discussions is posi-tive, but in the sustainability discussions of nature conservation it is negative. Particularly in Germany, the image of the public forestry administration as a culprit of problems is much higher than its image as a victim. This difference presents a big problem and a significant challenge for this actor, so as to improve the moral authority in the sustainability discussions of nature conservation.

There are two strategies for reaching this goal. The first aims at lowering the image of being a culprit causing problems, the second at nurturing the image of being a victim. The former could be called defensive and is very difficult to carry out. In democratical systems it is practically impossible to control the patterns of argumentation of a sustainability discussion in a way that forbids the image of an actor to be in the role of a culprit. Nevertheless in the actual sustainability discussion, forestry can often be noticed to try this hopeless strategy of defence. This strategy is even more hopeless since it is an integral aim of the action of NGOs to create a credibility gap ("Glaubwürdigkeitsdifferenz"; see Volz, 1997), a difference in credibility. The latter strategy, that of self-projection of being the victim of the problems by neutralizing the efforts of being stamped as a culprit, seems to be more effective and requires less resources. This strategy is rarely used by public forestry administration in the sustainability discussions conducted by the nature conservation sector.

In the forestry discussions when considering the sustainability concept, the image of being a victim is commonly accepted in all three countries. Public forestry administration suffers from problems caused by nature (game, climate etc.), by the market and by society, to illustrate just some of the patterns of argumentation used in the forestry sustainability discussions. It is incomprehensible why it should not be possible to introduce these patterns of argumentation into the sustainability discussions conducted by the nature conservation sector.

The changes in the position of the public forestry administration is shown in Tab. 4. Overall public forestry administration has suffered losses since the beginning of the actual surge in the sustainability discussion, despite the fact that this term has a long tradition in the forestry sector with the associated expertise in this sector. Indeed although the share of participation in the sustainability discussions increased in three cases, the image of being the solver of the problems decreased in four cases and the public forestry administration has suffered losses in moral authority in three cases.

The poorest development can be seen in the Spanish forestry sustainability discussion. Public forestry administration lost 17% of its share in sustainability discussion and consequently this actor suffered large losses in terms of image of being a problem solver, furthermore its moral authority is not far away from becoming negative. It must be assumed that the reason for these losses can be found in the change of political system after the Franco era and the general loss of confidence in traditional public authority.

Another problematical development for the public forestry administration can be seen in the German nature conservation sustainability discussion. In this discussion the involvement of public forestry administration has decreased to half that of former times and has been accompanied by a loss in image as a solver of problems and a severe loss in moral authority, with image changing from neutral to negative. If the public forestry administration really wants to solve this problem of loosing position within the German sustainability discussion of nature conservation, it is highly important to firstly re-intensify the involvement in this discussion in order to win back the former image as a solver of problems and secondly to follow the above described strategy in order to neutralise the accusation of being the culprit of problems by intensifying the image of being a victim of the problems.

	Germany				France				Spain			
	nature conservation		forestry		nature conservation		forestry		nature conservation		forestry	
	1960-89	1990-95	1960-89	1990-95	1960-89	1990-95	1960-89	1990-95	1970-89	1990-95	1960-89	1990-95
Share of all articles	- 2%		+ 13%		- 1%		+ 12%		+ 3%		- 17%	
	4% n=90 s=±0.016	2% n=90 s=±0.015	35% n=90 s=±0.051	48% n=90 s=±0.050	1% n=90 s=±0.005	0% n=90 s=±0	24% n=90 s=±0.040	36% n=90 s=±0.051	3% n=60 s=±0.011	6% n=90 s=±0.017	45% n=90 s=±0.061	28% n=90 s=±0.050
Image as culprit	+ 3%		- 1%		+ 2%		- 1%		- 3%		+ 3%	
	1% n=81 s=±0.009	4% n=94 s=±0.019	8% n=54 s=±0.044	7% n=66 s=±0.032	1% n=101 s=±0.009	3% n=137 s=±0.011	8% n=81 s=±0.040	7% n=76 s=±0.033	5% n=65 s=±0.027	2% n=147 s=±0.011	3% n=87 s=±0.016	6% n=127 s=±0.021
Image as victim	- 1%		+ 1%		+ 1%		+ 3%		- 1%		- 1%	
	1% n=77 s=±0.008	0% n=63 s=±0	19% n=63 s=±0.054	20% n=82 s=±0.043	0% n=121 s=±0	1% n=132 s=±0.006	15% n=103 s=±0.034	18% n=96 s=±0.051	2% n=57 s=±0.015	1% n=97 s=±0.006	8% n=77 s=±0.034	7% n=106 s=±0.026
Image as solver	- 1%		- 2%		+ 2%		+ 8%		- 3%		- 8%	
	2% n=83 s=±0.010	1% n=121 s=±0.007	37% n=54 s=±0.075	35% n=95 s=±0.049	1% n=114 s=±0.011	3% n=164 s=±0.016	24% n=97 s=±0.035	32% n=95 s=±0.051	11% n=80 s=±0.037	8% n=139 s=±0.023	26% n=119 s=±0.053	18% n=153 s=±0.033
Changes in position	▼		0		0		▲		0		▼	

Table 4. Changes in position of public forestry administration 1960-89 to 1990-95, ▲: improvement, ▼: loss, 0: little change

Only in the French forestry sustainability discussion was public forestry administration able to improve its position by strengthening its involvement in this discussion. Accordingly the image as a problem solver could be increased significantly and the moral authority improved too. In Germany the intensification of engagement of public forestry administration in the forestry sustainability discussion was unable to be transformed into an improvement in image. Indeed the image in the desired role of a solver of problems decreased and the moral authority increased only a little. Nevertheless the clear dominance of this actor in the sustainability discussion indicates high potential for dominating the patterns of argumentation so that an improvement of the image should obviously be easy to reach within this discussion.

Nature conservation NGOs

Nature conservation NGOs aim at conserving or reconstructing nature with biological, historical or recreational objectives. They are generally organized within regional or national institutions. In this analysis typical forestry or agricultural NGOs are excluded. The different aims show that this actor comprises a very heterogenous group, strongly influenced in organization and aim by the cultural differences between the three analysed countries. Information political instruments are the main instruments used by NGOs which usually do not have a sustainable financial base (that is except founds like the WWF), and who do not have the possibility of creating regulative instruments.

The position of the nature conservation NGOs is shown in Tab. 5. This group of actors has a big influence on the sustainability discussions in nature conservation, particularly in Spain where they are the dominant actor (32%). In Germany too, they are powerful actors in the nature conservation sustainability discussion. Only in France do they seem to have less influence.

In the forestry sustainability discussions on the other hand the voice of nature conservation NGOs does not carry much power. That does not mean to say that none of the authors are members of such an NGO, but just that it seems to be more acceptable to write in another capacity – as forestry officer, forestry scientist etc. The membership of a nature conservation NGO is mostly kept secret in forestry sustainability discussions. It is however possible to speak of camouflaged action by NGOs in the forestry sustainability discussions. This supposition is supported by the fact that nature conservation NGOs are represented as being the solver of problems in the Spanish and German forestry sustainability discussions, although they do not openly take part in either of these two sustainability discussions. The French forestry sustainability discussion does not perceive nature conservation NGOs at all.

Nature conservation NGOs are perceived to be in the desired role of the legitimised solver of problems in the German nature conservation sustainability discussion in a relative proportion to their activity in sustainability discussion. A very special case can be seen in France: despite a participation in sustainability discussion of only 8%, nature conservation NGOs are perceived as solvers of problems in 18% of all cases. This extraordinary fact shows that the confidence in the capacity of NGOs is much higher than their capacity to be active in a visible form. Probably in France the nature conservation NGOs still feel outnumbered and therefore need to conceal their action.

	Germany		France		Spain	
	nature conservation	forestry	nature conservation	forestry	nature conservation	forestry
Share of all articles	21% n=90 s=±0.044	0% n=90 s=±0	8% n=90 s=±0.030	0% n=90 s=±0	32% n=90 s=±0.055	0% n=90 s=±0
Image as culprit	2% n=94 s=±0.014	0% n=66 s=±0	0% n=137 s=±0	0% n=76 s=±0	0% n=147 s=±0	1% n=127 s=±0.007
Image as victim	5% n=63 s=±0.029	0% n=82 s=±0	2% n=132 s=±0.013	0% n=96 s=±0	0% n=97 s=±0	0% n=106 s=±0
Image as solver	19% n=121 s=±0.036	2% n=95 s=±0.013	18% n=164 s=±0.029	0% n=95 s=±0	22% n=139 s=±0.034	4% n=153 s=±0.016

Table 5. Position of nature conservation NGOs 1990-95, shaded: projection deficits.

Only in Spain do nature conservation NGOs show a projection deficit. Despite a share of 32% of all articles, the percentage of time that they are perceived as solvers of the problems is much smaller. Here lies the potential for the Spanish NGOs to improve their position in sustainability discussion.

NGOs are not very often perceived in any of the sustainability discussions as culprits or victims. This is in contrast to the oral sustainability discussion in forestry, which frequently sees nature conservation NGOs as the culprits of problems and particularly of economical problems and problems limiting the freedom of decision. Authors in written sustainability discussion certainly fear public opinion or the opinion of camouflaged members of NGOs, who perhaps occupy higher levels in the administrational hierachy. This is a phenomenon, that Noelle-Neumann (1991) calls "Schweigespirale" – the spiral of silence. In this way nature conservation NGOs are more often criticized in the published sustainability discussion by their own sector than from forestry.

Tab. 6 shows that nature conservation NGOs have profited from the new surge in nearly all sustainability discussions. The nature conservation NGOs are the most dynamical actors of all analyzed groups. Only in the French sustainability discussion of the nature conservation sector have they suffered losses. The published forestry sustainability discussion in France does not see them as factors of influence concerning the theme of sustainability.

A most interesting development can be seen in the sustainability discussion of nature conservation in Spain. Nature conservation NGOs were able to increase their share in sustainability discussion from nearly 0% to 32%. The journal "Quercus", founded in 1981, is very important for this development. With a growing influence on the sustainability discussion, the image of nature conservation NGOs as problem solvers grew from 11% to 22% and their moral authority has not been touched. Today therefore the nature conservation NGOs are the most important actors in the Spanish sustainability discussion in this sector.

In the German nature conservation sustainability discussion too this actor was able to remarkably improve its position: The share of participation could be enlarged by 8% and the image of being the solver of problems by 4%. Simultaneously they were able to improve their moral authority by changing from a perceived culprit to a perceived victim of the problems. Accordingly in the image of sustainability discussion today they are nearly as important as the nature conservation administration. Only in France did nature conservation NGOs suffer important losses. They lost 12% of their influence on the sustainability discussion and 10% of legitimacy to solve problems. Their moral authority remained unchanged, however today they are perceived more often as the victim of problems. They lost their powerful position and are perceived to be growing weaker.

In the forestry sustainability discussions nature conservation NGOs were unable to increase their share of participation, but their image as a problem solver increased. Based on a more or less neutral moral authority, this actor seems to have all prerequisites necessary for further development.

	Germany				**France**				**Spain**			
	nature conservation		**forestry**		**nature conservation**		**forestry**		**nature conservation**		**forestry**	
	1960-89	1990-95	1960-89	1990-95	1960-89	1990-95	1960-89	1990-95	1970-89	1990-95	1960-89	1990-95
	+ 8%		**- 1%**		**- 12%**		**0%**		**+ 32%**		**0%**	
Share of all articles	13% n=90 s=±0.039	21% n=90 s=±0.044	1% n=90 s=±0.011	0% n=90 s=±0	20% n=90 s=±0.041	8% n=90 s=±0.030	0% n=90 s=±0	0% n=90 s=±0	0% n=60 s=±0.005	32% n=90 s=±0.055	0% n=90 s=±0	0% n=90 s=±0
	- 2%		**0%**		**0%**		**0%**		**0%**		**+ 1%**	
Image as culprit	4% n=81 s=±0.027	2% n=94 s=±0.014	0% n=54 s=±0	0% n=66 s=±0	0% n=101 s=±0	0% n=137 s=±0	0% n=81 s=±0.004	0% n=76 s=±0	0% n=65 s=±0	0% n=147 s=±0	0% n=87 s=±0.003	1% n=127 s=±0.007
	+ 2%		**0%**		**+ 1%**		**0%**		**0%**		**0%**	
Image as victim	3% n=77 s=±0.026	5% n=63 s=±0.029	0% n=63 s=±0	0% n=82 s=±0	1% n=121 s=±0.007	2% n=132 s=±0.013	0% n=103 s=±0	0% n=96 s=±0	0% n=57 s=±0	0% n=97 s=±0	0% n=77 s=±0	0% n=106 s=±0
	+ 4%		**+ 1%**		**- 10%**		**0%**		**+ 11%**		**+ 4%**	
Image as solver	15% n=83 s=±0.047	19% n=121 s=±0.036	1% n=54 s=±0.008	2% n=95 s=±0.013	28% n=114 s=±0.043	18% n=164 s=±0.029	0% n=97 s=±0	0% n=95 s=±0	11% n=80 s=±0.034	22% n=139 s=±0.034	0% n=119 s=±0	4% n=153 s=±0.016
Changes in position	▲		▲		▼		0		▲		▲	

Table 6. Changes in position of nature conservation NGOs 1960-89 to 1990-95, ▲: improvement, ▼: loss, 0: little change

Private forestry

Private forestry in this study is defined as single private forest owners or organizations of private forest owners. In all three analysed countries, private forest owners possess a high percentage of the forested land, between 44% (Germany) and 72% (France; European Commission, 1997). Therefore private forest owners are of central importance for the implementation of forest policy and sustainable forest management.

	Germany		France		Spain	
	nature conservation	forestry	nature conservation	forestry	nature conservation	forestry
Share of all articles	0% n=90 s=±0	6% n=90 s=±0.025	0% n=90 s=±0	9% n=90 s=±0.024	0% n=90 s=±0	3% n=90 s=±0.019
Image as culprit	8% n=94 s=±0.029	3% n=66 s=±0.024	1% n=137 s=±0.005	8% n=76 s=±0.035	4% n=147 s=±0.020	3% n=127 s=±0.016
Image as victim	2% n=63 s=±0.019	25% n=82 s=±0.049	0% n=132 s=±0.005	26% n=96 s=±0.053	3% n=97 s=±0,021	14% n=106 s=±0.036
Image as solver	1% n=121 s=±0.007	9% n=95 s=±0.030	0% n=164 s=±0	15% n=95 s=±0.038	1% n=139 s=±0.009	5% n=153 s=±0.020

Table 7. The position of the private forestry sector 1990-95.

Taking into account the high percentage of privately owned forests, the participation of private forest owners in the sustainability discussion is very low (Tab. 7). In Spain only 3% of all articles are written by private forest owners or by one of their organizations and in France, in spite of the fact that French forest owners possess their own centralized journal, „Forêts de France“, something which German and Spanish forest owners do not have. Only 9% of all authors originate from the group of forest owners. This shows a very low level of interest by forest owners in taking part in sustainability discussion in all three countries. The reasons for this may be sought in the fact that a high percentage of the forest enterprises are only small scale and therefore do not generate a huge income or in the fact that those forest owners who own larger areas are able to implement their own visons on their own property and therefore do not have to influence others through sustainability discussion. Forest owners do not take part at all in the sustainability discussions conducted by the nature conservation sector.

Nevertheless, private forest owners are perceived more often as solvers of problems than it would be expected if conclusions were drawn from their percentage of participation. Their legitimacy for implementation therefore seems to be higher than their action in sustainability discussion. Nevertheless, normally public forestry authority and forestry science have a higher legitimacy to act.

The typical role of forest owners in forestry sustainability discussions is the image of a victim. In comparison, their image of being a culprit is very rare. Private forest owners suffer from different problems, for example from low prices for their products, from high costs for administration, from forest diseases, problems related to damage caused by game or storm. That may be the reason why their moral authority in forestry sustainability discussions remains positive.

In contrast, in the nature conservation sustainability discussions they are perceived to be mainly in the role of a culprit. Particularly in Germany, the image of being a victim is, in comparison to the image of being a culprit, of little relevance. The moral authority of this actor is negative. Obviously private forest owners are unable to show their problems within the sustainability discussions of nature conservation. It is not however the aim of this analysis to see whether the reason for the low engagement lies in a lack of interest concerning the forest owners or in filters of sustainability discussion, like for example the publishing institution, by resisting any intention of participation by forest owners. However here lies a major challange for forest owners and their organizations. They have to improve their moral authority and to enlighten nature conservation public both about their problems and their capacity to solve problems.

The visual representation of the dynamics in position of private forestry (Tab. 8) shows improvements and losses, with improvements taking place in the German and Spanish forestry sustainability discussions and losses in the German nature conservation sustainability discussion and in the French forestry sustainability discussion.

In the German forestry sustainability discussion private forest owners and their organizations increased their activity and accomplished a 10% increase in their image as a victim. Nevertheless their engagement in the forestry sustainability discussion remains very low (6%) and their image of being the solver of problems was not able to be increased.

The Spanish forest owners are beginning to get organized and take part in the sustainability discussion. By doing so they were able to improve their image in the published sustainability discussion in all fields: they are now perceived more often as solvers of problems, they increased the acceptance of being a victim of problems and they accomplished a decrease in their image of being the culprit of problems. Indeed the Spanish forest owners significantly changed their role in the sustainability discussion.

	Germany				France				Spain			
	nature conservation		forestry		nature conservation		forestry		nature conservation		forestry	
	1960-89	1990-95	1960-89	1990-95	1960-89	1990-95	1960-89	1990-95	1970-89	1990-95	1960-89	1990-95
	0%		**+ 4%**		**0%**		**- 8%**		**0%**		**+ 3%**	
Share of all articles	0% n=90 s=±0	0% n=90 s=±0	2% n=90 s=±0.013	6% n=90 s=±0.025	0% n=90 s=±0	0% n=90 s=±0	17% n=90 s=±0.044	9% n=90 s=±0.024	0% n=60 s=±0	0% n=90 s=±0	0% n=90 s=±0	3% n=90 s=±0.019
	+ 4%		**0%**		**0%**		**- 2%**		**- 1%**		**- 3%**	
Image as culprit	4% n=81 s=±0.027	8% n=94 s=±0.029	3% n=54 s=±0.020	3% n=66 s=±0.024	1% n=101 s=±0.005	1% n=137 s=±0.005	10% n=81 s=±0.039	8% n=76 s=±0.035	5% n=65 s=±0.027	4% n=147 s=±0.020	6% n=87 s=±0.034	3% n=127 s=±0.016
	0%		**+ 10%**		**- 1%**		**- 19%**		**+ 3%**		**+ 8%**	
Image as victim	2% n=77 s=±0.010	2% n=63 s=±0.019	15% n=63 s=±0.048	25% n=82 s=±0.049	1% n=121 s=±0.012	0% n=132 s=±0.005	45% n=103 s=±0.067	26% n=96 s=±0.053	0% n=57 s=±0	3% n=97 s=±0.021	6% n=77 s=±0.032	14% n=106 s=±0.036
	0%		**0%**		**0%**		**- 11%**		**- 3%**		**+ 4%**	
Image as solver	1% n=83 s=±0.008	1% n=121 s=±0.007	9% n=54 s=±0.045	9% n=95 s=±0.030	0% n=114 s=±0	0% n=164 s=±0	26% n=97 s=±0.049	15% n=95 s=±0.038	4% n=80 s=±0.024	1% n=139 s=±0.009	1% n=119 s=±0.008	5% n=153 s=±0.020
Changes in position	▼		▲		0		▼		0		▲	

Table 8. Changes in position of private forestry 1960-89 to 1990-95, ▲: improvement, ▼: loss, 0: little change

On the other hand; private forest owners suffered losses in the French forestry sustainability discussion and in the German nature conservation sustainability discussion. In France they lost nearly half their share in participation. As a consequence they lost acceptance as being solvers of problems, with 11% being a very considerable loss of legitimacy for action. Simultaneously they lost the image of being a victim of problems (19%) and along with this development they therefore lost both sympathy and help too. Besides these important changes, the small decrease in image as a culprit (2%) does not change the fact that the French forest owners can be considered one of the actors suffering most from the recent developments.

The increase of 4% in the image of the forest owners as culprits in the German sustainability discussion of nature conservation seems to be of minor importance in comparison to the losses in position experienced by French private forestry. Nevertheless forest owners are perceived in the German nature conservation sustainability discussion more and more as the culprits of problems. If they do not wish to remain in this role, they will have to begin to change their image, for example by strengthening their image of being a victim and a solver of problems.

Conclusions

By comparing and ranking the positions of all the discussed actors, the public forestry administration is seen to be in the best position. It is dominant in all three of the analysed forestry sustainability discussions and has access to one of the nature conservation sustainability discussions. It is perceived by a high percentage as a solver of problems in all of the forestry sustainability discussions and additionally it is also perceived as a problem solver in two of the nature conservation sustainability discussions. However, it carries a negative moral authority in the German nature conservation sustainability discussion. To achieve a neutral or positive image tn this specific sustainability discussion, it would need to balance out the image as a culprit and as a victim. A problem in constructing this image is its low involvement in this sustainability discussion. The public forestry authority has on the other hand major chances of improving its image in the forestry sustainability discussions in Germany and Spain. The analysis shows self-projection deficits because its participation in discussion is higher than its image as a solver of problems.

Nature conservation NGOs are seen to be placed in the second best postion. By being engaged in the German and the Spanish sustainability discussions conducted by the nature conservation sector to a high degree, they enjoy much confidence as the solvers of problems, even in the Spanish forestry sustainability discussion. In none of the sustainability discussions is their moral authority denied, which means that the nature conservation NGOs do not feel opposed in the sustainability discussions. This is a good basis for action. They have the best position in the Spanish nature conservation sustainability discussion. In this discussion, there is even greater potential for improving their position, since a self-projection deficit could be found.

Public nature conservation administration and forestry science are similarly positioned within the six analysed sustainability discussions. Forestry science shows slightly more involvement in the sustainability discussions but does not possess the full moral authority. In comparison to the public forestry administration and the nature conservation NGOs, both are not often perceived as being solvers of problems, but the public nature conservation administration does show the potential of improving its image as a problem solver in two sustainability discussions.

Private forestry and NGOs in forestry are badly positioned. Neither in the forestry sustainability discussions nor in the sustainability discussions conducted by the nature conservation sector do they participate with sufficient influence. Sometimes the percentage to which they are perceived as problem solvers exceeds their share in sustainability discussion, but nevertheless unlike the NGOs in the nature conservation sector they play a minor role in the sustainability discussions. This is a noteworthy fact because on the one hand the most important political instrument of NGOs is information, they have to communicate with their members through periodicals, they have to engage in sustainability discussion to awaken sympathy and to recruite new members and on the other hand the spirit of the last two decades has been a fertile soil for the development of NGOs.

Examination of the moral authority shows that all three analysed actors from the nature conservation sector possess the full moral authority in all sustainability discussions. Forestry however has more problems, particularly in Germany. Three actors from this sector show a negative moral authority - a major challenge in terms of further involvement in sustainability discussion.

By comparing the actor's positions during the time before and after 1990 (Tab. 9), all actors from the nature conservation sector show an overall improvement whereas all actors from the forestry sector show an overall deterioration. Particularly in terms of the sustainability discussions conducted by the other sector, actors from nature conservation show only improvements in forestry sustainability discussions whereas actors from forestry only show deteriorations in nature conservation sustainability discussions. The subject of sustainability obviously was unable to activate forestry expertise. Sectorinternally, in both sectors the developments are ambiguous.

	Sustainability discussions in the nature conservation sector			Sustainability discussions in the forestry sector		
	Germany	France	Spain	Germany	France	Spain
Public nature conservation administration	0	▲	▼	▲	0	▲
Nature conservation NGOs	▲	▼	▲	▲	0	▲
Public forestry administration	▼	0	0	0	▲	▼
Private forestry	▼	0	0	▲	▼	▲

Table 9. General overview on the changes in position of the analyzed actors 1960-89 to 1990-95, ▲: improvement, ▼: loss, 0: little change, shaded: profiting actors

Germany and Spain show a similar development, whereas the French sustainability discussions show a special tendency: In Germany and Spain the executive bodies and science, in the innersectoral nature conservation sustainability discussions as well as in the forestry sustainability discussions, were unable to improve their position, whereas nature conservation NGOs and private forestry are seen to hold a stronger position after 1990 as compared to before. In France the opposite tendency can be found: Private forestry and nature conservation NGOs lost ground in sustainability discussion, whereas the public executive and science were able to improve their position. The French forestry sustainability discussion seems to be completely blind to the image of nature conservation actors.

The results show that actors who are highly engaged in a sustainability discussion are normally perceived to be in a good position. Increasing involvement increases the possibility of introducing their own pattern of argumentation. A negative moral authority in a sustainability discussion normally results from too little participation of the respective actor in the sustainability discussion. If an actor wants to present himself as competent at solving problems or if he wants to be perceived as a victim of a problem who needs help by the transfer of resources or by the building of alliances, he has to take part in the sustainability discussion. Neither forestry nor nature conservation show an intensive engagement in the sustainability discussions conducted by the other sector. Spain seems to be the exception.

There could be two reasons for this fact. Firstly a lack of interest in acting in the sustainability discussion of the other sector or secondly, due to the fact that they are prevented from taking part in the discussions of the other sector. Probably both reasons hold true, but leaning towards the hypothesis of a filtered sustainability discussion – problematical in democratic systems – is the Spanish example: In Spain both sectors are engaged more often in the sustainability discussion of the other sector in comparison to Germany and France. In Spain the official institutions of forestry and nature conservation have not been separated as strictly as in Germany and France. The Instituto Nacional para la Conservación de la Naturaleza brought together both sectors with the result that authors from the forestry sector took part in the nature conservation sustainability discussion and vice versa.

If an actor is excluded from a sustainability discussion, he has to overcome the obstacles in a strategic way. There are various possibilities to act. He may act in the sustainability discussion in another role. This is a common strategy of NGOs, taking part for example in the forestry sustainability discussion in the role of an official who hides his membership in the NGO. This hidden participation of NGOs has the advantage that their patterns of argumentation can be introduced in spite of the restrictions. Another strategy is to co-operate with an actor from inside the filtered sustainability discussion. This actor may introduce the patterns of argumentation of his partner. This is a hopeful strategy for science, which has inter-disciplinary co-operation and for public administration with its branches in all sectors of society.

If an actor fears being misunderstood or perceived in the wrong way, the only way of changing the perception to try to influence the sustainability discussion where and when the wrong image is discussed. To lower activity in this discussion because of growing opposition, is not the right way to defend the own patterns of argumentation or to introduce new arguments. If the actors from nature conservation do not take part in the forestry sustainability discussions and vice versa, then their image is made by the other actors in the sustainability discussion.

Strangely, forestry science has not gained a noteworthy position as a problem solver in nature conservation sustainability discussions. Obviously, nature conservation sustainability presents a wide field for scientific investigation and for gaining public sympathy and support. To be perceived more often as a solver of problems in the nature conservation sustainability discussions would increase confidence in forestry science and could stimulate greater financial support for research projects. Increased involvement and the improvement of their own image in nature conservation sustainability discussions presents an important chance and a challenge for forestry scientists.

References

Alston, R.M.; 1992: *History of Sustained Yield in the United States (1937-1992)*, in: Gundermann, E. (ed.): Sustained Yield, Historical, Economic and Political Aspects, IUFRO Centennial, 31.8.-4.9.1992, S. 19-30

Bachem, R./Battke, K.; 1991: *Strukturen und Funktionen der Metapher "Unser gemeinsames Haus Europa" im aktuellen politischen Diskurs*, in: Liedtke, F./Wengeler, M./Böke, K.: Begriffe besetzen, Westdt. Verlag, Opladen, S. 295-307

Carlowitz, H.C.v.; 1713: *Sylvicultura oeconomica oder hauswirthliche Nachricht und naturgemäßige Anweisung zur wilden Baum-Zucht*, Leipzig

Duerr, W.A./Duerr, J.B.; 1975: *The Role of Faith in Forest Resource Management*, in: Rumsey, F./Duerr, W.A. (ed.): Social Sciences in Forestry, Saunders, S. 30-41

Eblinghaus, E./Stickler, A.; 1996: *Nachhaltigkeit und Macht*, IKO-Verl. für Interkulturelle Komm.

European Commission; 1997: *The European Union - Forestry*, Karte der europäischen Wälder, published by the Office for Official Publications of the European Communities, L-2985 Luxemburg

Hauff, V. (Hrsg.); 1987: *Unsere gemeinsame Zukunft*, Eggenkamp, Greven

Hütte, G.; 1999: *Nachhaltigkeit im europäischen naturschutz- und forstfachlichen Diskurs*, Cuviller-Verl., Göttingen

IUCN/UNEP/WWF; 1980: *Estrategía mundial para la conservación: La conservación de los recursos vivos para el logro de un desarrollo sostenido*, 1196 Gland, Schweiz

IUCN/UNEP/WWF; 1980: *Stratégie mondiale de la conservation: la conservation des ressources vivantes au service du développement durable*, 1196 Gland, Schweiz

IUCN/UNEP/WWF; 1980: *World Conservation Strategy: Living Resource Conservation for Sustainable Development*, 1196 Gland, Schweiz

Krott, M.; 1988: *Die ökonomieorientierte Politikfeldanalyse als Entscheidungshilfe in der Forst- und*

Holzwirtschaft, Forstarchiv, S. 127-131

Kuusela, K; 1994: *Forest Resources in Europe*, European Forest Institute

Noelle-Neumann, E.; 1991: *Öffentliche Meinung: Die Entdeckung der Schweigespirale*, Erw. Ausg., Ullstein

Prittwitz, V.v.; 1990: Das Katastrophenparadox - Elemente einer Theorie der Umweltpolitik, Leske + Budrich, Opladen 1990

Stahl, J.F. (ed.); 1780: *Onomatologiae forestalis-piscatorio-venatoriae Supplementum oder Beyträge und Verbesserungen des vollständigen Forst- Fisch- und Jagd- Lexicon*, Bd. 4, Stutgart 1780, S. 582

Statistisches Bundesamt; 1992: *Länderbericht - Europäischer Wirtschaftsraum 1992*, Metzler/Poeschel

Zürcher U.; 1965: Die Idee der Nachhaltigkeit unter spezieller Berücksichtigung der Gesichtspunkte der Forsteinrichtung, Schweizerische Anstalt für das forstliche Versuchswesen, Mitteilungen, Bd 41, Heft 4

Regulation of Timber Yield for Sustainable Management of Industrial Forest Plantations - Theory and Practice

Joseph P. Roise, Frederick W. Cubbage, Robert C. Abt, and Jacek P. Siry[1]

Abstract

The integration of forest plantations, sustained yield regulation of industrial wood fiber, and market responses of private forest landowners determine sustainable forest management for timber. The world has approximately 3.4 billion ha of forests; 230 million ha of all types of forest plantations; and 40 million ha of industrial wood plantations. A sample of private and public forestry analysts in the world indicated that they use a variety of timber harvest regulation approaches. These include Model I and Model II model formulations; Simplex and heuristic model solution approaches; sustained yield harvest constraints; and ecological constraints. An analysis of the forest plantation area and timber supply situation in the U.S. South indicated that intensive forest management can balance softwood timber inventories and timber harvests, at slightly increased real timber prices. The data on forest plantations and growth rates, as well as our survey of harvest regulation, indicate that the forest products industry and other forest owners have the means to achieve and modeling capabilities to measure and implement sustainable timber management. If this potential is realized, increased productivity in managed forests will lessen timber harvesting pressures and enhance the protection of natural forests.

Keywords: *timber, regulation, supply, sustainable, management*

---

[1] Associate Professor, Professor, Associate Professor, and Research Assistant Professor; Department of ForestryNorth Carolina State University, Raleigh, North Carolina 27695-8008; Facsimile: (919) 515-6193; joe_roise@ncsu.edu; fred_cubbage@ncsu.edu; bob_abt@ncsu.edu; jacek_siry@ncsu.edu

K. von Gadow et al. (eds.), Sustainable Forest Management, 217–255.

Introduction

Sustainable development and sustainable forest management have become the most widely accepted paradigms for resource management throughout the world. The expanding agenda for sustainable forest management has been developed via a host of international to local events. The 1987 Brundtland Report (World Commission on Environment and Development, 1987) was the modern catalyst for world sustainability. It defined sustainability as "development that meets the needs of the present generation without compromising the needs of future generations to meet their own needs..." Following that, the June 1992 United Nations Conference on Environment and Development (UNCED) was held in Rio de Janeiro. At this "Earth Summit", 144 countries developed and adopted a non-binding Statement of Forest Principles that recognized the importance of sustainable forest management for all types of forests.

Since then, sustainable forest management (SFM) Criteria and Indicators have developed rapidly, including the Montreal Process for most non-European temperate forests and the Helsinki Process in Europe. Earlier efforts by the International Tropical Timber Organization also were designed to enhance SFM (National Association of State Foresters [NASF], 1997). Sustainability is of course not a new criterion, nor the only important one for managing forests or for economic development. Classical criteria such as economic efficiency, social equity, administrative practicality, or biological feasibility (e.g., Clawson 1975) remain important for forest allocation and use. However, the new sustainable forest management literature and principles have encompassed many of these enduring criteria in broad sets of standards that have been agreed upon in international accords and are being implemented in various fashions throughout the world.

The United States NASF (1997) provides a succinct summary of the development of Sustainable Forest Management, as paraphrased below. A United Nations committee meeting on sustainable development of temperate and boreal forests was held in Montreal in 1993. That meeting included representatives from nine nations, who formed a Working Group on Criteria and Indicators for the Conservation and Sustainable Management of Temperate and Boreal Forests. The Montreal Process countries met in Santiago, Chile in February 1995 to endorse their commitment to the Montreal process. This "Santiago Declaration" accepted a comprehensive set of seven criteria and 67 indicators for the conservation and sustainable management of temperate and boreal forests. There are now 130 countries engaged in activities related to criteria and indicators. As of 1997, the 12 Montreal Process countries included Argentina, Australia, Canada, Chile, China, Japan, the Republic of Korea, Mexico,

New Zealand, the Russian Federation, the United States of America, and Uruguay. These countries cover five continents and contain 90% of the world's temperate and boreal forests and 60% of all forest on the globe. Member countries of the Helsinki Process cover most of the rest of the world temperate and boreal forests.

The Montreal Process developed broad *criteria* that were intended to represent a large-scale reflection of public values. *Indicators* were then developed to provide a means for measuring these forest conditions and tracking subsequent changes. The following seven criteria for sustainable forest management of forest ecosystems were developed by the Montreal Process:

1) Conservation of biological diversity.
2) Maintenance of the productive capacity of forest ecosystems.
3) Maintenance of forest ecosystem health and vitality.
4) Conservation and maintenance of soil and water resources.
5) Maintenance of forest contribution to global carbon cycles.
6) Maintenance and enhancement of long-term socio-economic benefits to meet the needs of societies.
7) Development of the legal, institutional, and economic framework for forest conservation and sustainable management.

In addition to the Montreal Process, several other efforts have been developed that relate to global or U.S. sustainable forest management. Similar criteria and indicators for measuring and assessing sustainable forest management were developed through the Helsinki Process in Europe. In 1994, the United States forest products industry has developed a Sustainable Forestry Initiative (SFI), which is mandatory for all its member companies (American Forest & Paper Association [AF&PA], 1998). SFI is a comprehensive system of principles, guidelines, and performance measures that integrates the perpetual growing and harvesting trees with the protection of wildlife, plants, soil, air quality, and water quality. The Internal Standards Organization also has developed an environmental auditing process that can consider forestry manufacturing under the ISO 14001 standards. These standards are similar to other corporate audits, such as ISO 9000, that ensure acceptable accounting or employment practices. Other programs such as the Forest Stewardship Council (FSC) standards, have been developed to certify that individual forests are managed sustainably and give a "green seal" to operations that meet those standards.

By 1997, the Montreal Process countries had developed and published their first review to determine their ability to gather the data needed to measure the 67 indicators. These were presented in each country's First Approximation Report (FAR) at the Ninth Meeting of the Montreal Process Working Group, held in July 1997 in Seoul, Korea (USDA Forest Service, 1997). In September 1997, the National Association of State Foresters endorsed the seven criteria established by the Montreal Process as a framework for integrating and measuring sustainable forest management, and asked the Chief of the Forest Service to work with NASF in developing a group to implement mutually determined actions to achieve this national policy. Since then, a sustainable forest management roundtable has been developed, which includes most major natural resource management and environmental protection agencies. The roundtable is assessing means to implement and refine the criteria and SFM criteria and indicators for national assessments and reporting. A national meeting of interest groups was held in November 1998 to review the status of implementing the criteria and indicators, and suggest future actions. The roundtable scheduled a series of three regional and national conferences that began in March of 2000, in order to develop protocols for implementing the U.S. Criteria and Indicators for sustainable forest management.

Of the 7 criteria and 67 indicators of the Montreal Process, Criterion 2--the Maintenance of Productive Capacity of Forest Ecosystems--covers most of the timber-related SFM measures (starting at the 10th of the 67 indicators):

- Area of forest land and net area of forest land available for timber production.
- Total growing stock of both merchantable and nonmerchantable tree species on forest land available for timber production.
- The area and growing stock of plantations of native and exotic species.
- Annual removal of wood products compared to the volume determined to be sustainable.
- Annual removal of non-timber forest products (e.g., fur bearers, berries, mushrooms, game) compared to the level determined to be sustainable.

Indicators 10 through 13 each measure factors that influence the sustainability and regulation of industrial forest plantations. In fact, of the 67 indicators, those four are among the most possible to quantify and monitor, and are among the indicators that have received the most focus from traditional forest management research and practice. However, even the status of regulation of timber yields from plantations has not been documented recently in the literature. Accordingly, this chapter surveys forest regulation and timber supply prospects in the world, and examines the practices and situation in the U.S. South closely.

The southern forestry sector has lagged in participation regarding international criteria and indicators, but has developed related efforts that make it poised to participate fully. The

South is particularly important for forest plantations and timber supply, comprising up to 17% of the total world industrial roundwood production; 60% of U.S. timber harvests; and a large share of the world's industrial wood plantations. The American Forest and Paper Association (AF&PA) Sustainable Forestry Initiative (SFI) has been fully developed on member company lands and with most of their logging contractors. AF&PA includes most major pulp and paper companies, whose land holdings are the largest in the South. This suggests that the forest industry has been following sustainable forest management principles at least since 1994. The AF&PA SFI program also has been moving toward second- or third-party certification of their SFI program, similar to that employed by the Forest Stewardship Council.

Forest plantations, timber regulation, and timber supplies are a critical factor in determining the amount of forest products manufacturing capacity that can be sustained. Based on derived demand, these factors also contribute significantly to decisions that firms and individuals will make in planting and growing trees. Timber supply and environmental concerns also influence what public policies might be pursued to encourage, subsidize, or require public and private landowners to manage their forest lands.

It has been suggested that intensive management of forest plantations will provide more wood fiber on less area, and gradually supplant the need to harvest and manage natural forest stands extensively (Binkley, 1997; Sedjo, 1995; Sedjo and Botkin, 1997; Mattoon, 1998; Sohngren et al., 1999). This premise, and the effects of intensive management of exotic plantations, usually in monocultures, is crucial in determining the sustainability if industrial and nonindustrial timber management and manufacturing. Critics of intensive management, however, suggest that we need more management of forest resources as complete ecosystems, balancing timber and nontimber goods and services (Abramowitz, 1998; Noble and Dirzo, 1997). This review focuses on means for measuring and achieving sustainable industrial timber production, which is at least necessary, if not sufficient, to even achieve broader multi-resource and multi-value forest resource sustainability.

Data for such analyses can be obtained from various sources. The World Resource Institute and the United Nations Food and Agriculture Organization (FAO) publish general data on many natural resource conditions, including forest land area, wood production, and trade by country. Several public institutions in the U.S. and abroad have examined timber supply situations in their country and in the world. And several consulting firms provide proprietary, or occasionally, public information about their views of the world's timber supply situation. Theoretical and historical information on forest regulation is pervasive in the forestry literature. However, little data are available on current practices. So we canvassed many companies in the U.S. South and in other countries to determine current practices and future directions. We will summarize and assess the data and conclusions from each of these sources.

World Forests, Plantations and Trade

The estimated world population as of 2000 was 6 billion people. The population is increasing by about 86 million people annually, or about 1.5% per year. Population levels are projected to reach 8.0 billion people by 2025, and 9.4 billion by 2050 (World Resources Institute, 1998). These population increases should prompt a proportional increase in the world wood fiber demands, either for industrial or nonindustrial uses, and perhaps greater increases as developing counties demand more paper and paperboard products and solid wood products.

Total Forest Area

World land, forest, and population statistics are summarized in Table 1, drawn from FAO (1998, 1999) data and many other sources described subsequently. Each of these relatively simple statistics is only moderately accurate and reliable, but they are the best available data. Even the FAO estimates of the closed forests of the world differ from others, depending on the definitions of forests and the measurement method employed. Williams (1994) summarizes many prior estimates of world forest area, and notes that even recent estimates in the last decade ranged from about 3.7 to 4.6 billion hectares (ha), depending on the source. For simplicity, we will employ the FAO (1999) estimate of 3.4 billion ha of forests as of 1995. Estimating planted forest areas and industrial wood plantation areas is at least as challenging, and we rely on many sources for that purpose, as described below.

According to the FAO (1999) data, South America and the Russian Federation have the largest total forest cover in the world, at 871 million ha and 764 million ha, respectively. Brazil alone has 551 million ha. Africa (520 million ha), led by the Democratic Republic of the Congo (109 million ha), follows next in total area covered by forests. Asia has 503 million ha, including Indonesia (133) and China (133). North America has 457 million ha of forest land, with 245 million ha in Canada and 212 in the United States.

Forest Plantation Area

Data on world plantation coverage are sparse and incomplete. The FAO (1999) data enumerate "total forest area" and "natural forest area." The difference between these data, when available, was computed to estimate the "total plantation area" reported in Table 1.

Region/ Country	Total Land Area (000 ha)	Total Forest Area (000 ha)	Natural Forest Area (000 ha)	Total Plantation Area (000 ha)	Fast-Grown Industrial Plantation Area (000 ha)	Total Population, 1997 (million)	Population Density, 1997 (people per sq. km.)
Africa	2.936 960	520 237	515455	4 782	2 080	778	26
Congo, Dem Rep.	226 705	109 245	109 203	42		48	21
Ethiopia	100 000	13 579	13 439	140		60	60
Kenya	59 914	1 292	1 174	118		28	50
Nigeria	91 077	13 780	13 629	151		118	130
South Africa	122.104	8.499	7.204	1.295	1.450	43	36
Other Southern Africa					630		
Sudan	237.600	41.613	41.410	203		28	12
Zimbabwe	38.685	8.710	8.626	84		12	30
Asia	3.073.436	503.001	434.278	68.723	4.214	3.589	117
China	932.641	133.323	99.523	33.800		1.244	133
India	297.319	65.005	50.385	14.620		960	323
Indonesia	181.157	109.791	103.666	6.125	3.063	204	112
Japan	37.652	25.146	14.946	10.200		126	334
Korea, Republic	9.873	7.626	6.226	1.400		46	463
Malaysia	32.855	15.471	15.371	100	50	21	64
Myanmar	65.755	27.151	26.875	276		47	71
Philippines	29.817	6.766	6.563	203	102	71	237
Thailand	51.089	11.630	11.101	529	265	59	116
Turkey	79.963	8.856				63	82
Viet Nam	32.549	9.117	7.647	1.470	735	77	235
Oceania	849.084	90.695	88.161	2.534	2.558	46	5
Australia	768.230	40.908	39.828	1.080	1.080	18	2
New Zealand	26.799	7.884	6.406	1.478	1.478	4	13
Papua N.G.	45.286	36.939	36.909	30		5	10
Europe	2.260.128	933.326	807.949	125.377	5.352	729	32
Northern	112.329	53.246	9.220	44.026			
Finland	30.459	20.112	2.190	17.922		4	17
Norway	30.683	8.697	3.578	5.119		4	14
Sweden	41.162	24.437	3.452	20.985	500	9	21
Western	245.569	59.479	35.849	23.630	4.852		
Austria	8.273	3.877	1.347	2.530		8	99
Belgium & Luxemb.	3.282	709	375	334		11	323
France	55.010	15.034	10.122	4.912	1.087	59	106
Germany	34.927	10.740	4.300	6.440		82	235
Italy	29.406	6.496	5.364	1.132	280	57	195
Spain	49.944	8.388	5.581	2.807	1.900	40	80
United Kingdom	24.160	2.390	803	1.587	750	58	242
Eastern	1.902.230	821.309	763.588	57.721			
Belarus	20.748	7.372				10	50
Czech Republic	7.728	2.630				10	132
Poland	30.442	8.732	2.082	6.650		39	127
Russian Federation	1.688.850	763.500	719.230	44.270		148	9
Slovakia	4.808	1.989				5	112
North America	1.838.037	457.086	436.086	21.000	18.000	304	17
Canada	922.097	244.571	241.571	3.000		30	3
United States	915.912	212.515	194.515	18.000	18.000	272	30
C. Amer/Caribbean	264.774	79.443	78.958	501	250	131	49
Mexico	190.869	55.387	55.278	109	109	94	49
South America	1.751.708	870.594	863.515	7.079	7.710	332	19
Argentina	273.699	33.942	33.395	547	840	36	13
Brazil	845.651	551.139	546.239	4.900	4.300	163	19
Chile	74.880	7.892	6.877	1.015	1.600	15	20
Colombia	103.870	52.988	52.862	126	100	37	36
Ecuador	27.684	11.137	11.092	45	70	12	43
Paraguay	39.730	11.527	11.518	9		5	13
Peru	128.000	67.562	67.378	184	150	24	19
Uruguay	17.481	814	658	156	150	3	18
Venezuela	8.205	43.995	43.742	253	500	23	26
WORLD	13.048.410	3.442.369	3.212.373	229.996	40.164	5.930	45

*Table 1. World Land, Forest, and Population Statistics by Region and Selected Country, circa 1995. Sources: Base data from Food and Agriculture Organization [FAO] (1999), Tables 1 and 2 Europe data drawn from Brown (1998) and authors' estimates in Table 2 Japan and Canada plantation data taken from Brooks (1993) Asia plantation data: FAO (1999); Brooks (1993) for Japan; Yin (1998) for China Selected Asian and Central America fast-grown industrial plantation data: authors U.S.A. plantation data estimated by authors based on USDA Forest Service survey data Latin America industrial wood plantation data from Flynn (1996) Other southern hemisphere industrial wood plantation data from Edwards (1996) Country population data reported for 1997 in FAO (1999) Continental population data reported for 1998 in World Resources Institute (1998) Empty cells indicate no reported area or that data were not available FWC; NCSU Forestry; *30\s\worldarea.wb3; 18 January 2000; V#5*

However, the "natural forest area" figure was not reported for most major developed countries, including all of Europe, Canada, the United States, Japan, New Zealand, and Australia. Data for plantation area for these countries was thus taken from other sources, including Brooks (1993), Flynn (1996), Edwards (1996), Brown (1998), Cubbage et al. (1996), and our own current timber supply estimates.

The amount of industrial wood plantations also is difficult to quantify exactly, based on the general shortcoming on plantation data and the definition of "industrial." Nevertheless, for this analysis we assumed that "industrial" plantations included all those grown on industrial or nonindustrial private or on some public forest lands that could be used to grow and harvest industrial roundwood for manufacture of forest products. We further added a subset of these to include fast-grown industrial wood plantations that were based on exotic or native species with growth rates of about 5 m^3/ha/yr or more, and common rotation lengths of less than 30 years. Tables 1 and 2 summarize our different estimates of total forest area, natural forest area, and fast-grown industrial plantation area according to these criteria.

Table 2 summarizes our estimates of forest area, industrial wood plantations, and exotic plantations for Europe separately, which includes a large share of the planted stands in the world, managed either extensively or intensively. In several countries in temperate and boreal regions, especially in Europe, it is difficult to distinguish between plantations composed of native species and natural forests. Historically, only areas planted with exotic and/or fast growing species are called plantations in Europe. There are about 5 million ha of this type plantations. However, most of European forests have been planted and then clearcut to provide wood for industrial purposes for decades, if not centuries. They cannot be considered natural forests and they mostly resemble a hybrid of plantations and natural forests. What distinguishes them from a "pure" plantation is a rather long rotation period. In order to assess the area of those forests, we assume that planting, clear cutting, and industrial wood production confer a plantation status on forests composed of native species. We approximated that area by calculating the number of hectares of exploitable coniferous forests. This is because most of coniferous stands are planted monocultures and produce industrial wood. While some of those stands may be naturally regenerated or have nonindustrial uses, they would be more than just offset by deciduous plantation forests. Regarding the former Soviet Union (including European and Asian parts), we were more conservative due to low planting survival rates and assume that only 15% of exploitable coniferous forests are plantations. This approach indicates that there are about 125 million ha of plantations in Europe, with 81 million ha in European countries excluding the former Soviet Union, which accounts for another 44 million ha.

European Region/Country	Total Forest	Total Exploitable Forest	Total Exploitable Coniferous Forest	Estimated Total Plantation Forest	Exotic/fast Growing Species
Northern					
Finland	20112	19511	17922	17922	
Norway	8697	6638	5119	5119	
Sweden	24437	22048	20985	20985	500
Subtotal	53246	48197	44026	44026	500
Western					
Austria	3877	3330	2530	2530	
Belgium	620	620	303	303	
Denmark	466	466	316	316	
France	14230	13551	4912	4912	1087
Germany	10490	9852	6440	6440	
Greece	2512	2289	935	935	
Ireland	396	394	356	356	
Italy	6750	4387	1132	1132	280
Luxemburg	85	82	31	31	
Netherlands	334	331	202	202	
Portugal	2755	2346	1322	1322	835
Spain	8388	6506	2807	2807	1900
Switzerland	1130	1093	757	757	
United Kingdom	2207	2207	1587	1587	750
Subtotal	54240	47454	23630	23630	4852
Eastern					
Bulgaria	3386	3222	1147	1147	
Hungary	1675	1626	260	260	
Former USSR	754958	414015	295130	44270	
Former Yugoslavia	8370	7768	976	976	
Former Czechoslovakia	4491	4491	2869	2869	
Poland	8672	8460	6650	6650	
Romania	6190	5413	1549	1549	
Subtotal	787742	444995	308581	57721	0
TOTAL	895228	540646	376237	125377	5352

Table 2. Forest Area and Plantations in Europe, 1995 (thousand ha). Note: Former USSR includes both European and Asian parts. Source: European Forest Institute Databank, Kuusela (1992), Nilsson et al., 1992, UN/ECE Timber Committee; Jacek Siry; NCSU Forestry; #30\s\europlanted.wb3; updated 18 January 2000; V#3

Flynn (1996) and Edwards (1996) provide relatively accurate estimates of industrial plantation areas for the southern hemisphere; Cubbage et al. (1996) do the same for North America. Brooks (1993) estimates broad classes of industrial plantation area for Japan, as does Yin (1998) for China, but probably only a fraction of this could be considered wholly for industrial production purposes. Most of the Japanese, Chinese, Indian, and African forest plantations probably cannot be considered industrial wood plantations because they were established for watershed protection, fuelwood, or desertification protection. In addition, the plantations often were not very successful; industrial timber harvests may be excluded by higher recreational or development uses; and the accuracy of the data is modest. For selected countries in Asia major and in Central America, we assumed that about half of the total reported forest plantation area would be devoted to fast-grown industrial wood production. Most planted stands in Europe, however, have such long rotations that we excluded them from the "Fast-grown" category.

In total, then, we estimate that there about 230 million ha of forest plantations in the world. Europe contains the largest area of planted stands, with 125 million ha or about 55% of the world total, but only about 5 million ha are in fast-growing exotic species. The United States contains about only about 8% of the world's planted stands, but about 45% of the fast-grown industrial wood plantations. The U.S. South alone would contain about 33% of the fast-grown industrial wood plantations per our definitions, with 13.5 million ha. The Southern Hemisphere of the world contains the most fast-growing exotic plantations, with approximately 16.8 million ha. This comprises about 42% of all industrial plantations, but a far greater, and increasing, share of industrial roundwood production.

The industrial wood plantation area of about 40 million ha, or 1.2% of the world's forests, represent the minimum area that is actually subject to traditional or modern forest harvesting regulation techniques. Probably much of the other planted stands in the world, many of which are in Europe, are also managed actively and subject to sustained yield timber regulation and management. The 229 million ha of planted stands (6.7%) are probably the most actively managed forests in the world. The remaining 93% of the world's forests are managed in natural stands, left to grow naturally, or are reserved from commodity production entirely.

World Wood Production and Trade

Table 3 summarizes the world wood production and trade statistics by region and selected countries as of 1996 (FAO, 1998). Asia and Africa produce the greatest quantities of fuelwood and charcoal in the world, at 905 million m^3 (49% of world fuelwood production) and 520 million m^3 (28%), respectively. Asia also produces a large amount of sawnwood (144 million m^3) and pulp and paper (125 million m^3).

Region/Country	Fuelwood & Charcoal Production (000 cu m)	Industrial Roundwood (000 cu m)				Sawnwood + Wood Panels (000 cu m)				Pulp + Paper, & Paperboard (000 cu m)				Total Forest Products Trade Value (thousand $US, 1996)		
		Production	Imports	Exports	Net Exports	Production	Imports	Exports	Net Exports	Production	Imports	Exports	Net Exports	Imports	Exports	Net Exports
Africa	519.884	67.931	676	7.439	6.763	10.969	5.143	1.698	-3.445	4.826	1.536	1.248	-288	2.589.057	3.163.056	573.999
Congo, Dem Rep	45.142	3.433	0	178	178	121	0	49	49	3	2	0	-2	1.979	81.005	79.026
Ethiopia	46.982	2.373	0	2	2	112	0	44	44	7	8	0	-8	5.884		-5.884
Kenya	41.003	1.961	0	0	0	237	5	2	-3	195	25	0	-25	22.351	398	-21.953
Nigeria	105.832	8.479	2	11	9	2.838	20	25	5	64	62	0	-62	46.465	16.707	-29.758
South Africa	7.156	18.176	77	2.473	2.396	2.227	524	62	-462	3.418	411	944	533	552.997	1.258.672	705.675
Sudan	14.600	2.274	14	0	-14	47	74	0	-74	3	3	0	-3	27.248		-27.248
Zimbabwe	6.269	1.859	0	15	15	329	7	80	73	126	24	4	-20	24.377	24.850	473
Asia	905.236	280.153	68.509	14.336	-54.173	143.458	40.660	16.407	-24.253	124.795	24.856	8.560	-16.296	44.172.867	17.651.458	-26.521.409
China	204.239	108.718	7.169	3.580	-3.589	42.318	7.296	1.236	-6.060	55.004	5.701	978	-4.723	6.898.915	1.490.413	-5.408.502
India	279.350	24.989	336	23	-313	17.808	37	47	10	4.895	615	9	-606	453.024	25.028	-427.996
Indonesia	153.540	47.245	178	683	505	17.466	80	8.731	8.651	7.021	836	2.342	1.506	865.424	5.206.522	4.341.098
Japan	360	22.897	47.860	7	-47.853	31.541	18.232	44	-18.188	41.079	4.935	1.027	-3.908	18.890.397	1.781.177	-17.109.220
Korea, Republic	4.497	1.994	9.066	2	-9.064	5.576	2.805	144	-2.661	8.299	2.918	1.384	-1.534	4.425.527	1.258.793	-3.166.734
Malaysia	10.035	35.771	224	7.152	6.928	15.152	585	8.991	8.406	777	872	41	-831	881.539	4.161.279	3.279.740
Myanmar	20.612	3.023	0	498	498	376	2	191	189	24	13	0	-13	9.678	205.067	195.389
Philippines	37.280	3.394	636	13	-623	909	845	186	-659	762	412	25	-387	1.069.313	99.977	-969.336
Thailand	36.894	2.818	939	388	-551	818	2.438	226	-2.212	2.744	814	335	-479	2.048.397	476.359	-1.572.038
Turkey	8.534	10.745	820	64	-756	5.409	542	180	-362	1.589	720	37	-683	737.487		-737.487
Viet Nam	31.250	4.487	1	350	349	760	47	51	4	183	204	37	-167	83.641	59.332	-24.309
Oceania	8.756	41.461	14	18.347	18.333	8.922	1.221	1.880	659	4.881	1.508	1.305	-203	2.182.344	3.044.006	861.662
Australia	2.904	19.813	7	7.895	7.888	4.432	941	261	-680	2.825	1.295	241	-1.054	1.848.257	774.192	-1.074.065
New Zealand	50	16.999	3	6.275	6.272	4.094	51	1.549	1.498	2.156	185	1.064	879	272.360	1.566.420	1.294.060
Papua N.G.	5.533	3.239	0	3.039	3.039	233	2	17	15	0	6	0	-6	6.169	526.298	520.129
Europe	82.439	369.650	44.700	46.074	1.374	154.452	48.521	57.624	9.103	118.983	48.036	50.526	2.490	59.100.191	62.642.567	3.542.376
Northern	8.388	102.798	16.189	3.210	-12.979	30.964	2.096	21.187	19.091	43.298	1.548	22.378	20.830	3.426.119	23.357.176	19.931.057
Finland	4.094	42.503	7.323	749	-6.574	10.808	299	8.198	7.899	20.117	243	10.064	9.821	699.632	10.301.017	9.601.385
Norway	470	7.695	3.123	530	-2.593	10.823	2.575	1.034	-1.541	4.365	622	2.222	1.600	1.402.551	2.059.960	657.409
Sweden	3.824	52.600	5.737	1.931	-3.806	4.233	2.205	438	-1.767	18.816	683	10.092	9.409	1.323.936	10.996.199	9.672.263
Austria	3.059	11.346	5.708	2.236	-3.472	9.765	1.378	5.374	3.996	5.049	1.456	3.317	1.861	1.988.878	4.149.678	2.160.800
Belg & Luxmbrg	550	3.635	3.503	1.315	-2.188	3.818	2.716	2.840	124	1.814	2.605	1.208	-1.397	3.544.574	2.180.694	-1.363.880
France	10.460	30.980	2.753	3.650	897	13.420	3.435	3.191	-244	11.047	5.134	2.882	-2.252	5.356.351	4.193.414	-1.162.937
Germany	3.795	35.543	1.639	4.148	2.509	24.763	8.517	3.959	-4.558	16.549	10.226	6.651	-3.575	11.926.822	9.428.751	-2.498.071
Italy	4.970	4.163	5.398	9	-5.389	5.572	6.346	603	-5.743	7.517	5.889	1.810	-4.079	6.148.593	2.486.782	-3.661.811
Spain	2.707	12.997	2.113	266	-1.847	5.997	2.356	695	-1.661	5.229	2.668	1.170	-1.498	3.552.249	1.523.810	-2.028.439
United Kingdom	261	7.021	772	68	-704	4.801	8.688	491	-8.197	6.764	8.086	1.382	-6.704	8.476.689	1.957.907	-6.518.782
Belarus	809	9.206	0	516	516	1.919	30	204	174	162	14	7	-7	19.956	55.634	35.678
Czech Republic	689	12.317	217	3.239	3.022	4.465	414	1.861	3.019	1.106	372	474	102	506.753	796.677	289.924
Poland	1.688	17.783	486	484	-2	7.470	481	1.513	6.438	2.373	770	485	-285	954.600	709.295	-245.305
Russian Fed	29.250	67.000	964	16.312	15.348	24.636	79	5.662	5.583	6.937	68	2.459	2.391	115.030	2.995.568	2.880.538
Slovakia	454	4.887	2	545	543	941	100	502	402	562	100	269	169	143.829	316.628	172.799
North America	94.029	589.708	9.288	23.584	14.296	221.482	54.579	67.507	12.928	186.462	18.695	38.050	19.355	25.180.739	42.273.054	17.092.315
Canada	5.319	183.113	7.022	2.607	-4.415	72.795	2.793	57.408	54.615	42.817	1.851	23.526	21.675	2.622.203	25.333.157	22.710.954
United States	88.710	406.595	2.266	20.977	18.711	148.687	51.786	10.099	-41.687	143.645	16.844	14.524	-2.320	22.558.536	16.939.897	-5.618.639
C. Amer/Caribbean	61.408	10.739	145	117	-28	5.377	2.120	652	-1.468	3.833	2.101	361	-1.740	2.349.450	525.798	-1.823.652
Mexico	16.731	5.914	27	28	1	3.149	960	529	-431	3.558	1.150	328	-822	1.336.908	396.706	-940.202
South America	193.007	129.890	40	10.066	10.026	34.373	780	5.591	4.811	19.047	3.317	5.542	2.225	22.558.536	16.939.897	-5.618.639
Argentina	4.498	6.220	6	968	962	1.590	239	260	21	1.930	684	216	-468	3.077.539	5.356.500	2.278.961
Brazil	135.652	84.711	8	2.215	2.207	22.649	335	2.804	2.469	12.110	1.118	3.396	2.278	1.154.971	3.233.476	2.078.505
Chile	9.984	21.387	2	6.451	6.449	4.646	44	1.505	1.461	2.720	162	1.797	1.635	185.188	1.424.751	1.239.563
Colombia	18.062	2.703	2	8	6	820	32	428	396	983	421	63	-358	373.970	55.015	-318.955
Ecuador	5.474	5.514	0	20	20	2.266	6	173	167	86	324	7	-317	215.192	71.054	-144.138
Paraguay	6.524	3.877	6	1	-5	496	2	484	482	13	26	1	-25	21.930	105.382	83.452
Peru	7.315	1.546	4	6	2	776	16	18	2	188	148	4	-144	138.107	17.114	-120.993
Uruguay	3.050	1.043	2	298	296	275	60	54	-6	115	62	10	-52	64.012	47.314	-16.698
Venezuela	918	1.366	7	48	41	1.520	250	7	-243	900	307	50	-257	211.327	41.894	-169.433
WORLD	1.864.760	1.489.530	123.372	119.963	-3.409	579.030	153.022	157.344	4.322	462.926	99.490	105.686	6.196	138.652.187	134.656.439	-3.995.748

Table 3. World Forest products production and trade statistics by region and selected country, 1996. Sources: Food and Agriculture Organization [FAO] (1999), Table 4; FAO (1998) Table on Forest Products Trade Values, p.224-226. Fred Cubbage; NCSU Forestry; #30\s\worldtrade.wb3; updated 18 January 2000; V#4.

Leading roundwood producers include the United States (407 million m^3 (27% of world production), with about 250 million m^3 (17%) in the South); Europe (370 million m^3 (25%), including Scandinavia (103 million m^3 ; 7%)); Canada (183 million m^3; 12%); and South America (130 million m^3; 9%, including 85 million m^3 in Brazil). The same countries tend to dominate in pulp and paper production, the principal output using industrial roundwood. Brown (1998) estimates that about 25% the world's industrial roundwood now comes from forest plantations, with as much as 75% of industrial roundwood coming from plantations in Oceania and 50% in Africa, and less than 20% in the former USSR, Europe, and North and Central America. These statistics appear to agree with our subsequent estimates of U.S. South timber production, but seem low for European forests areas. At least they provide some bounds for the contributions of industrial plantations to world timber production.

In the volume of forest products trade, the United States and Canada have by far the largest volumes of exports and imports, but Canada has large net exports, and the U.S.A. is a large net importer. Europe also conducts a large volume of trade and is a net exporter, based mostly on Scandinavia. Asia as a whole is a large importer, except for Indonesia and Malaysia. For the total value of forest products trade, Canada ($23 billion in 1996) and Scandinavia ($20 billion) dominate net export receipts. Indonesia ($4.3 billion), Malaysia ($3.3 billion), the Russian Federation ($2.8 billion), Austria ($2.2 billion), Argentina ($2.3 billion), Brazil ($2.1 billion), New Zealand ($1.3 billion), and Chile ($1.2 billion) are other major net exporters of forest products by value. Asia is by far the largest importer for forest products by value, at $27 billion, and $17 billion for Japan alone. The United Kingdom ($6.5 billion), the United States ($5.6 billion), and China ($5.4 billion) were the world's next largest net importers of forest products by value.

Forest regulation

The preceding discussion illustrates the importance of sustainable forest management (SFM) as a new paradigm, and the extent of industrial forest plantations in contributing to the world's wood fiber needs. Sustained yield as the basis for forest regulation is probably the original timber (and forest) sustainability criterion. This has been implemented through classical timber harvest scheduling approaches, and modernized with the use of computers and extension to multiple use optimization models.

The nexus between SFM, traditional and modern forest regulation, and timber supply becomes crucial in determining whether we can indeed satisfy the precept of SFM with existing

Forest regulation

The preceding discussion illustrates the importance of sustainable forest management (SFM) as a new paradigm, and the extent of industrial forest plantations in contributing to the world's wood fiber needs. Sustained yield as the basis for forest regulation is probably the original timber (and forest) sustainability criterion. This has been implemented through classical timber harvest scheduling approaches, and modernized with the use of computers and extension to multiple use optimization models.

The nexus between SFM, traditional and modern forest regulation, and timber supply becomes crucial in determining whether we can indeed satisfy the precept of SFM with existing forests and timber management practices. This will help us determine whether we are meeting the implicit objectives stated in the Montreal Process SFM Criteria 2, indicators 10 through 13. It also will greatly affect whether we can achieve the other SFM criteria and indicators, or whether wood production from natural stands will have to continue to provide a large share of industrial and fuel wood purposes.

Classical Forest Regulation

It is instructive to briefly review the history of forest regulation to provide perspective for modern concerns about sustainable forest management. Not long ago by forestry standards, when human population became sufficiently dense, the notion of the forest as commons for hunting and gathering wood shifted to the notion of a forest as property which needed regulation. With that shift came the profession of forestry. Before there was a need for timber management there was a need to control wildlife harvesting. The professional forester's first duties were to act as game warden and hunting guide (Fernow, 1911). Later as wood consumption increased enough the need to protect and regulate wood harvesting became evident. In 15th century Germany the first recorded attempts to bring order into forest harvesting occurred by dividing the whole area into a number of felling areas. This was not exactly area control, but close. In the next century, 1547 to be exact, the first recorded attempts to regulate harvest by volume control occurred. The earliest use of the term "sustained" yield we could find was in 1713 by Von Carlowitz (Wiebecke and Peters, 1984). In 1740, Frederick the Great, who understood the importance of a steady wood supply to the state, ordered the use of area control for the German state pineries. Forty-eight years later (1788), an anonymous Austrian official first wrote about the concept of a normal forest, with a normal stock, distributed in normal age classes, so as to ensure sustained yield management.

About 1820, new theories were advanced which led to the formulation of methods based upon the idea of the normal forest. It was in this time that Huber, Hundeshagen, Heyer, and Koenig developed their rational formula methods of volume regulation (Fernow, 1911). These mathematical techniques were generally considered too complex for all but the most educated foresters. Thus the need for higher education forestry schools became a necessity. The need for sustained yield of forest products was the major impetus in the creation of the forestry profession, separate from being a game warden or hunting guide.

The same forest management paradigm shift, from forest as a commons to forest as a property, occurred slowly in the United States in the early part of the twentieth century. The possibility of a timber famine was popularized by Pinchot (1919, 1947), and became a national political issue. Wood products companies started to realize that the supply of timber was exhaustible and that regenerating cutover lands was an investment that had the possibility of a positive payoff. This was a radical thought at the time since natural timber was still relatively abundant and cheap. A few far-sighted people saw that in the not so distant future this situation would change. The first organized attempt in the U.S.A. to invest in regeneration that we found was in 1904, when Henry Hardtner of Urania Lumber Company in Louisiana, began buying and regenerating cutover land and practicing sustained yield harvesting of the company lands. Hardtner's success was emulated by F.O. Bateman at the Great Southern Company in Bogalusa, who managed the planting of thousands of acres of pine plantations (Burns, 1984). By the end of the 1920s, about 64,000 acres of forest land had been planted with trees, with about 44,000 on industry lands. From 1936 to 1942, more than 100,000 acres per year were planted in the South, mostly on nonindustrial private forest land, in response to the depression and dust bowl conservation issues (USDA Forest Service 1988).

In 1941, Weyerhaeuser dedicated the Clemons Tree Farm in Washington state, which perhaps heralded the modern era of large-scale industrial plantation forestry in the United States (Twining, 1984). Beginning in 1948 and extending to the present, both nonindustrial owners and industrial private forestry firms in the South began substantial pine plantation activity, beginning with planting of more than 186,000 acres in 1948 and increasing to more than 2 million acres per year by 1985 (USDA Forest Service 1988). Exact data don't exist, but we estimate that the southern share has probably comprised about 75% of all forest plantations in the United States.

Modern Approaches

The problem which classical scheduling methods were designed to solve was to regulate timber harvests at a sustainable level. This classical question remains essentially the same in modern forestry, however, the social context and our technological capabilities are much different. In the past simple formulas stressing biological maturity seemed sufficient for the task, until people started to wonder about economic efficiency, about alternatives for management of future stands and social responsibility of such regulation approaches. What are the impacts on future yields, and therefore current cutting levels, of precommercial thinning, commercial thinning, fertilization, and other silviculture activities? These questions could not be answered well until modern computer technology developed. Without computers, it would take a life time to do routine calculations required in even a simple formula with immense data matrices. Along with the addition of computers, the mathematics of forest modeling progressed beyond the scope of early work in timber harvest regulation. This progress continues through today in efforts to answer increasingly complex forest resource problems.

With the advent of the computer, full-scale multiple-use planning became possible (Iverson and Alston, 1986). A series of progressively more sophisticated forest planning models were developed using classical regulation methods. The first of many forest simulation models was ARea VOLume check method (ARVOL, Chappelle, 1966)--a computerized simulation model using area-volume check. This was followed shortly by ShOrt Run Allowable Cut (SORAC, Chappelle, 1968), which was a designed to look beyond the current rotation toward how intensively managing the regenerated stands might affect current harvest levels. The next step in complexity came with Simulating Intensively Managed Allowable Cut (SIMAC, Sassaman et al., 1972), which permitted a wide range of intensive management simulations.

Model I and Model II Formulation.

In the late 1950s linear programming (LP) formulations for forest planning were proposed and their use grew through the 1980s. All the documented programs consist of a matrix generator that takes the forest data and generates a linear programming matrix. Then solution software takes the matrix and solves for an optimal solution; and a report writer interprets the solution into a readable format. The Timber Resource Allocation Model (Timber RAM, Navon, 1971) was an early LP matrix generator and report writer widely used within the U.S. Forest Service. The plans themselves were developed to address the questions of biological sustainability of the cut on a forest wide basis. Timber RAM was designed to only solve timber management questions. At about the same time, scientists at the University of Georgia developed MAXMILLION (Clutter et al., 1978), a Model I linear programming harvest scheduling package that works directly with forest inventory data. MAXMILLION was widely adopted by forest products firms throughout the South. A variety of other systems such as CAPS, a Model II formulation of

Crown Zellerbach, and a Model I system used by Champion International were also being employed (Davis and Johnson, 1987).

In the middle to late 1970s, planning issues were beginning to shift from harvest scheduling to multiple use questions. The Multiple Use-Sustained Yield Calculation technique (MUSYC, Johnson and Jones, 1979) was an attempt to improve the limited orientation of Timber RAM and provided for integration of other forest uses into timber planning. Out of this effort emerged the Model I/Model II nomenclature and timber scheduling formulations (Johnson and Scheurman, 1977). The difference between these two formulations is how the decision variables are defined. Both TIMBER RAM and MAXMILLION were Model I simulations. In Model I, a decision variable represents the amount of land allocated to a set of management activities from the present to the planning horizon. The set of management activities may include actions on the existing stand through perhaps several harvests on future regenerated stands. The model I formulation is:

$$MAX \quad C^T X$$

$$S.T. \quad AX \leq b$$
$$X \geq 0$$

Where C is a (N) vector of per unit area contributions to the objective function, C_{IJ} is an individual contribution for management regime i and strata j, X is a (N) vector of decision variables representing area assigned to management regime/strata combinations. *b* is a (M) vector of resource constraints, including area per strata and relations between outputs and strata. A is a (NxM) matrix of coefficients.

In Model II, there are decision variables for existing stands and different ones for regenerated stands. Each time a stand gets harvested it gets transferred to another decision variable. The model II formulation is:

$$MAX \quad C^T X \; + \; D^T Y$$

$$S.T. \quad A\begin{bmatrix} X \\ Y \end{bmatrix} \leq b$$
$$\begin{bmatrix} X \\ Y \end{bmatrix} \geq 0$$

Where X is a (N_1) vector of existing stand decision variables; Y is a (N_2) vector of regenerated stand decision variables; D is the (N_2) vector of contributions to the objective function from regenerated stands.

Both Model I and II formulations have merits for timber harvest and activity scheduling. When ecological considerations need to be incorporated, the Model I formulation has a slight advantage because a unit of land can be easily tracked from beginning to end of the planning horizon. Another advantage is that integer constraints on management units can easily be tracked from start to finish. Both formulations have a fundamental weakness in terms of ecological forest planning and analysis. They both are essentially strata based models, while most ecological problems require spatial integrity.

The required spatial integrity can be incorporated by making the Model I formulation an integer program with each stand receiving only one management regime, but this leads to very large integer formulations which are usually not solvable even with the most powerful computers. A common solution to this problem usually consists of a heuristic procedure to find a good or simply feasible solution. An early example of this is the Hoganson and Rose (1984) approach, which uses characteristics of the dual formulation and a decomposition procedure to find a simultaneous solution to the stand management and harvest scheduling problem.

FORPLAN and SPECTRUM

MUSYC did not live up to its name. However, it did provide a stepping stone to the development of the FORest PLANning (FORPLAN) model, which was used by the U.S. Forest Service for most of the 1980s and early 1990s. FORPLAN, besides being a highly flexible modeling program, incorporated both the strata-based information required for per acre yields and an innovative area-based formulation required by recreation, wildlife, water and other environmental outputs needing area-wide coefficients. The areas were called zones and they could receive totally different sets of management prescriptions depending on the type of management emphasis which was allocated to each zone. FORPLAN was unique among planning models within the United States. It could handle a wide variety of problems in forest management. It could evaluate area specific projects. It could evaluate groups of projects connected in space and time. For the first time forest planners had the opportunity to shift the focus of planning away from functional concerns toward truly integrated planning (Iverson and Alston 1986). It was used extensively throughout the U.S.A. as a forest planning tool by the Forest Service, private industry, and state planning personnel.

However, by the early 1990s FORPLAN had received a reputation of being very difficult to learn, understand, and interpret. The U.S.D.A. Forest Service redirected its sustainable forest management planning software effort to SPECTRUM in the mid-1990s. The most visible difference between SPECTRUM and FORPLAN was a "relatively easy to use" graphical user interface (GUI). The underlying matrix generator was essentially the same with several modeling enhancements and some deletions of unused components. The enhancements included model data storage in relational database format allowing for easy linkages with other programs and geographic information systems (GIS), expanded objective function capability, direct rollover capability, new data checking procedures, database formatted reports and several other small but useful changes.

As flexible and comprehensive as SPECTRUM is, it still has at least one major deficiency. It is a Linear Programming based analysis tool and nonlinearities presented by spatial arrangement cannot be modeled. Spatial analysis is often referred to as area-based while the analysis represented by SPECTRUM and it's forerunners is referred to as strata-based. SPECTRUM does allow for formulations that are area-based (or a mixture of strata-based/area-based) but the available solution software limits the number of areas that can be practically modeled.

A more critical deficiency is that integer linear programming does not provide a modeling environment sufficient to answer several important spatial problems. LP can be used to piecewise approximate highly nonlinear relationships between inputs and outputs over time. When different management prescriptions are included they involve different levels of intensity of some input. The different A-matrix coefficients under these different management prescriptions can reflect highly nonlinear responses to changing input intensity. A common example is the use of nonlinear growth and yield models to provide volume estimates. For nontimber outputs, the most important nonlinearity is probably the impact of spatial configuration of management actions (Hof, 1993). In LP management prescriptions are typically defined on a per unit area basis, and LP determines the amount of area to which each management prescription applies. The problem is the nonlinear response to different sizes and shapes of the management action is lost in a fixed per unit area production coefficient.

Hierarchical Levels of Management and Planning and Analysis

Forest management decisions are made over broad spectrums of social organizations, space, and time. In forest industry there is a wide range of management decisions, from the overriding decision of whether or not to be part of forest industry to the detail of where to set skid trails in harvest operation. Along with organizational decision levels, there is also a wide range of spatial scales in decision making ranging from decisions over millions of hectares to sub-meter placement of a drainage structures. Also there is a wide spectrum of time scales which forest management deals with ranging from centuries to minutes. Given the task of managing over ranges of space and time, as well as organizational hierarchy, management always breaks down the decision problem into sets of more manageable sub-problems. These breakdowns are commonly categorized as strategic, tactical, and operation divisions of management.

Strategic level planning is the process of envisioning the organization's future goals and developing the necessary procedures and operations to achieve those goals. It usually involves upper levels of the organization's management considering wide time and space scales. In forestry, strategic level plans usually deal with one or more forests covering a time horizon of 1½ or more typical rotations.

Tactical level planning is the process of taking the goals, procedures and operations developed at the strategic level and developing the actual process for achieving them. It usually involves midlevel management looking over limited time horizons and space scales. In forestry, tactical level plans usually deal with sub-forest areas (districts, watersheds, etc.) covering a time horizon of 1 to 10 years. A typical tactical level decision would define which stands will be harvested each year for the next 10 years.

Operational level planning is the process of taking the tactical level decisions and developing the methods for carrying them out. It usually involves midlevel management looking over short time and small space scales.

An analytical problem that developed by the 1980s was caused by the failure of analysts to recognize different levels of forest planning. In 1992, the University of Toronto hosted a workshop on hierarchical approaches to forest management in public and private organizations, where many of these analysis problems were clarified, as paraphrased below. Forest management planning crosses all levels of an organization from top to bottom and many ecological scales from regional to the micro. Failure to recognize these different levels at the outset can lead to infeasible and possibly non-sustainable forest operations, as well as solutions that cannot be implemented. Within a firm the planning hierarchy is commonly categorized into strategic, tactical and operational. Using a hierarchy, decisions can be analyzed in separate, manageable levels. A difficult problem with an analytic hierarchy is to adequately link decisions between levels. Obviously, the decisions are interrelated. The decisions at the strategic level should be reflected at the tactical level and so on down the hierarchy (Weintraub, 1992). It is well known that the details and reality at lower levels in the hierarchy can often make upper level decisions infeasible.

Some procedure to link analytic levels is used by all forest industry landowners. However, the level of explicitness of procedure varies greatly between companies. A major strategic decision is how much volume can be harvested each period. The major set of tactical decisions is what areas should be harvested each period. Operation planning carries out the tactical decision.

The way that forest industry deals with the hierarchy can be grouped into at least three different analytical approaches (see Tables 4 and 5). One approach, and perhaps the most ambitious, is to integrate all levels into one very large model. Using this approach to hierarchical planning stands are treated as units, planning periods are one year long, the planning horizon is at least one timber rotation long, and all the detail and decisions needed at each level are included. This approach was attempted in the 1970s and 1980s but lost favor due to: (1) difficulty in obtaining feasible results; (2) if results were found they were difficult to interpret; and (3) the amount of detail needed up-front to solve the models was immense. However, this approach has found some renewed interest in the late 1990s due to advancements in data handling capabilities and heuristics solution software.

A second approach, and the one with most widespread current use and number of analytic variations, is to solve the strategic forest management problem using a traditional linear programming system and then allocate the solution out to different areas on the forest. The allocation of the strategic solution can take on a variety of different forms. Some of the more sophisticated firms use some analytic procedures to allocate solution to stands and are then mapped in a GIS. The analytic procedure varies greatly between companies and may be an integer linear program, taboo search, simulated annealing, genetic algorithms or some other heuristic.

Less sophisticated firms leave it up to the forest managers to do a manual allocation. Instead of emphasizing quantitative analytical techniques, some companies place more emphasis on the experience base of the forest manager. In fact, even the "sophisticated" mathematical techniques are always used in conjunction with the experience of the forest manager, who often has much more information about what is on the ground than can be stored in a database.

The last approach, one with a growing number of users, is to solve the hierarchy in sequence with iteration. The basic links express the flow of information as data move up, to a strategic level, and flow of directives as decisions move down, to tactical and operational levels.

In respect to all the different systems being employed, it appears to be a very common practice, if not universal, to take the solutions of even the most sophisticated analytical system to the land management foresters for their approval. Foresters are still the final arbiters of what will or will not work on the ground.

Current Industrial Practice of Sustainable Forest Management Scheduling

Forest industry uses a wide variety of quantitative methods to ensure sustainability of harvest. In a survey of 13 analysts, we ascertained the timber harvest scheduling/sustainable forestry practices for 69 private firms and 12 public organizations. Tables 4 and 5 summarize current strategic and tactical planning methods, respectively, for the surveyed firms. Most, but not all, forest products companies divide the sustainable management-scheduling problem into strategic and tactical levels.

Two large companies that integrate strategic and tactical planning models feel that the decision levels are too intricately linked to be separated. In addition, they feel that computer and mathematical programming technology has advanced far enough to eliminate most problems encountered in early attempts to solve the strategic and tactical problems simultaneously.

Strategic Level Planning (Table 4)

Not surprisingly for industry, maximization of net present value is the most common objective, but some companies use other and multiple objectives to coordinate more closely with the larger integrated company strategy. Model formulations are dominated by Model I closely followed by Model II. One company reported use of the classic technique of area control with volume check. Most of the Model I and II formulations are linear programs using some version of the Simplex Method for solution. The notable exceptions are Company F, which has a Model I integer formulation using a heuristic and Company I, which uses area control-volume check with straightforward rules for harvest. Both F and I solve the combined strategic-tactical problem.

Company/Contact (# rep.)	Objective	Model Formula-tion	Solution Technique	Commercial Program Name	Planning Horizon	Period Length	Analysis Unit Definition	Spatial Integrity	Sustained Yield Constraints	Update Cycle	Ecological Constraints
A (1) Forest Products	Net Present Value (NPV)	Model I	Simplex Method	FORPLAN 15.5 (Spectrum no shell)	100 years	5 years	Stand type aggregates	Not in strategic model	Sequential Lower Bound between periods	5 years, moving to annual	Riparian zones, T and E Habitat, HSI, Ecosystem Diversity
B (1) Forest Products	NPV	Model II	Simplex Method	Woodstock (Remsoft)	30-100	variable	Stands	stands can be split	Volume flow	Continuous	SFI
C (3) Forest Products	NPV	Model I	Simplex Method	Epics	100 or more years	5 to 10 years	Stand type aggregates	Dissag-gregation heuristic	Nondeclining yield (NDY), flow leveling	variable	Many – Habitat Conservation Plan one of the outputs
D (1) Forest Products	Min Cost of fiber to mill	Model I	Simplex method	FORPLAN (Spectrum with no shell)	45 years in south USA	3 years	Stand type aggregates	Not in strategic model	NDY of cash flow	yearly	Not included at this level
E (USDA Forest Service)	Multiple (NPV, Volume,etc)	Model I and II	Simplex Method	Spectrum (C-WHIZ Solution software)	variable	variable	variable	variable	NDY, LTSY, perpetual timber harvest	variable	Variable
F (1) Forest Products	Multiple but always including NPV	Model I	weak heuristic, simulated annealing and taboo search	In house	40 years	yearly	stands	Complete, stands treated as integer values	Harvest flow	Yearly	SFI
G Forest Product (21,13,11)	Volume and/or NPV	Model II	Simplex Method	Woodstock (Remsoft)	variable	variable	stands	stands can be split	Even-flow	variable	SFI and others
H (1) Forest Products	NPV	Model I	Simplex Method	In house	SE 40-50 NW 50-100 years	1 year for first five years and 3 years after	Stands	stands can be split	Harvest <= growth	3 to 5 years	Not included at this level
I (1) Forest Products	Achieve Cutting Budget	area control with volume allotment	Harvest oldest and slowest growing	In house	10 years	1 year	Stands	Stands are treated as units	Harvest <= growth	yearly	SFI
J (21) NZ Forest Products	NPV	Model II	Simplex Method	FOLPI	60-90 years	3-6 years	Crop types	Not in strategic model	volume flow, NDY, LTSY	yearly	Riparian zones restrictions
K (3) Chile Forest Products	NPV	Model I	Simplex Method	MEDFOR	50 years	3 years	Stand type aggregates	Not in strategic model	volume flow	Several times per year	Constrain out critical areas, Riparian zones restrictions
L (1) LM ORG	NPV	Model I	Simplex Method	In house	50 years	5 years	Stands	Not included	Logistical constraints	5-8 years	Not included at this level
M (1) Forest Products	NPV	Model I	Simplex Method	In house	30-45 years	3-5 year periods	Stand type aggregates	Not at this level	volume control, NDY, LTSY	4-6 years	Not included at this level

TABLE 4. Strategic Planning Methods in use by Industry. Contact column is usually a single person, which could represent a single company, a software development company or a consulting company that represents many forest products firms. Information presented in both table roise1 and roise2 was gathered by informal survey of industry, industrial consultants and government agencies conducted in July 1999. Information summarizes scheduling practices from 32 North American, 3 Chilean, and 21 New Zealand and Australian companies. In addition 13 consulting firms and 12 government agencies are represented. The number in parenthesis in column 1 represents the number of companies reported by this contact. Contact G reported 21 firms, 13 consultants and 11 government agencies.

The planning horizon varies among and within companies depending on their geographic location. Period length also varies, with 5 years and under being common and with three companies reporting yearly periods. The majority of companies report that they do not aggregate stands within their analysis. Several reported that they do not aggregate because it helps at the tactical level and the two companies (F and I) with simultaneous strategic-tactical solutions must use stands to make their approach work.

It is interesting to note that many companies update the strategic plan yearly. Many industry contacts reported that strategic planning in their company was a continuous process and that they may update their strategy several times per year. Others felt that strategy was a little more static and they only update when needed. The largest update interval reported was 5 to 8 years. This difference between companies may simply reflect differences in the planning hierarchy between companies. One company's strategic decision is another's tactical decision.

Constraints to ensure sustainability also vary between companies. Nondeclining flow, volume flow, long term sustained yield and constraining harvest to be less than growth were all reported in use. Only the private nonindustrial landowner (L) reported no sustained timber yield constraint. All other industrial and public land owners were concerned about sustained yield. While the nonindustrial private land owner was concerned about maximizing return to investors, sustained yield was not quantified. Regeneration after harvest and use of best management practices was a given for all organizations surveyed.

Tactical Level Planning (Table 5)

There seems to be much more variation at the tactical level of management then at the strategic level. While NPV is still the primary objective function in use, target values provided from strategic planning come into use also. Model I formulations are most commonly used, though with the short planning horizon the problem becomes more of a land allocation problem and less of a scheduling problem. Our contact persons for some of the model formulation responses listed as Model I in Table 5 were ambivalent about what the formulation actually should be considered. Some felt that because the nature of the different levels was different the Model I-Model II dichotomy does not apply. Companies that use the same model for both tactical and strategic analysis usually use strategic results to set tactical goals or constraints and use a more localized formulation.

The most common tactical solution technique is a spatial heuristic. The second most common tactical solution method uses people to allocate the computer-generated strategic decision. The simplex method was not used by any firm at this level. Most firms reported that the results were mapped into a GIS. However, the degree of automation in allocating the harvest solution into a GIS varied from totally manual to totally computerized (with management approval).

Contact	Objective	Model Formulation	Solution Technique	Commercial Program Name	Planning Horizon	Period Length	Analysis Unit Definition	Spatial Integrity	Constraints	Update Cycle
A (1) Forest Products	Net present value	Model I	Heuristic: Analysis Units Disaggregation Model into Stands	In house	100 years with first period being of most interest	5 years	Stands	Allocated to the stand level	Opening size, adjacency	5 years now moving to annual update
B (1) Forest Products	NPV	Not clear	Heuristic	Stanely (Remsoft)	Variable	1 year	harvest units	yes	Opening size, adjacency green up period	Yearly
C (3) Forest Products	Both tactical and strategic plans use same model									
D (1) Forest Products	Targets from strategic model	Experience based allocation	Manual		8 years	1	Stands	Allocated to the stand level	Opening size, adjacency	yearly
E (USDA Forest Service)	Tactical model can take on many different forms including spatial disaggregation, project planning using a variety of tools including SNAP. Each forest has flexibility to do what they feel is most appropriate for them.									
F (1) Forest Products	Strategic and tactical planning are integrated into one large model, which avoids disaggregation problems.									
G Forest Product (21,13,11)	Targets from strategic model	Not clear	Heuristic	Stanley (Remsoft)	variable	1 year	Polygons	yes	SFI	Yearly
H (1) Forest Products	Strategic and tactical planning are integrated into one large model, which avoids disaggregation problems. Results are given to district foresters who allocate results to ground checking for SFI and other spatial concerns. Use of GIS to map out solutions but a manual check is made to insure compliance.									
I (1) Forest Products	No differentiation between strategic and tactical planning									
J (21) NZ Forest Products	The IFS simulator and the FOLPI optimizer developed by NZ Forest Research are used by about 20 organizations in New Zealand and Australia. They are both modeling systems in which the user builds their own models. They provide the system, which the user can apply in any number of ways. Tactical planning is even more company specific.									
K (3) Chile Forest Products	NPV	Model I	Heuristic rounding, Lagrangian relaxation, other heuristics in development	OPTIMED	3 to 5 years	biyearly	Stands	stands treated as integer	Access, roads, long range targets	biyearly
L (1) LM ORG	NPV	Model I	Manual	-	3 to 5 years	1 year	stands`	stands treated as integer	Opening size and adjacency	Rolling plan with yearly updates
M (1) Forest Products	NPV	Model I	Spatial Heuristics	In house	15 years	1 year	GIS Polygons	Cutting unit Polygons are integer	SFI	4-6 years

TABLE 5. Tactical planning methods being used in Forest Industry.

The tactical planning horizon ranged from 3 years to 100 years. Period lengths ranged from ½ year to 5 years. All contacts except one reported that the unit of analysis was a stand, harvest unit, or polygon, which were treated as integers so that spatial integrity could be maintained. One contact (G) reported use of a substand GIS layer. These small land units are recombined in the analysis to define land units to be treated over time. This technique is particularly useful in the southern United States where large homogenous plantations might be subdivided in to smaller units to satisfy SFI requirements. All U.S.A. companies reported the use of SFI constraints at this level, while not all used them at the strategic level. The tactical plans are updated most commonly on a yearly basis but range from ½ to 5 years for different companies.

Summary

Forest products companies from North America, South America, New Zealand and Australia are managing their lands with a concern about sustainability. Foremost in their efforts was ensuring a sustainable supply of timber expressed through use of constraints within their harvest scheduling analysis. Several people expressed the view that the investment in regeneration and growth today demonstrates a commitment to sustainable forestry. The companies plan to be around in 30 years when the investment pays off, or at least have the value of that investment reflected in their current stock prices. Most firms also profess a commitment to sustaining the forest ecosystem. Most U.S. firms report use of voluntary American Forest and Paper Association Sustainable Forestry Initiative (SFI) regulations. Many firms go beyond the practices required by law and regulation. Riparian zone protection, wildlife corridors, special habitat protection, bottomland protection, and use of best management practices were reported as being used by several companies.

The methods and technology of harvest scheduling vary among companies. Some things are similar. The Model I-II classification of models is useful for describing different mathematical formulations. The Simplex Method is still in common use in strategic planning, but only of minor use in tactical planning. Computer technology has explained our ability to handle the large amounts of data required of a forest plan.

Southern Timber Supply – sustainable forest regulation?

The preceding two sections summarize the importance of plantations in world timber supply and the means that forest products firms and public organizations use to ensure sustained yield of timber products. It is the interaction of timber plantations and industrial and nonindustrial firm decisions about sustained yield that actually determine whether a region or a country achieve sustainable timber management, and the even broader purposes of sustainable forest management. Now we will examine that interaction of plantations, intensive management, and sustained timber supply. The U.S. South will be used as a case study of regulation of timber yield for sustainable management of industrial timber.

Plantation Trends in the U.S. South

Since the turn of the century softwood and hardwood inventories in the South have been increasing. Softwood lumber production peaked just after the turn of the century leaving large areas of harvested and unmanaged timberland. Lower production, natural regeneration and advances in fire control had the biggest impact on increasing inventories until the middle of the century.

With the advent of sulfate pulping technology and the second growth forest, the southern pulp and paper industry began expanding in the 1930s. This was followed by the emergence of industrial forestry and pine plantation management in the 1940s and 1950s, as noted earlier. The dramatic increase in plantation area and volumes can be seen in Figures 1 and 2, per data from the USDA Forest Service Forest Inventory and Analysis (FIA) surveys.

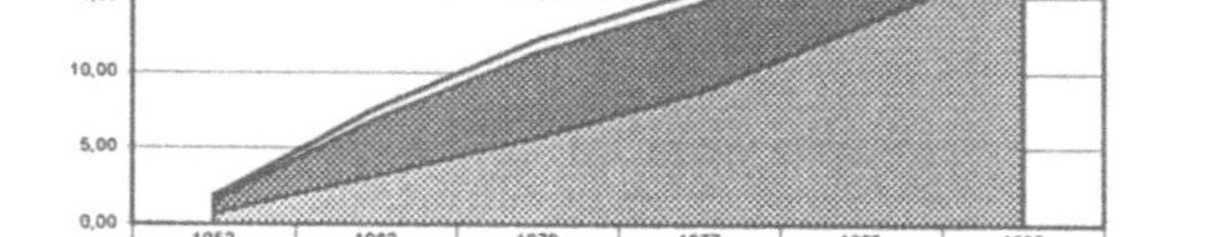

	1952	1962	1970	1977	1985	1995
Public	336	720	814	857	1163	1979
NIPF	850	3740	5616	5843	6568	14200
FI	660	3127	5714	8723	13153	17700

Figure 1. Development of Southern U.S. Plantation Acres By Owner.

From 1952 to 1985, forest plantations in the South increased from 1.8 million acres (0.7 million ha) to 20.8 million ac (8.5 million ha) (USDA Forest Service, 1988). Estimates of 1995 data are based on the latest FIA surveys in each state. The official FIA definition of pine plantations is based on the stand origin (planted) and the forest type (pine or mixed pine). In 1995 there were 33.9 million acres (13.7 million ha) of planted pine in the South. Industrial lands (7.2 million ha) are owned by companies with processing facilities and include lands leased by industry. Nonindustrial lands comprised 5.7 million ha of plantations, and public lands 0.8 million ha in 1995.

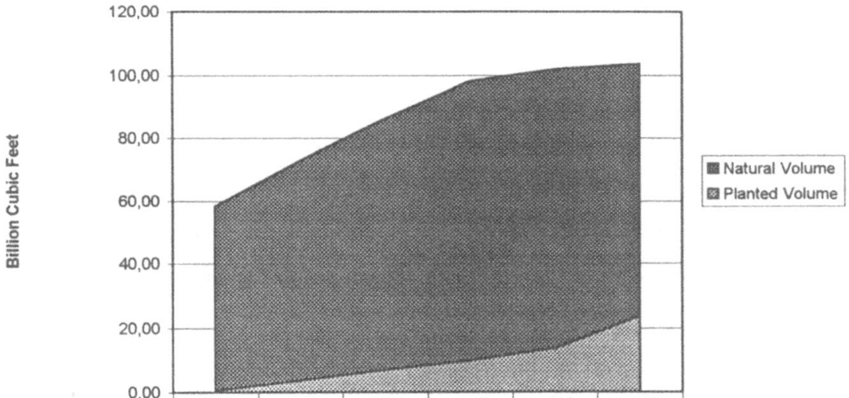

	1952	1962	1970	1977	1985	1995
Natural Volume	57589	69171	79270	88159	88269	79715
Planted Volume	658	3736	7027	9822	13567	23818

Figure 2. Development of Southern U.S. Softwood Growing Stock Volume.

In the 1950s and 1960s plantations were split relatively equally among industrial and nonindustrial private forest (NIPF) ownerships. The increase on NIPF lands was driven in part by the Soil Bank program, which subsidized plantation establishment. During the 1980s and 1990s the Conservation Reserve Program and higher stumpage prices led to another jump in NIPF plantations, while industry acreage continued to increase at historical rates. Currently about 52% of southern U.S. plantation area is owned by industry (Figure 3). With the recent growth in timber investment management organizations (TIMOs), some former forest industry land has been reclassified as other corporate.

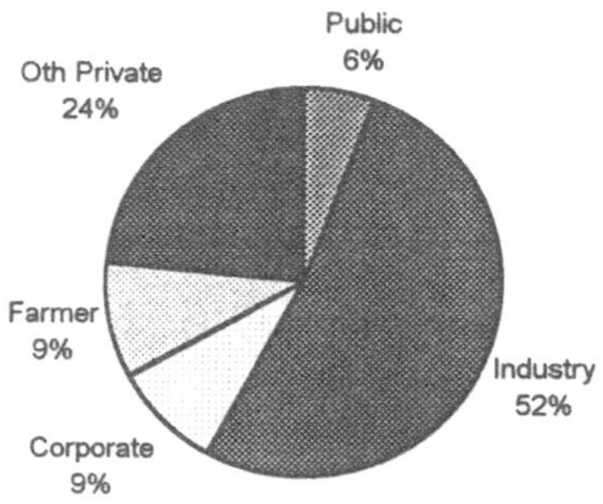

Figure 3. Current distribution of Southern U.S. plantation ownership.

The impact of this area on volume is realized 15 to 20 years after planting, as shown in Figure 2. Total softwood volume in the South increased from 58 billion cubic feet (1.66 billion m^3) of wood in 1952 to 104 billion cubic feet (2.96 billion m^3) in 1995. The higher growth and harvest rates associated with pine plantations are crucial in determining whether we can achieve sustainable timber supplies in the U.S. South, or indeed the world. Figure 4 shows estimates of growth per acre on coastal plain plantations in the southern U.S. based on FIA data.

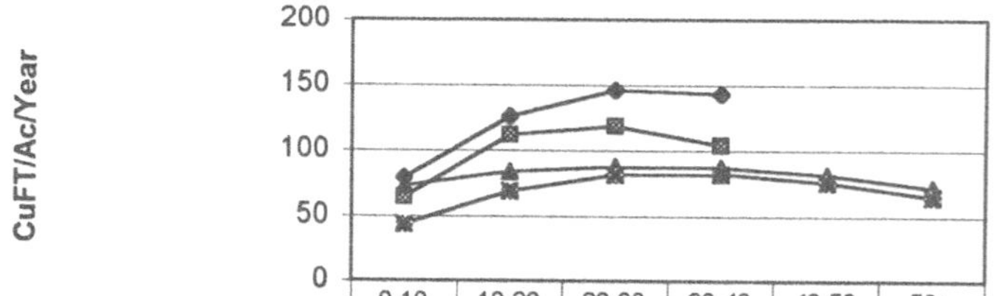

	0-10	10-20	20-30	30-40	40-50	50+
Industry Planted	79	126	146	143		
Oth Private Planted	64	112	119	104		
Industry Natural	72,5	83,5	87,5	87	81,5	71,5
Oth Private Natural	44	69	81	82	75,5	65

Figure 4. Growth rates in coastal plain pine plantations in the Southern U.S.

On average, plantations have 50% greater growth rates than natural pine stands, and industry plantations have growth rates 20% greater than non-industrial landowners. Based on the 1995 data, softwood removals from plantations have increased to nearly 40% of harvest from industry lands or 25% of total softwood removals (Figure 5). The conversion to plantations has continued slowly. To date, however, the U.S. South is still harvesting proportionately more wood volume from natural stands than plantations, compared to the respective forest areas in each forest management type.

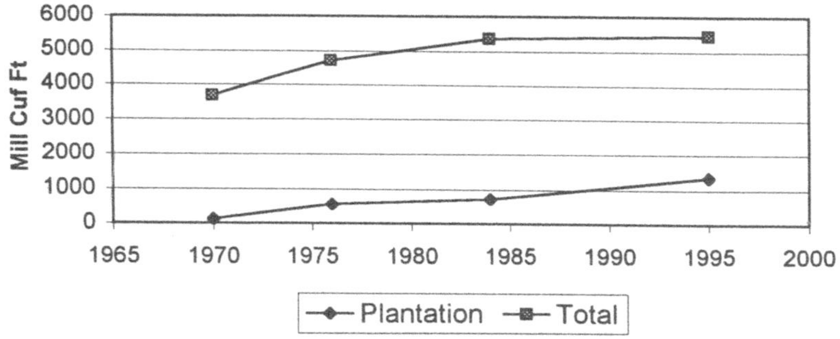

Figure 5. Southern softwood removals by type.

Hardwood plantations in the U.S. South are currently not a factor in timber supply. While a significant research effort is underway, current plantings tend to be strategic wet-weather accessible sites near mills. Eucalyptus plantations in South America will likely have greater short-term effect on southern hardwood markets than domestic plantations.

Role of Plantations in Future Sustainability

The historical trends discussed above show an increasing role of plantations in southern U.S. softwood timber resource sustainability. The role of intensive management and plantations in the future of the softwood resource have been explored using the Sub-Regional Timber Supply (SRTS) model (Abt et al., 1995, 1999). SRTS combines an inventory module based on the FIA data. Growth, removal and inventory data are integrated with an economic model that describes price responsiveness.

Baseline

The baseline simulation described below is a 27 year projection from 1993 to 2020 for private lands in the South, using US Forest Service RPA projected harvest trends (Haynes et al.,

1995). The base year of 1993 reflects the approximate midpoint of the current set of FIA surveys available for the 12 southern states. The baseline run consists of 40% increase in softwood harvest during the 1993 to 2020 period, and a 50% increase in hardwood harvest. To reflect the large increases in harvest for both softwoods and hardwoods experienced in the last decade, the majority of the projected harvest increase was assumed to occur in the beginning of the projection period (Figure 6) by using a logarithmic trend.

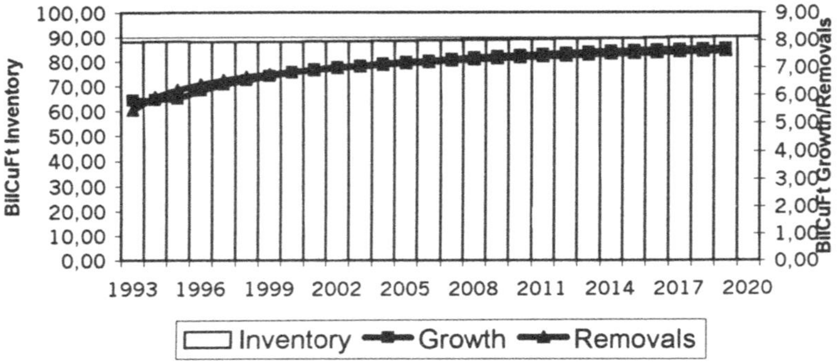

Figure 6. Southern U.S. softwood inventory projection.

We assumed a 50% increase in forest industry pine plantation growth rates over the FIA-based data during the projection period. A 25% increase was assumed for nonindustrial plantations. Other southern growth and yield analyses have suggested that such growth increases should be attainable (Yin et al., 1998; Siry et al., 2000). This growth rate change is applied with a logarithmic function over time so that most of the increase is experienced in the first half of the projection. The growth rate change is assumed to affect all age classes. Pine plantation areas were assumed to increase from 1993 to 2000 based on baseline rates, but they were held constant from that point forward at about 26% of the private land base. Areas were removed from natural and mixed pine types based on their relative abundance in the sub-region. Together, these assumptions imply that total timberland area is constant during the projection period.

Figure 6 shows the South-wide softwood inventory results for private forest lands only. Public lands were excluded from the analysis, since they are assumed to be unaffected by timber market activity. Private forest inventory increases slightly, from about 90 billion cubic feet (2.57 billion m^3) in 1993 to 93 billion cubic feet (2.66 m^3) in 2020. The increase in plantation area and in plantation growth rates (50% for industry, 25% for NIPFs) largely offset the 40% increase in harvest. Figure 7 shows the increasing harvest and inventory trends as indices along with the price and demand trends that are consistent with the harvest assumption. Since harvest is increasing faster than inventory, there is upward pressure on prices, which more than double in real terms by the end of the projection. Note that the results indicate that most of the harvest and price increases occur in the 1990s, with cnly a 35% increase in real prices from 2000 to 2020. Increasing harvest rates even with higher prices during the last two decades of the projection correspond with an implicit 30% demand increase.

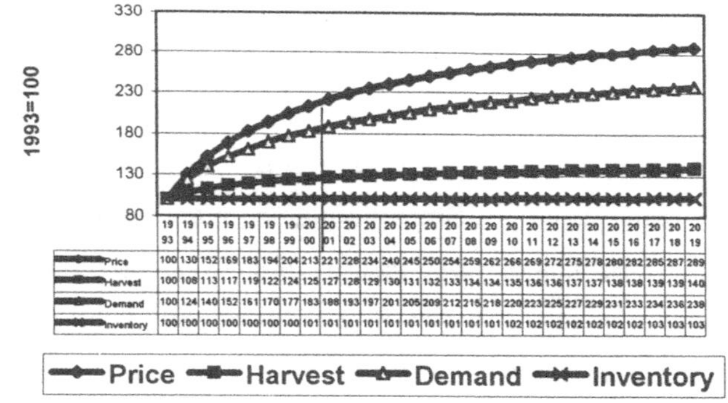

	1993	1994	1995	1996	1997	1998	1999	2000	2001	2002	2003	2004	2005	2006
Price	100	130	152	169	183	194	204	213	221	228	234	240	245	250
Harvest	100	108	113	117	119	122	124	125	127	128	129	130	131	132
Demand	100	124	140	152	161	170	177	183	188	193	197	201	205	209
Inventory	100	100	100	100	100	100	100	101	101	101	101	101	101	101

	2007	2008	2009	2010	2011	2012	2013	2014	2015	2016	2017	2018	2019
Price	254	259	262	266	269	272	275	278	280	282	285	287	289
Harvest	133	134	134	135	136	136	137	137	138	138	139	139	140
Demand	212	215	218	220	223	225	227	229	231	233	234	236	238
Inventory	101	101	101	102	102	102	102	102	102	102	103	103	103

Figure 7. Southern U.S. baseline softwood market shifts.

The spatial distribution of plantations affects the distribution of harvest. Those states with the largest concentrations of industry plantations (Georgia and Florida) had the largest harvest increases. States which had less favorable growth/drain ratios based on the last available survey, for example Louisiana, lost harvest relative to other states.

Figure 8 shows that plantations soon become the dominant source of pine harvest. By the year 2004, less than half of softwood harvest comes from natural stands. The trend continues so that by 2020, over two thirds of softwood harvest comes from plantations.

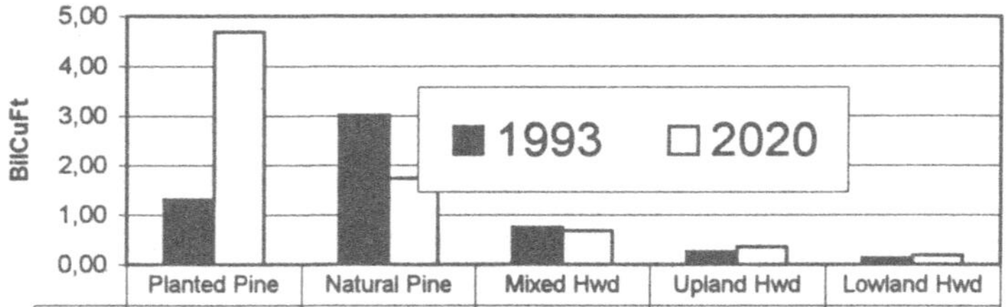

	Planted Pine	Natural Pine	Mixed Hwd	Upland Hwd	Lowland Hwd
■ 1993	1305342	3014696	744387	247234	133220
□ 2020	4670129	1734482	675989	358222	180465

Figure 8. Baseline softwood harvest shifts by management type.

Intensive Management

Recent data suggest that large gains in plantation growth can be realized through the use of fertilization and herbicide treatments. An alternative intensive management scenario was developed using the same demand as baseline scenario. The model was used to estimate the price and harvest consequences of altering plantation management relative to the baseline. Instead of the base assumptions, forest industry plantations were assumed to have growth rates 75% higher than current FIA growth rates at the end of the projection period; non-industrial plantations were assumed to show 37.5% greater than current FIA annual growth rates. Plantation areas increased from 20% to 36% of the private timberland base from 1993 to 2020 using 1993 RPA trends. The increase in growth rates will have an immediate effect on inventory, but the changes in area beyond the year 2000 only begin to influence supply by the end of the projection period.

Figure 9 shows the market impacts of substantially increasing plantation growth on a plantation area base that is assumed to continue expanding after 2000. The same demand trend applied to the faster growing softwood resource leads to lower prices and higher harvest levels, especially at the end of projection period. Real prices increase only about 15% after the 1990s and then decrease as the new plantations affect supply.

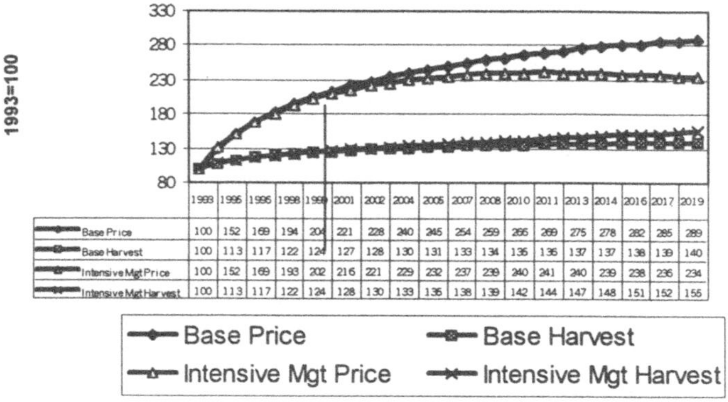

	1993	1995	1996	1998	1999	2001	2002	2004	2005	2007	2008	2010	2011	2013	2014	2016	2017	2019
Base Price	100	152	169	194	204	221	228	240	245	254	259	266	269	275	278	282	285	289
Base Harvest	100	113	117	122	124	127	128	130	131	133	134	136	136	137	137	138	139	140
Intensive Mgt Price	100	152	169	193	202	216	221	229	232	237	239	240	241	240	239	238	236	234
Intensive Mgt Harvest	100	113	117	122	124	128	130	133	136	138	139	142	144	147	148	151	152	155

Figure 9. Softwood baseline and intensive management scenario.

All of the trends toward increased dependence on plantations were accelerated in these scenarios, with plantation-rich states having slightly more harvest and less dependence on natural stands relative to the baseline.

Sustainability Policies and Plantations

Because of concerns related to erosion and timber supply there have been a succession of policies that subsidized the establishment of plantations. In the 1950s the Soil Bank program led to the establishment of 1.9 million acres (770,000 ha) of plantations in the South, with 700,000 acres (280,000 HA) in Georgia alone (Cubbage et al., 1993). Since that time, the Conservation Reserve Program (about 2.2 million acres in the U.S.; 900,000 ha), the Forestry Incentives Program (about 4 million acres in the U.S.; 1.6 million ha), the Agricultural Conservation program (7 million acres; 2.8 million ha) have been successful in encouraging plantation establishment (Gaddis et al., 1995). So have private efforts such as certification of Tree Farms (more than 50 million acre; 20 million ha, but not all plantations). These efforts to ensure soil and timber sustainability, however, have generated concern about biodiversity and sustainability of nontimber forest resources. Recent efforts aimed at broader sustainability questions include the certification programs being promoted by the Forest Stewardship Council (FSC) and the industry-sponsored Sustainable Forest Initiative (SFI). Both programs impose additional costs on plantation management, and the FSC program actively discourages further plantation establishment.

Murray and Abt (1999) analyzed the costs of forest certification and the financial incentives that would be required to offset those costs. They found that a large proportion of the cost impact would be on industry plantations. Passively managed hardwood stands would meet certification standards without much added management efforts. Increased softwood forest management costs, especially for industrial owners, would require price premiums that averaged about 10% but ranged up to 31%.

Conclusions

This chapter has examined the regulation of timber yield for sustainable management of industrial forest plantations. As one can surmise after reading the chapter, even this fairly elementary forestry concept of sustained yield of industrial timber plantations--for a firm, forestry sector, or region--is fraught with challenges. These include determining the world forest area; defining industrial timber plantations; ascertaining means that firms regulate timber on a sustained yield basis; and estimating aggregate effects of forest plantations on regional timber supply. Like most economists, we hopefully have made defensible assumptions about each of these questions, and developed reasonable results about plantations, sustainability, forest management, and timber supply. However, all of these issues remain subject to debate, so our analysis may provide a benchmark, but certainly not a capstone regarding such questions.

It is worth noting that these timber-related questions are probably some of the most data-rich and most easily analyzed issues that fall under the broader sustainable forest management criteria and indicators. This does not auger well for quick or facile efforts to measure, monitor, and enhance the world's forests. In fact, it instead suggests that SFM will be a laudable goal that will indeed take continuous improvements in science, practice, and monitoring forever.

These interactions of natural and planted forest management and protection; intensity of forest management; population and demands; and science and practice will in the end determine whether we will have sustainable forest management and adequate timber supplies to meet burgeoning world wood fiber needs at reasonable prices. We of course bring some bias to this analysis, as purveyors of forest science, technology, education, and practice. But we will need significant advances in the development and application of technology in order to meet world demands.

Our review of industrial and public analytical methods indicates that all major corporations and United States national forests intend to achieve sustained yield of their timber plantations, and use elegant models to ensure that they achieve that goal. The survey of those firms also indicates that they seriously consider the broader mandates of the private Sustainable Forestry Initiative or public SFM criteria and indicators. In aggregate, industrial forestry firms, nonindustrial private forest owners, and public lands produce the southern U.S.A. timber supply. Our study of sustainability in southern timber supplies indicates that current forest inventory levels are not sustainable at current plantation management intensities, but that enhanced productivity can indeed help us achieve the objective of sustained yield of southern U.S. softwood supplies.

Even the merits of intensive fiber production on some lands, while leaving other forests more natural, is subject to considerable debate. At the very least, this dominant use or zoning approach rejects the classical multiple use approach, which tries to provide a balanced set of goods and services on all lands. Zoning land uses may be controversial because it concentrates potentially more adverse environmental effects in a small area, affecting fewer people. It also raises more fears about maintaining ecological and site sustainability, since natural processes are altered more by management (e.g. Baumol and Oates, 1979).

Intensive food and fiber production are now moving toward use of genetically modified plants, which have been opposed by an increasing array of interest groups. Tree selection has been used for decades for intensive softwood plantations in the northern hemisphere, and more recently for hardwoods in the southern hemisphere. Biotechnology promises to help select, or even insert, genetic material into trees that can convey desirable properties such as fast growth, good form, or disease resistance. However, the latent dislike of tree farms or forest plantations, coupled with the fear of genetically modified plants, is increasing rapidly (e.g. Halweil, 1999). These opinions implicitly suggest that the public prefers achieving broader ecological sustainability that relies on less intensive, but far more widespread, management of natural systems.

These tradeoffs between forest management intensity, environmental effects of intensive uses, and public acceptance will determine the extent of the application of advanced technology in achieving sustainable industrial timber management. Our economic analyses do indicate that we have the technical means--both biological and analytical--to measure and achieve sustainable management of industrial forest plantations, and indeed perhaps sustainable management of timber supply in most large regions of the world. This still leaves the question of broader Sustainable Forest Management for all forest resources (goods and services) unanswered. This analysis does provide more insight about sustainability for a set of four of the existing Montreal Process SFM Criteria and Indicators, and broader world timber supply as well.

We are shifting from wood-based manufacturing based on extraction of natural timber supplies to more intensive management of planted or even second-growth natural stands. For firms, controlling one's own wood fiber production will be important for success in this new market situation. Timber plantation research investments being made today--including improving seed sources, better genetics and breeding programs, improved matching of species to sites, better planting and competition control, and better intermediate stand management--must be implemented to ensure the viability of individual firms and their competitiveness as global fiber demands increase and available supplies decrease. Providing public incentives for implementation could also benefit nonindustrial private forest landowners, resulting in countries or regions developing a comparative advantage in timber growing and processing.

The data on forest plantations and growth rates, as well as our survey of harvest regulation, indicate that the forest products industry and other forest owners have the means to achieve and modeling capabilities to measure and implement sustainable timber management. If this potential is realized--which will require investments by forest landowners and acceptance by the general public--increased productivity in managed forests will lessen timber harvesting pressures and enhance the protection of natural forests. If investments by landowners are lacking, or public opposition to plantation forestry and genetic manipulation dominate, sustainable timber productivity may be more difficult to achieve.

Literature

Abt, R. C., Cubbage, F. W., and Pacheco, G., 1995: The timber supply situation in the Southeast: implications for intensive management. In: Proceedings, 23rd Southern Forest Tree Improvement Conference, Asheville, NC. Sponsored Publication No. 45 of the Southern Forest Tree Improvement Conference. Raleigh, NC: North Carolina State University Tree Improvement Cooperative. p. 1-6.

Abt, R. C., Cubbage, F. W. and Pacheco, G., 1999: *Southern Forest Resource Assessment Using the Sub-Regional Timber Supply Model.* Working Paper No. 9. Research Triangle Park, NC: Southern Forest Resource Assessment Consortium. 21 p.

Abramovitz, J. N., 1998: Forest decline continues. In: Vital Signs, 1998. Linda Starke, Editor. New York: WorldWatch Institute, W.W. Norton & Co. p. 124-125.

American Forest & Paper Association, 1998: Sustainable forestry for tomorrow's world. Washington, D.C.: American Forest and Paper Association. 20 p.

Baumol, W.J. and Oates, W.E., 1979: *Economics, Environmental Policy, and the Quality of Life.* Englewood Cliffs: Prentice Hall.

Binkley, C. S., 1997: Ecosystem management and plantation forestry: the case for allocation with illustrations from British Columbia. *Forestry Chronicle* 73(5):553-559.

Brooks, D., 1993: U.S. Forests in a Global Context. General Technical Report RM-228. Fort Collins, CO: U.S.D.A. Forest Service, Rocky Mountain Range and Experiment Station. 24 p.

Brown, C., 1998: Global forest products outlook study: thematic study on plantations. Consulting Report for: Food and Agriculture Organization of the United Nations. Rome. mimeo.

Burns, A. C., 1984: The Kisatchie Story: From Wasteland to Sustained Yield. In: History of sustained yield: a symposium. H.K. Steen, Editor. Durham, NC: Forest History Society. 469 p.

Chappelle, D.E., 1966: A Computer Program for Calculating Allowable Cut Using the Area-Volume Check Method. Research Note PNW-33. Portland, OR: U.S.D.A. Forest Service, Pacific Northwest Forest and Range and Experiment Station. 4 p.

Chappelle, D.E., 1968: A computer program for scheduling allowable cut using either area or volume regulation during sequential planning periods. Research Note PNW-93. Portland, OR: U.S.D.A. Forest Service, Pacific Northwest Forest and Range and Experiment Station. 9 p.

Clawson, M., 1975. *Forests for Whom and For What?* Baltimore, MD: Johns Hopkins Press. 175 p.

Clutter, J., Fortson, J. C., and Pienaar, L. V., 1978: MAXMILLION II: A computerized forest management planning system, user's manual. Athens, GA: School of Forest Resources, University of Georgia.

Cubbage, F. W., O'Laughlin, J., and Bullock, C. S., III., 1993: *Forest Resource Policy*. New York: John Wiley & Sons. 562 p.

Cubbage, F. W., Dvorak, W. S., Abt, R. C., and Pacheco, G. 1996: World timber supply and prospects: models, projections, plantations, and implications. Paper prepared for speech presented at: Central American and Mexican Coniferous Resource Cooperative Meeting. Bali, Indonesia. October 21-23.

Davis, L. S. and Johnson, K. N., 1987: *Forest Management, 3rd Edition*. New York: McGraw-Hill. 655 p.

Edwards, M., 1996: The South African forestry and forest products industry: a synopsis. In: Proceedings of the International Woodfiber Conference. Omni Hotel at CNN Center. Atlanta, GA. May 13-14, 1996. Published by the Pulp & Paper Report, Miller Freeman. looseleaf; 36 figures.

European Forest Institute (EFI). Forestry Databank. Http://www.efi.fi/Forestry_Databank.

Fernow, B. E., 1911: *A Brief History of Forestry in Europe, the United States, and Other Countries*. Cambridge, MA: Forestry Quarterly, Toronto University Press.

Flynn, B., 1996: Latin America: the future of Fiber Exports. In: Proceedings of the International Woodfiber Conference. Omni Hotel at CNN Center. Atlanta, GA. May 13-14, 1996. Published by the Pulp & Paper Report, Miller Freeman. looseleaf; 6 p, 10 figures.

Food and Agriculture Organization, 1999: State of the World's Forests, 1999. Obtained from World Wide Web site, July 1999: http://www.fao.org/fo/sofo/sofo99/pdf/sofe_e/coper_en.pdf. Rome: Food and Agriculture Organization of the United Nations.

Food and Agriculture Organization, 1998: Forest Products 1996, FAO Yearbook. Rome: Food and Agriculture Organization of the United Nations. p. 224-226.

Gaddis, D.A., New, B.D., Cubbage, F.W., Abt, R.C., Moulton, R.J. 1995: Accomplishments and economic evaluations of the Forestry Incentives Program: a review. Working Paper No. 78. Research Triangle Park, North Carolina: Southeastern Center for Forest Economics Research. 52 p. + appen.

Hagler, R. W., 1996: The global wood fiber equation--a new world order? *TAPPI Journal* 79(1):51-54.

Halweil, B., 1999: The emperor's new crops. *WorldWatch* 12(4): 21-29.

Haynes, R. W., Adams, D. M. and Mills, J. R., 1995: The 1993 RPA Timber Assessment Update. General Technical Report RM-GTR-259. U.S.D.A. Forest Service, Rocky Mountain Forest and Range Experiment Station. Fort Collins, CO. 66 p.

Hof, J., 1993: *Coactive Forest Management*. New York: Academic Press, Inc.; Harcourt Brace Jovanovich. 189 p.

Hoganson, H. H. and Rose, D. W., 1984: A simulation approach for optimal timber management scheduling. *Forest Science* 30(2):220-238.

Iverson, D. C. and Alston, R. M., 1986: The Genesis of FORPLAN: A Historical and Analytical Review of Forest Service Planning Models. General Technical Report INT-214. Boise, ID: U.S.D.A. Forest Service, Intermountain Forest and Range and Experiment Station. 32 p.

Johnson, K. N. and Jones, D. B., 1979: A user's guide to multiple use sustained yield resource scheduling calculation (MUSYC). Washington, D.C.: U.S.D.A. Forest Service, Timber Management Staff. 242 p.

Johnson, K.N. and Scheurman, H. L., 1977: Techniques prescribing optimal timber harvest and investment under different objectives-discussion and synthesis. *Forest Science*, Monograph No. 18. 31 p.

Kuusela, K. 1992. Forest Resources in Europe. *Cambridge University Press.*

Mattoon, A. T., 1998: Tree plantations taking root. In: Vital Signs, 1998. Linda Starke, Editor. New York: Worldwatch Institute, W.W. Norton & Co. p. 126-127.

McNutt, J. A., 1996: The global fiber resource situation: Challenges for the 1990s. In: Proceedings, 1995 Southern Forest Economics Workers Annual Meeting. Atlanta, GA: Wachovia Timberland Management. p. 1-32.

Murray, B. C. and Abt, R.C., 1999: Forest certification in the U.S. South: some empirical evidence on price premium requirements. In: Proceedings of the Southern Forest Economics Workers, Part II. Research Triangle Park, NC: U.S.D.A. Forest Service, Southern Research Station. p. 220-227.

National Association of State Foresters, 1997: Forests for a sustainable future: the use of criteria and indicators in sustainable forest management. NASF White Paper. Washington, D.C. 9 p.

Navon, D. I., 1971: Timber RAM...A long-range planning method for commercial timber lands under multiple-use management. Research Paper PSW-70. Berkeley, CA: U.S.D.A. Forest Service, Pacific Southwest Forest Experiment Station. 22 p.

Nilsson, S., Sallnas, O., Hugosson, M., Shvidenko, A. 1992. *The Forest Resources of the Former European USSR*. IIASA, Laxengurg, Austria.

Noble, I. R. and Dirzo, R., 1997: Forests as human-dominated ecosystems. *Science* 277(5325):522-525.

Pinchot, G., 1919: The lines are drawn. *Journal of Forestry* 17(8):899-900.

Pinchot, G., 1947: *Breaking New Ground.* New York: Harcourt, Brace, and Company. 522 p. Reprinted by Island Press. Covela, California.

Powell, D. S., Faulkner, J. L., Darr, D. R., Zhu, Z, and MacCleery, D., 1993: Forest Resources of the United States, 1992. General Technical Report RM-234. Fort Collins, CO: U.S.D.A. Forest Service, Rocky Mountain Range and Experiment Station. 132 p. + map. [Revised, June 1994].

Sassaman, R.W., Holt, E., and Bergsvik, K., 1972: User's manual for a computer program for simulating intensively managed allowable cut. General Technical Report PNW-1. Portland, OR: U.S.D.A. Forest Service, Pacific Northwest Forest and Range and Experiment Station. 50 p.

Sedjo, R. A., 1995: The potential of high-yield plantation forestry for meeting timber needs: recent performance and future performance. Discussion Paper 95-08. Washington, D.C.: Resources for the Future. 30 p.

Sedjo, R. A. and Botkin, D., 1997: Using forest plantations to spare natural forests. *Environment* 39(10):14-20.

Siry, J. P, Cubbage, F. W., Malmquist A., 2000: Potential impacts of increased management intensities on planted pine growth and yield and timber supply modeling in the South. In: Proceedings of the 28th Annual Southern Forest Economics Workshop. Edited by: Ian Munn. Available from: Department of Forestry, Mississippi State University, MS 39762. In press.

Sohngen, B., R. Mendelsohn, and R. Sedjo, 1999: Forest management, conservation, and global timber markets. *American Journal of Agricultural Economics* 81(February 1999): 1-13.

Twining, C. E., 1984: Weyerhaeuser and the Clemons Tree Farm: Experimenting with a Theory. In: History of Sustained Yield: A Symposium. H.K. Steen, Editor: Durham, NC: Forest History Society. 469 p.

UN/ECE Timber Committee, 1999: Forest and forest industry country fact sheets. Http://www.unece.org/trade/timber/

U.S.D.A. Forest Service, 1988: The South's fourth forest: alternatives for the future. Forest Resource Report No. 24. Washington D.C.: U.S.D.A. Forest Service. 512 p.

U.S.D.A. Forest Service, 1997: Report of the United States on the Criteria and Indicators for the Sustainable Forest Management of Temperate and Boreal Forests. Washington D.C.: U.S.D.A. Forest Service.

Wiebecke, C. and Peters, W., 1984: Aspects of sustained yield history: forest sustention as the principle of forestry--idea and reality. In: History of Sustained Yield: A Symposium. H.K. Steen, Editor. Durham, NC: Forest History Society. 469 p.

Weintraub, A., 1992: Data and decision aggregation processes in forest hierarchical planning. In: Proceedings, Workshop on Hierarchical Approaches to Forest Management in Public and Private Organizations. Toronto, Canada: University of Toronto.

Williams, M., 1994: Forests and tree cover. In: William B. Meyer and B.L. Turner II (editors): Changes in Land use and Land Cover: A Global Perspective. Cambridge, England: Cambridge University Press. p. 97-124.

World Commission on Environment and Development, 1987: *Our Common Future*. Oxford, England: Oxford University Press.

World Resources Institute ***et al.***, 1998: *World Resources 1998-99*. Joint publication by the World Resources Institute, The United Nations Environment Programme, The United Nations Development Programme, and The World Bank. New York: Oxford University Press. 369 p.

Yin, R., 1998: Forestry and the environment in China: the current situation and strategic choices. *World Development* 26(12):2152-2167.

Yin, R., Pienaar, L. V., and Aronow, M. E., 1998: The productivity and profitability of fiber farming. *Journal of Forestry* 96(11):13-18.

Theory and Practice of Yield Regulation Systems for Sustainable Management of Tropical and Subtropical Moist Natural Forests

Armin H.W. Seydack[1]

Abstract

The broader context and prerequisites for sustainability of timber harvesting in tropical and subtropical natural forests are outlined. With sustainability as guiding principle, four components of yield regulation systems were identified: yield optimisation, stand regeneration strategies, tactical components of harvesting cycle, and harvest tree selection criteria. Yield regulation systems occupy positions on a gradient of high to low impact on the forest ecosystem: from manipulation to mimicking of forest structure and dynamics. Most systems occupy high impact positions on this gradient in attempting to achieve productivity enhancement, stand regeneration and favourable operational economics. The complexities of these issues are discussed, suggesting the opportunities and advantages of rather selecting low impact yield regulation approaches. Existing yield regulation systems are critically assessed. An example of a naturalistic yield regulation system, as practised in a South African natural forest, is described.

Keywords: *Forest regulation, Sustainability, Tropical and Subtropical forests.*

[1] Dr. Armin Seydack (Ph.D Forestry) Forest Ecologist, Department of Water Affairs and Forestry, Private Bag X12, KNYSNA 6570, South Africa, Tel: +27 44 3825466, Fax:+27 44 3825461

K. von Gadow et al. (eds.), Sustainable Forest Management, 257–317.

1. Sustainability as guiding principle

Sustainable timber yield regulation is here considered to include the scientific and technical components of harvest quota regulation, silvicultural measures and logging approaches. Sustainability is taken as the guiding principle for timber harvesting under a professionally sound and ethically defendable yield regulation system. Following the definition by Goodland et al. (1990), sustainable use of moist subtropical and tropical forests is here taken to mean that use of natural forest which indefinitely maintains the forest substantially unimpaired both in the environmental services which it provides, as well as in its biological quality. Thus any harvest must not exceed the regeneration rate of the resource, nor impair the potential for similar harvests in the future".

The success of implementation of sustained yield regulation is however most often determined within the broader socio-economic and political context (Palmer, 1975, Weidelt, 1989; Bruenig and Poker, 1991; Schmidt, 1991). The technical challenge of sustainability; i.e. optimally matching harvest rates with biological productivity rates, has to be pursued within the framework of three externalities (*vide* Goodland et al., 1990; Bruenig, 1996): human population pressure; national resource policies and sustainability of effective management institutions. As phrased by Goodland et al. (1990): „Sustainability has two fundamentals: stabilising populations to live within the carrying capacity of their resources, and stabilising economic systems so that they do not increase strains on the finite environment". Following from these, success in the achievement of sustainability depends on a resource policy which integrates forestry equitably with agricultural and economic development (Bruenig,1996). Under conditions of high population growth and rural poverty the management of tropical and subtropical forests for economic production is considered a key element in their conservation (Schmidt, 1991); although Dickinson et al. (1996) caution against the general validity of the „use it or lose it" principle. The successful management for economic productions requires pre-conditions such as long-term security, operational control, a suitable financial environment and adequate information (Poore et al.,1989).

The successful implementation of sustainable forestry directives requires effective managing agencies. Some important factors making for effective managing agencies are (Seydack,1997):

- Political backing. When attempting to enforce forestry regulations, forestry officials require the backing of political authorities.

- Professional autonomy. Professional considerations must take precedence over political convenience.
- Corporate specialist focus. Managing agencies should be specialised in and as far as practically possible confined to indigenous forest management.
- Strategic leadership. To be effective, managing agencies require persons with strategic leadership qualities in key positions in order to focus managerial activity towards the achievement of long-term objectives.
- Competent staff. This is a function of adequate recruitment, training and experience.
- Participative management. This allows the authorities to remain in touch with the needs and opinions of external stakeholders and ensures their support for any initiatives to be carried out.

From the above, strategic leadership must be singled out as particularly important. „Leaders are people who do the right thing; managers are people who do things right. Both roles are crucial, but they differ profoundly“ (Bennis, 1989). The issue here is that people with an affinity towards long-term strategic thinking as against those predisposed to short-term operational activity, have quite different inherent personality profiles. To focus the operational activists onto the achievement of planned objectives requires purposeful leadership. Strategic leadership therefore is the integrative force of long term strategic objectives with short-term operational efficiency. Both are important in their own right, but leadership is the key link between information and implementation.

If the principle of sustainability is to function effectively in guiding timber harvesting, sustainable timber harvesting must be defined explicitly in terms of its pre-requisites and constraints so as to match resource use with the productivity potential of the ecosystem. Sustainable timber harvesting accordingly implies that similar amounts and types of products (dimensions, quality, species) continue to be harvestable at periodic intervals in perpetuity. Harvesting must therefore be organised in such a way that it remains within the renewability capacity of the forest system both in respect of growth as well as the success and type of regeneration. Wherever relevant such harvesting should strive to be compatible with local forest dependent communities (Bruenig and Poker, 1991) and biodiversity conservation (*vide* Struhsaker, 1998).

The components of a yield regulation system which are involved in order to achieve these objectives are dealt with in the next section.

2. Components of yield regulation systems

A yield regulation system can be perceived as consisting of four inter-linked components, represented by two strategic facets and two tactical elements:

- Yield optimisation strategy
- Stand regeneration strategy
- Harvesting interval (felling cycle)
- Harvest tree selection criteria

These four inter-linked components are also seen as making up the contents of silvicultural systems. The term yield regulation system is however preferred here since it is used in a broader sense which includes options devoid of any active silvicultural intervention.

Both the yield optimisation as well as the stand regeneration strategies revolve around the optimal balance between the amount and type of the growing stock to remain (RGS) as opposed to that portion of the growing stock to be removed by harvesting and/or release thinning operations. In practice the quantity perceived to be the minimum economic cut most often resulted in a distortion of this balance in favour of too heavy removals and thereby resulting in unsustainable harvesting.

In addition to logging economics, unsustainably heavy removals were often carried out in an attempt to induce regeneration or enhance productivity. These issues are discussed in the third section.

Yield optimisation strategies

Selecting a yield optimisation strategy requires decisions regarding I.) the stocking of the residual stand and II.) at what stage of maturity trees are to be harvested. Individual tree diameter growth may either be maximised, most often by concentrating growth onto pre-selected crop trees (low residual stand density / stocking: fewer faster-growing trees) or maximum productivity is pursued per unit of growing space (high residual stand density / stocking: more, slower-growing trees). Furthermore, trees may either be harvested at a relatively premature stage (low target diameter) or harvestable trees may be allowed to continue with growth whilst already having attained comparatively large sizes.

Depending on the yield optimisation strategy, growing stock reductions in pursuit of productivity enhancement may be implemented through one or more selected interventionist treatments (*vide* Baur, 1964b; Hutchinson, 1988; Uebelhör et al., 1989; Parren and De Graaf, 1995; Bruenig, 1996).

- Conversion-geared canopy opening. Removal of the tolerant understorey in order to release regeneration or removal of non-utilizable overstorey trees may be undertaken.
- Climber cutting. This operation results in some opening of the canopy, but its main purpose is to reduce felling damage.
- Exploitation. Harvesting of trees and the associated damage to residual trees normally results in the most substantial canopy opening. Depending on the clumping of harvest trees, the effect is often spatially irregular.
- Post-logging thinning treatments. *Refining* involves the removal of undesirable elements all over the stand (usually non-commercial species down to a prescribed diameter). *Liberation thinnings* are carried out to remove competing trees around selected potential final crop trees.

Any particular yield optimisation strategy is positioned on a high to low impact continuum; from systems characterised by severe growing stock reductions and low minimum harvestable diameters (MHD) to selective harvesting of mature trees from fully stocked stands. The position of the RGS to be selected on the low to high residual growing stock gradient depends on I.) the nature of the forest and its associated natural disturbance regime; II.) the proportion and temperament of the commercially valuable species and III.) any conservation constraints limiting the degree of growing stock reduction. A severely reduced residual growing stock would, for example, be inappropriate in a forest with a fine-grained disturbance pattern dominated by small-gap species. On the other hand, the light requirements for the regeneration and recruitment of gap opportunist species may require substantial canopy opening which would then be associated with relatively low residual growing stocks.

The harvest maturity threshold (HMT) is the selected stage of maturity above which particular trees are to be harvested and can be defined either by diameter or maturity condition criteria (Seydack, 1995; Seydack et al., 1995). The HMT is determined in relation to I.) the type of product pursued (log dimensions); II.) the diameter, volume or value increment culmination pattern and III.) thc pattern of wood quality deterioration with age.

It is often not taken into account sufficiently that most volume and value increment of particular trees takes place once they have already achieved substantial dimensions. Even a relatively slow diameter increment rate will, in trees of large dimensions, result in substantial volume productivity (e.g. Vooren, 1992). Most non-pioneer trees, and particularly the gap opportunist emergents, sustain extended growth at high diameters and premature harvesting does not permit the realisation of the value accretion potential as yield. The optimum yield realisation strategy is therefore to stipulate relatively high harvest maturity thresholds, but before substantial wood quality deterioration sets in (e.g. heart rot: Parren and De Graaf, 1995).

The optimum HMT diameter does not only depend on the diameter growth pattern but also the mortality rate (e.g. Alder, 1992). The diameter at which volume growth of individual trees is maximised (retention limit) is usually considerably higher than the cutting limit which takes mortality into account (Vanclay, 1989). This is so because the volume productivity of a tree progressively increases with size, but so do the number of trees lost to mortality during the time required to reach the higher diameters. If a premium is paid for large logs, maximum value production would favour even larger cutting and retention limits (Vanclay, 1989). The dilemma of relatively low optimum diameter cutting limits is that a certain proportion of trees are accordingly harvested prematurely in terms of their productive potential which is only realised at larger dimensions (Uebelhör et al., 1990). This dilemma can be partly or largely overcome by making use of maturity condition based harvest tree selection criteria (2.3).

Considering the RGS and HMT parameters in combination leads up to a fundamental yield optimisation dichotomy.

- Yield realisation from accumulated net growth: achieved by reducing the growing stock and age structure to an appropriately early developmental stage characterised by relatively free-growing trees and low levels of mortality.

- Yield realisation through selective mortality pre-emption: Growing stocks are relatively high in accord with stands fully stocked with predominantly mature-phase trees. More trees are growing at moderate rates and the relatively high mortality levels which would be applicable are converted into yield through mortality pre-emption by selectively harvesting mature trees of declining vigour.

The latter strategy is considered more congruent with the natural dynamics of tropical/subtropical moist forests and may hold more scope to reconcile biodiversity conservation and timber harvesting. Considerable development work is however required before such an approach may be adopted under divergent circumstances of forest dynamics and economic constraints.

Stand regeneration strategies

Successful regeneration depends on the proximity of seed sources (seed trees) to spaces suitable for seedling establishment (establishment space) and suitable recruitment conditions for established seedlings to survive and reach adult canopy positions (recruitment space). The regeneration space (combined establishment / recruitment spaces) refers to the variable and changing shelter and light requirements of recruits along the time axis in accordance with the metabolic type of the species concerned (traditionally narrowly seen as a shade tolerance / intolerance gradient). Stand regeneration strategies pertain to decisions aimed at the maintenance or creation of regeneration spaces in the effective vicinity of seed trees thereby sustaining conditions expected to result in successful regeneration.

Two types of silvicultural systems may be recognised. Monocyclic systems remove all timber at one cut and rely on seedlings to form the next crop. Polycyclic systems involve selective removal of a few trees on several occasions per cycle (Whitmore, 1991). Stand regeneration strategies are here grouped into four categories:

Monocyclic systems:

- Uniform systems (a)
- Tropical Shelterwood systems (b)

Polycyclic systems:

- Manipulistic selection cutting systems (c)
- Naturalistic selection cutting systems (d)

Which of these four stand regeneration approaches is favoured depends on regeneration status (actual representation of regeneration) and perceived regeneration requirements in terms of seed source distribution and regeneration space (Section 3.2).

a) Uniform systems

A typical example is the Malayan Uniform System (MUS). The objective of management is a more or less even-aged forest with a high proportion of commercially desirable speci-es.

The system was generally operated on a 70 year rotation and involved a series of operations (Table 1). It required that advance seedling regeneration, present on the ground at the time of felling, would survive undamaged, develop rapidly and grow through to crop dominance without assistance. The system was successfully applied in the lowland diptero-carp forests of Malaysia (Appanah and Weinland, 1992), but was not considered suitable for hill dipterocarp forests mainly due to a lack of seedlings in virgin stands (Thang, 1987).

Year	Operation
E – 1½	Linear sampling (2m x 2m) of regeneration, and enumeration of merchantable trees
E to E +1	Exploitation, followed by poison-girdling down to 5 cm dbh
E + 3 to E + 5	Linear sampling (5m x 5m) of new crop, followed by cleaning, climber cutting and poison-girdling as required
E + 10	Linear sampling (10 m x 10m) of new crop, followed by treatment as required or passed as regenerated
E + 20, E + 40	Sampling and thinning as required
E: Exploitation (70 year rotation)	

Table 1. Sequence of operations of the Malayan Uniform System (Thang, 1987)

b) Tropical Shelterwood Systems

The aim of Tropical Shelterwood Systems (TSS) is to produce a more or less even-aged forest by establishing sufficient regeneration of economic species within a limited regeneration period starting before and extending beyond full canopy logging. This is pursued through the formation of a shelterwood of mother (seed) trees which are subsequently removed when regeneration has become established.

The natural regeneration of commercially valuable species (mainly Meliaceae) in West African tropical rainforests is often considered inadequate. As a result TSS were chosen in an attempt to induce the regenerations of such species. As an example, the Nigerian TSS comprises a series of operations designed to open the canopy in order to induce regeneration and enhance the growth and development of existing regeneration. When the regeneration is rated successful, the temporary shelterwood, functioning as shelter and source of seeds, is removed leaving only young trees in the stand (Table 2 ex Parren and De Graaf, 1995). Parren and De Graaf (1995) caution that these need to be tended if they are expected to survive in the face of exuberant spread of climbers and failure to grow adequately.

Year	Operation
E –4	a) Demarcation of compartment b) 1st climber cutting and seedling assistance c) 1st poisoning (= 1st canopy opening)
E –3	a) 2nd climber cutting and 1st regeneration count b) 2nd poisoning
E – 2	a) 1st and 2nd cleanings (=under-storey opening)
E – 1	a) 3rd and 4th cleanings and 2nd regeneration count
E	a) Exploitation in the case of sufficient regeneration, otherwise the whole procedure shifts accordingly
E + 1	a) 5th cleaning and 3rd regeneration count b) Repair of exploitation damage c) Removal of overwood depending on light demand of regeneration
E + 6 or up to E +10	a) 6th cleaning b) Removal of overwood c) 4th regeneration count
E: Exploitation (±100 year rotation)	

Table 2. Sequence of operations of the Tropical Shelterwood System in Nigeria (ex Parren and De Graaf, 1995)

c) Manipulistic Selection Cutting Systems

Selection cutting systems amount to determining minimum harvestable diameter (MHD) felling limits for harvesting and retaining advanced residuals (e.g. 30 – 35 cm DBH) for the next harvest at relative short felling cycles. Within this category those selection cutting systems are grouped where the harvesting impact is deemed to require regeneration enhancing measures. The term selection cutting systems used here should not be viewed as equivalent to the selection silvicultural system practised in northern hemisphere forests.

Harvesting intensities of commercial species are such that low MHDs result in relative premature harvesting; requiring specific stipulations to retain residual seed trees. This category also includes selection cutting systems involving measures / operations geared to the enhancement of regeneration: enrichment planting, tending and canopy opening through the culling of non-harvest trees in order to favour the regeneration of commercially desirable species. Examples of manipulistic selection cutting systems are the Selection

Management System (SMS) operated in Peninsular Malaysia (Thang, 1987), the Queensland Selection System (Baur, 1964a) and the selection system practised in Ghana (Parren and De Graaf, 1995).

The selection cutting option under SMS involves that all commercial species above 45 cm DBH for non-dipterocarps and 50 cm for dipterocarps are felled at felling cycles fixed at 25 to 30 years. A certain minimum number of trees per hectare are stipulated to be left behind as residual stocking. The sequence of operations in the SMS as shown in Table 3 (Thang, 1987). Climber cutting to reduce logging damage and line planting of areas inadequately stocked of young regeneration may be prescribed. Although considered necessary for sustained stand regeneration over many felling cycles, the stipulated regeneration enhancing measures are not necessarily effectively implemented in practice (Appanah and Weinland, 1992).

Year	Operation
E2 to E1	Pre-felling forest inventory using systematic-line-plots and determination of cutting regimes
E1 to E	Climber cutting to reduce damage during logging. Tree marking incorporating directional felling. No marking of residual trees for retention.
E	Felling of all trees as prescribed
E+2 to E+5	Post felling inventory using systematic-line-plots to determine residual stocking and appropriate silvicultural treatments
E: Exploitation (25 – 30 year felling cycles)	

Table 3. Sequence of operations in the Selection Management System (Thang, 1987)

The Queensland Selection System which was used in northern Queensland, Australia, was considered particularly successful (Bruenig, 1996). The third example, the (Ghana) Selection System involved the identification of residual seed bearers and improvement thinnings. The system was rated as well suited to aiding the regeneration in forests that were well stocked with commercial species (Parren and De Graaf, 1995). Small-scale examples of selection cutting systems at experimental level include the Puerto Rico Selection System and the polycyclic CELOS Silvicultural System in Surinam (De Graaf, 1986; Wyatt-Smith, 1987a).

d) Naturalistic Selection Cutting Systems

Any selection cutting system where the intensity and kind of harvesting is in a manner permitting successful regeneration to take place naturally without any additional regeneration enhancing measures, is taken to fall into this category. Harvesting and natural regeneration are fully integrated (*vide* Bruenig, 1996). This requires that the spatial concentration and extent of felling is in accordance with the grain of disturbance naturally experienced by the forests involved. This implies that the degree of canopy opening associated with harvesting must result in that range of canopy gaps which is congruent with the regeneration space requirements of the species required to regenerate sustainably. An example of such a system is the Senility Criteria Yield Regulation System practised in Afromontane forests in South Africa (Seydack et al., 1995).

Tactical elements of yield definition

The sustainable yield ultimately depends on the area under forest and its productivity in respect of commercially suitable species. The harvesting interval (felling cycle for polycyclic systems; rotation for monocyclic systems) is linked to the harvest quota retrievable per unit of area (the harvest intensity). Longer harvesting intervals are associated with higher harvesting intensities to be obtained from proportionately smaller areas. The maximum area to be harvested annually should be limited to that proportion of the total forest area which allows for an approximately equal annual harvest in perpetuity (= total area ÷ harvesting interval in years). Whereas the harvesting interval is primarily determined by factors relating to the stand regeneration strategy and operational economics, the selected yield optimisation strategy has an overriding influence on which trees are to be harvested; i.e. the harvest tree selection criteria.

The harvesting interval should be the outcome of balanced compromise and careful consideration of the following three factors:

- Desired degree of canopy opening
- Impact of the harvesting maturity threshold
- Operational economics

The harvesting interval should be selected in accordance with the above requirements within a multi-cycle time frame and not subject to opportunistic changes for short-term harvest output manipulation.

As the intention is to harvest the growth accrued between harvesting events, longer harvesting intervals are associated with higher harvest intensities and thus with higher degrees of canopy opening. Severe canopy opening predominantly favours light demanding species whereas intermediate canopy opening is suitable for gap opportunist species (3.2). Small gap species which predominate in forests with fine grained disturbance patterns require correspondingly low degrees of canopy opening associated with shorter harvesting intervals.

If the yield optimisation strategy requires high harvesting maturity thresholds, long harvesting intervals may result in reduced success of mortality pre-emption and increased timber losses due to wood quality deterioration with age. For monocyclic systems high harvest maturity thresholds, in order to optimise value increment, would require long harvesting intervals (rotations).

Longer harvesting intervals, which involve relatively high rates of timber removals per hectare from proportionately smaller areas, are usually superior in terms of operational economics. A relatively high yield output is then achieved per unit of operational input as a result of cost effective logging and harvest tree selection measures (tree marking, stock mapping, inventories). However, compromises have to be struck whenever the stand regeneration strategy requires relatively low degrees of canopy opening, i.e. requiring short felling cycles.

Lastly, in addition to the felling cycle and scheduling of annual harvesting coupés, a yield regulation system must also define which trees are harvestable (selection criteria). Trees to be harvested may be defined (I) on an area basis, i.e. all trees on a particular area are harvested, (ii) through minimum harvestable diameters (MHD), or (iii) based on features of tree condition (maturity condition criteria).

I) In the case of a monocyclic system all trees can be harvested from a particular area.

II) Most yield regulation systems applied to tropical and subtropical forests define trees to be harvested by minimum harvestable diameters (MHD).

III) Harvest tree definition may also be according to maturity condition criteria (MCC) based on indications of advanced maturity and short remaining life expectancy.

IV) A combination of II and III: the harvesting of all trees above the retention limit (a relatively high MHD) as well as those of profitably merchantable dimensions below the retention limit indicated by maturity condition criteria (Fig. 9).

In the case of maturity condition criteria the HMT threshold must be defined in a manner to permit the full value accretion, which takes place overwhelmingly when already large trees continue to grow, to be realised before substantial wood quality deterioration with age sets in. Being based on individual tree assessments, harvest tree selection by maturity condition criteria places higher demands on professional sophistication and in the field involvement of personnel than harvest tree definition by MHDs. On the other hand MHD criteria are associated with a number of serious disadvantages.

I) The use of MHDs normally results in the premature removal of fast-growing trees with high value increment (e.g. Rietbergen, 1989), since the fastest growing trees are generally in the higher diameter classes (Seydack et al., 1995). The reaction to release and general growth performance of residuals, on the other hand, is uncertain.

II) The repeated removal of the fastest growers may have a dysgenic effect (*vide* Palmer, 1975; Weidelt, 1986; Jonkers, 1987; Appanah and Weinland, 1992; Whitmore, 1991).

III) A proportion of trees below the minimum harvestable diameter but above the minimum utilisable diameter are relatively old, slow-growing and succumb to mortality. This mortality is not retrieved as harvest and represents lost growth. Depending on the diameter and mortality distribution, many trees may be involved.

IV) If the minimum harvestable diameter is set relatively high, the losses under (III) are increased. If the minimum harvestable diameter is too low in relation to sexual maturation and seed production, the reproduction of the species may be negatively affected (cf. Weidelt, 1989).

V) It is plausible to assume that site differences affect the optimal minimum harvestable diameter (see Vanclay, 1989). To institute such a site differentiation would however greatly increase the pre-harvesting inventory requirements.

The allowable yield is often determined according to some method (FAO, 1998) and then forced onto the situation by moulding the felling cycle and the harvest tree selection criteria to meet this calculated yield. Accordingly, the emphasis in yield regulation has hitherto been on quantifying or predicting the allowable and sustainable yield. However, here the emphasis has been placed on the development of yield regulation systems which stipulate the treatment of forests in such a way that maximum sustained yields are forthcoming from the forests. Yield quantification or prediction then is of relevance only for economic considerations and operational planning.

Fundamental issues of yield regulation systems

Productivity enhancement

The potential for and limitations of productivity enhancement through thinnings or release operations is one of the central issues in yield regulation systems. Thinning is generally taken to have an increment-promoting effect in most cases. This is deduced from the apparently favourable reaction of the remaining trees. However if the canopy consists primarily of commercial species it is often not the increment of individual trees which is of decisive importance, but the increment per unit area (Assmann, 1970). In this regard, the evidence available to Synnott (1980) remained compatible with Dawkins' (1958) conclusion that in the longer term silvicultural treatment would not in most cases have greatly affected total yield; i.e. that the annual production per hectare will be similar over a fairly wide range of basal area stocking and reduction in stocking. This is in keeping with Assmann (1970), who states that within wide limits, the basal area can shift without any marked change in the basal area or volume increment. Apart form the first favourable reactions to thinnings, the optimum basal area over a period on favourable sites very quickly approaches the maximum (Assmann, 1970).

Often, as in the Afromontane moist evergreen forests at Knysna (South Africa), the more highly stocked forest stands also have the highest basal area increment (Figure 1).

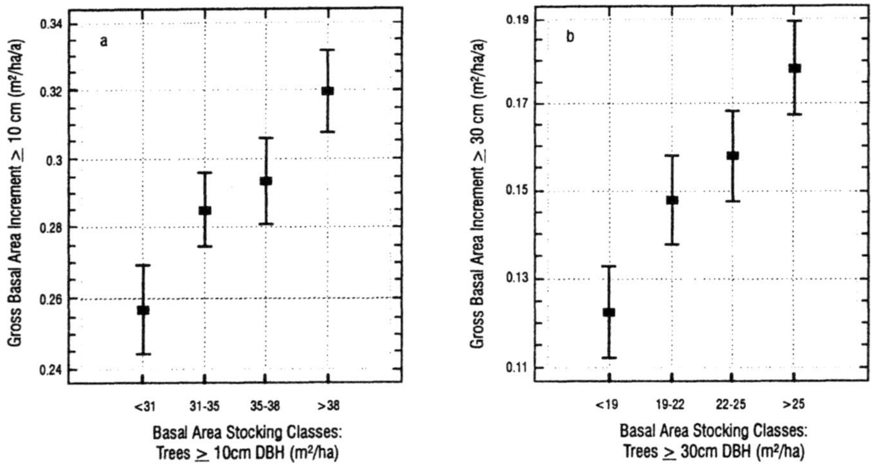

Figure 1. Gross basal area increment (m²/ha/annum) over basal area stocking classes (m²/ha) for 131 Afromontane forest stands (Southern Cape; South Africa) betweeen 1-3 ha in size. 1a: Data for all trees ≥ 10 cm DBH; 1b: Data for all trees ≥ 30 cm DBH (Seydack, in prep.).

This implies that the positive effect of many and large stems growing per unit of area overrides any negative effect of crowding on individual tree diameter increment (Figure 2). This is congruent with the statement of Assmann (1970) that high absolute productivity of mixed stands is usually associated with high growing stocks.

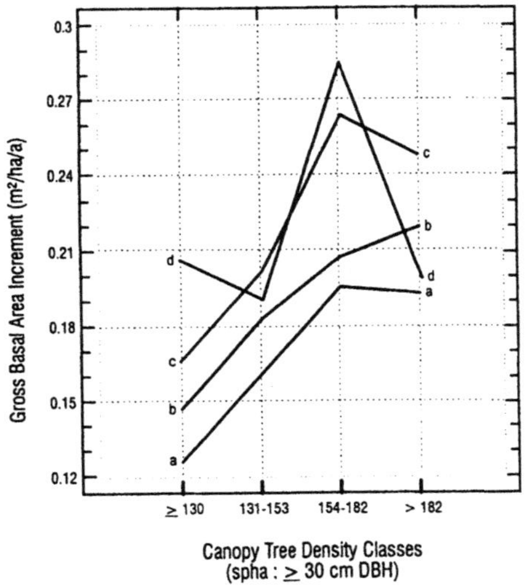

Figure 2. Gross basal area increment (m²/ha/annum) over canopy tree density classes (sph; ≥ 30 cm DBH) and stand mean diameter Classes (cm; ≥ 30 cm DBH): a: ≤ 41,0; b: 41,1-42,8; c: 42,9-44,3; d: ≥ 44,4 cam DBH; (Data for 131 southern Cape Afromontane forest stands, South Africa; Seydack, in prep.)

In multi-aged, multi-species moist forests, canopy trees are relatively widely spaced, associated with a prevalence of large crowns. Unless senile, these large-sized canopy trees generally have the highest growth rates (Jonkers, 1987; Manokaran & Kochummen, 1987; Primack et al. 1987a; Swaine et al., 1987a; Weidelt, 1989; Appanah & Weinland, 1992). The lower mean diameter increment rates of trees in subcanopy diameter classes of canopy and emergent species (Fig. 3) can be attributed to the subdominant crown positions of most of these trees in subcanopy positions. Whenever low diameter trees have good crown positions, they often have increment rates similar to larger diameter trees (e.g. Korsgaard, 1992; Geldenhuys, 1997). The importance of light as a limiting factor for sub-canopy tree cohorts (10 – 20 cm DBH) relative to canopy occupants (>30 cm DBH) clearly emerges from the growth data analyses depicted in Figures 4a-f.

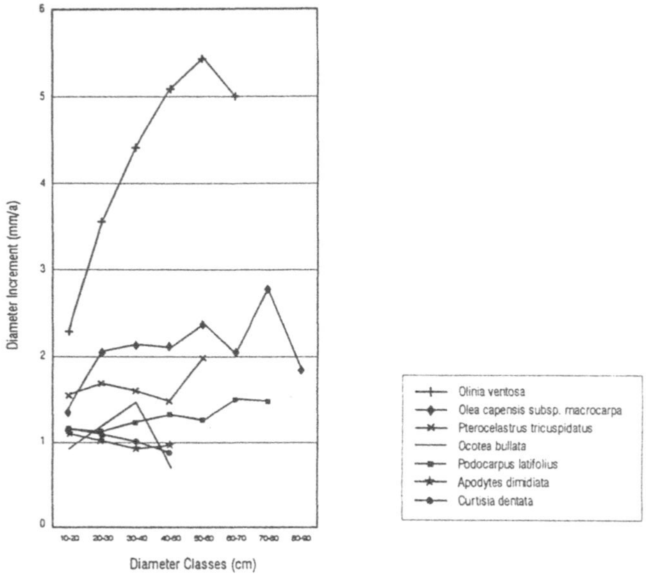

Figure 3. Diameter increment pattern (mm/annum) over diameter classes for selected species in a southern Cape Afromontane forest (South Africa). Gap opportunist species: Olinia ventosa; canopy dominants: Olea capensis subsp. macrocarpa and Podocarpus latifolius; forest matrix occupants: Pterocelastrus tricuspidatus; Ocotea bullata; Curtisia dentata and Apodytes dimidata.

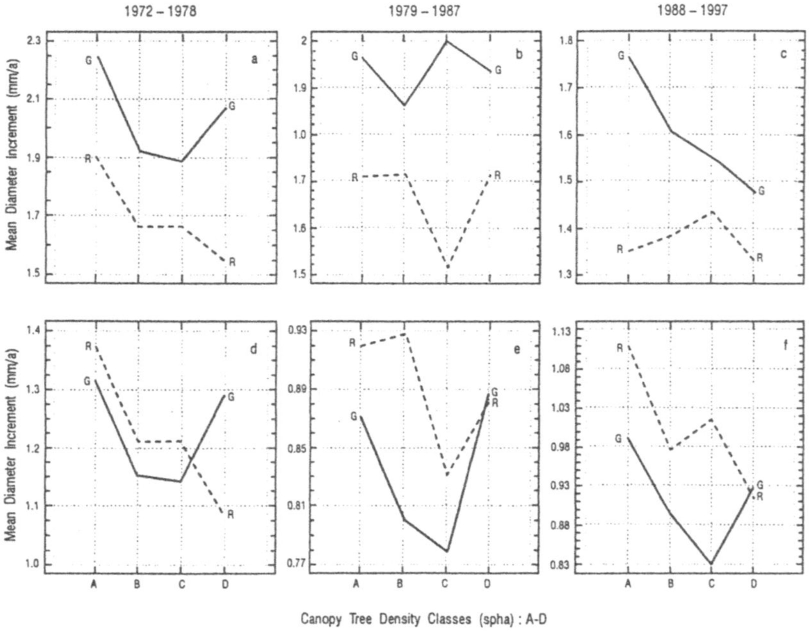

Figure 4. Mean diameter increment (mm/a) for canopy trees (≥ 30 cm DBH: a-c) and pool-sized trees (10-20 cm DBH: d-f) over three successive increment periods; differantiated according to canopy tree density classes (A: ≤ 130; B: 131-153; C: 154-182 and D: >

182 spha) and two site types (G: guelley sites; R: ridge sites). Data for 131 southern Cape Afromontane forest stands, South Africa. (Seydack, in prep.)

Whereas canopy trees are relatively unaffected by canopy stem density but respond favourably to moisture availability (gully versus ridge sites), sub-canopy tree growth is negatively affected by the density of overtopping canopy trees. Also, faster canopy tree increments prevail on ridge sites, which presumably experience higher levels of solar radiation.

Thus, unlike larger-sized canopy occupants which usually remain unaffected by spacing operations (e.g. Dawkins, 1959; Palmer, 1975, Weidelt, 1986, 1989), this implies that favourable growth responses are to be expected when subcanopy trees are released from being overtopped (see below). However, the bulk of the canopy trees are usually particularly productive in terms of volume and value increment (in their grand periods of increment *re* Bruenig, 1996) and nothing is therefore to be gained in terms of enhanced productivity by felling these prematurely in order to release subadult recruits. The importance of viewing a natural uneven-aged mixed species forests as a two-layered system (faster growing canopy occupants over partially waiting sub-canopy trees) needs to be stressed in this context.

Productivity gains through silvicultural thinning operations are often expressed in terms of the full growing stock ≥ 10 cm DBH. Most of any such gains pertain to sub-canopy trees of which the survival to harvestable dimensions is rather unpredictable. Such gains established over relatively short observation periods are accordingly difficult to extrapolate to the time of harvesting.

As trees grow larger over time the number of trees in the cohort declines (thinning rule). Westoby (1984) states that it would seem that the survival of individuals in mixed stands is more determined by which species the individual belongs to, rather than by its size relative to others, as in single-species stands. Particularly in the case of even-aged monospecific cohorts the rate of self-thinning may be retarded in the face of increasing resource deficits and increment rates decline. Thinnings are then useful to overcome such growth retardation. This self-thinning retardation is postulated to be more pronounced when members within the tree cohort compete for the same resources at the same time, as in even-aged monospecific stands. For example, in the southern Cape Afromontane forests (South Africa) the greater negative impact of intraspecific competition relative to overall crowding seems to be evident in *Olea capensis* ssp *macrocarpa* and *Podocarpus latifolius* canopy and subcanopy trees respectively (Figures 5 and 6).

Hasse and Ek (1981) noted that uneven-aged mixed stands have a greater tendency to concentrate growth in larger, fewer individuals. This assertion was also made by Assmann (1970), suggesting less potential for self-thinning retardation to occur in uneven-aged mixed species stands. The tendency in most tropical/subtropical forests to avoid conspecific neighbours (Sterner et al., 1986; Hubbell and Foster, 1990; Hubbell et al., 1990; Wills et al., 1997); usually resulting in relatively wide intraspecific spacing, in conjunction with continuous or periodic regeneration through the vegetation matrix, leads to a differentiated species / age mixture. The more this species / age intermixture manifests, the less scope there seems to be for much gain from lateral thinnings.

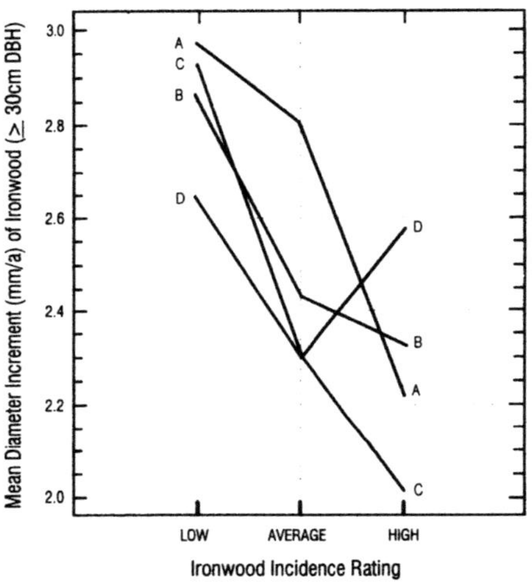

Figure 5. Mean diameter increment of Ironwood (Olea capensis subsp. mactocarpa (≥ 30 cm DBH) in relation to conspecific incidence: Proportion of canopy stem frequency: Low: ≤ 0,38; Average: 0,39-0,58; High: ≥ 0,59; and canopy tree density classes, all species pooled ((A: ≤ 130; B: 131-153; C: 154-182 and D: > 182 spha). Data for 131 southern Cape Afromontane forest stands, South Africa. (Seydack, in prep.)

The intermingled species mixture generally encountered in tropical and substropical forests is presumably brought about by some mechanism mitigating against conspecific neighbourhood. For example, suggestively following from the above, Lübbe and Geldenhuys (1991) found that seedlings in a arboretum trial of monospecific stands of natural forest species were largely absent below conspecific canopies. These patterns and interactions should give rise to caution against creating monospecific stands through silvicultural operations if regeneration and growth performance is to be sustained over successive tree generations.

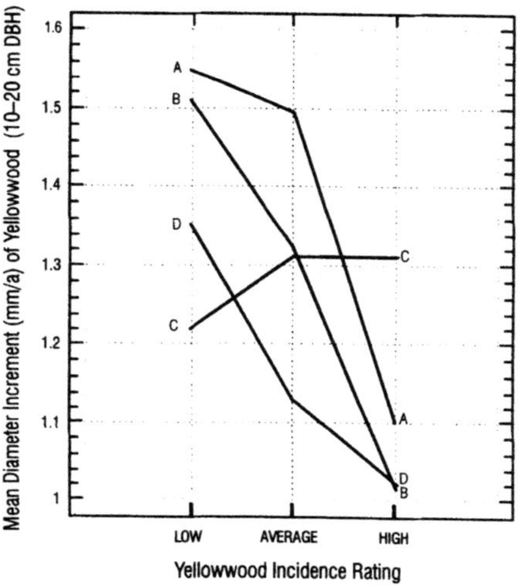

Figure 6. Mean diameter increments of pole-sized Yellowwod trees (Podocarpus latifolius: 10-20 cm DBH) in relation to conspecific incidence: Proportion of canopy stem frequency: Low: ≤ 0,16; Average: 0,17-0,29; High: ≥ 0,30; and canopy tree density classes, all species pooled (A: ≤ 130; B: 131-153; C: 154-182 and D: > 182 spha). Data for 131 southern Cape Afromontane forest stands, South Africa. (Seydack, in prep.)

While it may seldom be possible to appreciably increase volume production per hectare with thinning operations, the growth potential of a site can be concentrated on desirable species through the removal of competing and overtopping non-commercial species (*vide* Boxman et al., 1985; Weidelt, 1986, 1989; Schmidt, 1987, 1991; Hutchinson, 1988; Bruenig, 1996). However, as more of the stand can be used, the less scope there is for this (*cf* Synnott, 1980). The general trend is that more and more species are considered utilisable (Schmidt, 1987; Primack et al., 1987b; Poore et al., 1989) and this trend therefore de-emphasises the scope of silvicultural release operations. Furthermore, true sustainability requires that future options remain open. This constraint counter indicates any drastic transformation of the forests involved.

While some workers suggest that improvement fellings in primary forests is not an effective way to increase the long-term growth rates of trees (Primack et al., 1987a, 1987b), good reactions to liberation thinnings in terms of increased growth of small to medium-sized trees is commonly reported (Synnott, 1980; Kio, 1980 *op. cit.* FAO, 1988; Weidelt, 1986;

Maitre, 1987 and 1992; Catinot, 1986 *op. cit.* FAO, 1988; De Graaf, 1986; Woell, 1989; Korsgaard, 1992). Increased diameter increments, especially of young trees who have retained their ability to react when freed from overtopping, of between 50 and 100% are recorded by these authors. Such positive growth responses are primarily found in gap opportunist species (mainly *Meliaceae* and *Dipterocarpaceae*) which feature as gregarious or individual emergents in the forest matrix. Gap opportunist species, with an initial capacity for shade tolerance, establish in gaps or below-canopy spaces much smaller than they finally need for full development. Their regeneration space must therefore be progressively extended to allow good growth, as circumstantially evidenced by the trend of an increasing growth rate with larger diameters. This happens naturally through death of neighbours. If not, the recruits stagnate, lose their capacity to react and are subject to higher rates of mortality. Under this scenario, intervention in the form of limited crown release of subadult and pole-sized individuals of gap opportunists is perceived to result in sustainable growth increases.

Within the context of yield optimisation strategies, decisions are required as to whether or not release thinnings are to be applied and if so, details concerning which type of trees are to benefit and the type and degree of release must be stipulated. In this decision process it is desirable to be aware of potential release response complexities and constraints. Two issues are of importance here: the sustainability and the life strategy related consequences of growth accelerations in respect of tree survival, timber quality and growth culmination shifts.

The upsurge of growth following the first release operations is often misleading, because this may be mainly due to the acceleration of growth which is later cancelled by setbacks (Assmann, 1970). It therefore remains to be seen whether the accelerated increment can be sustained over a long felling cycle, or repeated in the case of short felling cycles (FAO, 1988). Diminishing magnitudes of reaction are realistically expected with repeated release operations.

Growth acceleration responses to canopy opening release operations are often relatively short-lived, extending for between three and a dozen years in reported cases (e.g. Weidelt, 1986; Silva et al., 1995). This is generally attributed to competitive re-encroachment or renewed overtopping of subject trees. However, in many cases, particularly for trees already occupying good crown positions, such an explanation does not always appear plausible. According to an alternative interpretation, temporary growth accelerations may only last as long as newly available crown and root zone space is being extended into. Thereafter growth continues according to rates as applicable before the release event and in line with the metabolic type of the subject trees. Thus, after a particular spatial crown and root occupation has been achieved, no further release reactions are to be expected.

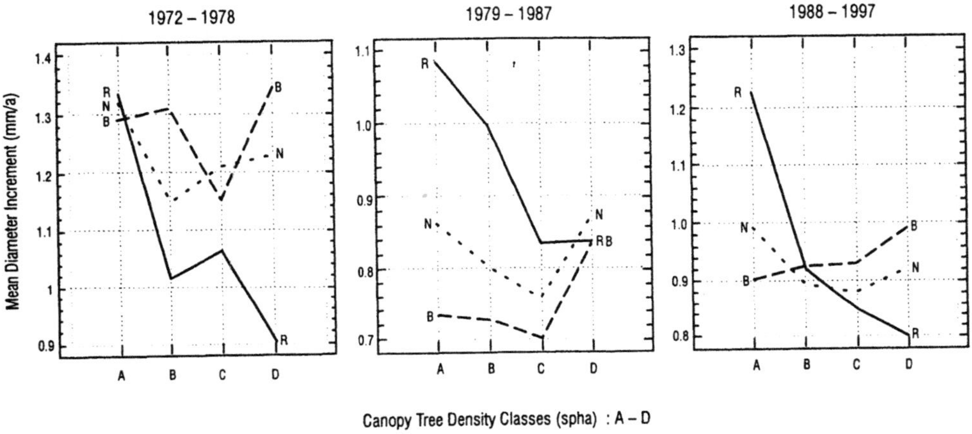

Figure 7. Mean diameter increment of pole-sized trees (10-20 cm DBH) of three different growth response types R, N and B (see text) and over canopy tree density classes A: ≤ 130; B: 131-153; C: 154-182 and D: > 182 spha. Data for all species pooled; southern Cape Afromontane forests, South Africa. (Seydack, in prep.)

This raises the question: do trees with good crown positions grow fast because of associated favourable light resource levels or do they attain good crown positions because they are able to sustain growth in spite of crowding? Both scenarios seem to exist. In a South African evergreen moist forest, the growth patterns of three different growth response types of pole-sized trees (of all five major canopy species pooled) of three successive measurement periods and in dependence to canopy tree density are shown in Figure 7.

The dominant growth response modes of stands were identified according to whether their canopy tree increments were higher in period 1 (1972 – 1978) or period 2 (1979 – 1987) as B and R types respectively (N intermediate). Although the data are for all species pooled, these spatiotemporal growth responses over 131 forest stands tended to be parallel over species and growth response types therefore designate intraspecific response type differentiations. It can be seen that the growth of the R-types is sensitive to overhead crowding, whereas that of B and N types is not. The underlying metabolic profiles and trade-offs are being investigated (Seydack, in prep.).

Growth response types R perform best at high light resource levels whereas B types exhibit sustained growth under resource limitation. This presumably reflects a metabolic trade-off between the capacity for maximum performance at the cost of sustained performance (resource use efficiency). R-types are accordingly expected to particularly benefit from release operations whereas B (and N) types have higher growth efficiencies per unit of crown space occupied. The former require favourable light conditions to grow fast towards good crown positions whereas the latter sustain growth towards improving crown positions and progressively grow faster as these are attained.

The life strategy pattern of a plant is a matter of economics (Bloom et al., 1985). Growth, reproduction and survival represent competing demands for finite resources and the material produced by a plant must be allocated preferentially to these, since there is never enough for optimal allocation to all three (Barbour et al., 1987). According to Loehle (1988), increased longevity requires investment in chemical and structural defences to resist decay, herbivory and wind. These defences have costs in terms of direct synthesis and photosynthate not available for growth. According to this growth versus survival or stress tolerance trade-off, a well-defended tree will be less competitive relative to neighbours and might be more easily overtopped.

The results of the interspecific analysis by Loehle (1988) support the hypothesis that trees protect themselves against mortality factors via energy investments and that these investments are at the cost of growth rate (e.g. Coley, 1987). Tree species which specialise in the capture of resource surpluses or pulses tend to be shade intolerant, fast growing, early maturing, short-lived and have an abrupt decline phase (of the ephemeral type according to the terminology of Loehle, 1988). Tree species specialising in resource use efficiency tend to be shade tolerant, slower growing, especially early in life, later maturing, longer lived and with a persistent adult stage (persistent types; Loehle, 1988). Thus, an early growth peak, where maximum resource capture overlaps with a phase of juvenile metabolic capacity, results in high turnover which is at the cost of the maintenance capacity (i.e. at the cost of long-term growth performance, defence, structural strength and longevity).

The described patterns apparently also apply intraspecifically. These patterns are particularly relevant for monocyclic systems and second generation polycyclic systems where the canopy was opened up extensively. Some evidence suggests that trees with uninhibited juvenile growth may suffer from premature senility, as manifested by a rapidly declining increment and susceptibility to decay (cf. Leibundgut, 1960). Fast grown *Shorea* species will produce timber of reduced strength and poorer quality (Wyatt-Smith, 1987a). The same has been reported *of Ocotea usambarensis* (Willan, 1965). The incidence of brittle-heart may be increased if the early growth of certain dipterocarp species is unimpeded (Appanah and Weinland, 1993). Fast juvenile growth appears to favour insect attacks and root rots in certain species of the Meliaceae (cf. FAO, 1988).

With full light available to crowns at an early stage, trees reach the stage of full vigour not only at an earlier age, but also at a relatively lower tree height (Assmann, 1970). The associated earlier onset of senility may, depending on the species, result in lower final dimensions. Leibundgut (1978) also pointed out that fast-growing stands of the same species break down sooner than slow-growing stands, whereas individuals experiencing an extended juvenile stage can attain exceptional sizes and may retain unusual vitality into old age (Mayer, 1979). These patterns demonstrate the growth law *vide* Backman (1943) and relate to the role of nurse stands.

The long-term role of natural forests is likely to lie predominantly in producing high value timbers (Leslie, *op. cit.*, Poore et al., 1989), both in terms of timber quality and dimensions. High value production requires that growth is concentrated onto the larger individuals. If this is to be achieved without compromising timber quality, the described life history strategy trade-offs should be taken cognisance of in terms of the effect that excessive canopy opening or the absence of a nurse stand could have on the second rotation tree crops.

The growth law manifests clearly in Blackwood (*Acacia melanoxylon*), a forest matrix invading gap opportunist species which tends to occur as gregarious emergent in certain parts of Afromontane evergreen moist forests in the southern Cape, South Africa (Seydack, 2000a, b, c). Depending on the amount and stage of solar radiation available during the establishment and recruitment phases, Blackwood differentiates into growth strategists P, SPP, SP (Figure 8). S strategists are independent of conditions relating to solar radiation (Seydack, 2000a).

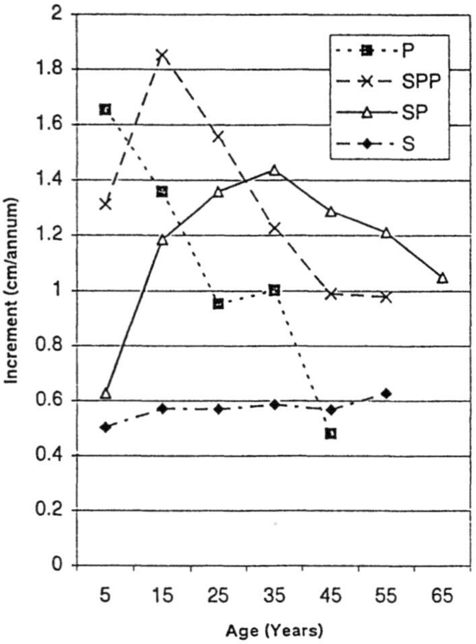

Figure 8. Mean annual increment (cm) of blackwood growth strategists P, SPP, SP and S (as defined in text) over age (Seydack, 2000a).

As expected according to the growth law high levels of solar radiation during the establishment phase results in early growth peaks and subsequent growth rate declines. P strategists experienced substantial solar radiation levels at establishment and may be taken as reminiscent of growth responses expected to prevail during second rotation monocyclic management. SPP strategists could reflect the response to overhead release during early recruitment under polycyclic management, whereas the SP strategists would represent virgin forest conditions. A productivity performance analysis of these divergent growth response scenarios revealed or verified certain patterns (Table 4):

- The periodic annual volume growth rate for P strategists is higher than for SPP and SP strategist types only in the growth period up to 10 cm DBH. The current volume growth rate of SPP strategy types is superior to SP strategists at diameters below 40 cm DBH, whereas the latter maintains superior current volume growth rates > 60 cm DBH.

- The mean annual volume production per tree rises continuously with size. For all three strategy types, allowing growth to the highest attainable diameter is associated with the highest lifespan volume growth productivity (within the data range represented in Figure 8). However, variable mortality rates of individuals within and between populations would result in the optimum minimum harvestable diameters to be lower in order to pre-empt losses due to mortality (2.1). Optimum MHDs would accordingly differ between growth strategists (P < SPP < SP).

- The productivity up to reference diameters (m^3/a over growth period to reference diameter) of SPP and SP strategists draws even only at 90 cm DBH after about eight decades of growth. Below this diameter the productivity of SPP strategists is superior to that of SP strategists. The latter could thus only make up for the slower growth during the first two decades with superior growth in decades four to nine.
- Target diameters for harvestable trees of between 50 and 70 cm appear achievable about a decade earlier when release operations are applied (SPP pattern) compared to the virgin forest situation (SP pattern). This is similar to the results on the effects of thinning treatments for gap opportunist dipterocarp species in the Philippines (Woell, 1989).
- Whenever high levels of solar radiation at establishment under a monocyclic system results in a preponderance of the P strategic mode, short rotations will be required in the face of early onset of mortality which is expected to parallel early declining of diameter growth rates. This outcome may prove to be characteristic of some second rotation monocyclic management situations.

DBH (cm)	Standing Volume at DBH (m^3)	Growth strategists *								
		P			*SPP*			*SP*		
		Age (years)	Mean annual volume growth rate to DBH (m^3/a)	Periodic annual volume growth rate per DBH class (m^3/a)	Age (years)	Mean annual volume growth rate to DBH (m^3/a)	Periodic annual volume growth rate per DBH class (m^3/a)	Age (years)	Mean annual volume growth rate to DBH (m^3/a)	Periodic annual volume growth rate per DBH class (m^3/a)
10	0,027	6	0,0045	0,0045	8	0,0034	0,0034	13	0,0021	0,0021
20	0,151	12	0,0121	0,0207	13	0,0116	0,0248	22	0,0067	0,0138
30	0,416	20	0,0208	0,0331	19	0,0219	0,0442	29	0,0143	0,0379
40	0,854	29	0,0294	0,0487	25	0,0342	0,0730	36	0,0237	0,0626
50	1,490	41	0,0359	0,0530	32	0,0458	0,0909	43	0,0347	0,0909
60	2,349				41	0,0573	0,0954	51	0,0461	0,1074
70	3,451				50	0,0690	0,1224	59	0,0585	0,1378
80	4,521				62	0,0729	0,0892	68	0,0665	0,1189
90	6,462				78	0,0828	0,1213	79	0,0818	0,1765

* Growth strategists: P: full sunlight during establishment; SPP: sufficient light early in the recruitment phase; SP: limited overhead gap space during recruitment to emergent status (Seydack 2000)

Table 4. Productivity performance analysis of divergent growth strategists of Blackwood (Acacia melanoxylon) as gap opportunist invaders in Afromontane forests (South Africa)

When trees are harvested according to maturity condition criteria (2.3) the constraint of having to harvest trees prematurely, as when MHD are used as selection criteria, is relaxed because tree mortality is (partly) pre-empted by harvesting trees at variable diameters when-ever individual trees approach some advanced stage of maturity (see Figure 9). Thus maturity condition harvest tree selection criteria permit that trees can be allowed to individually grow to relatively high diameters when volume productivity is highest. As a general statement it appears that productivity is superior for released gap opportunists (as represented by the SPP strategy type) with yield optimisation strategies stipulating low harvesting maturity thresholds associated with optimised MHD, whereas high harvest maturity thresholds optimally possible with maturity condition based selection criteria would result in superior yields associated with virgin forest growth patterns (as represented by SP strategy types).

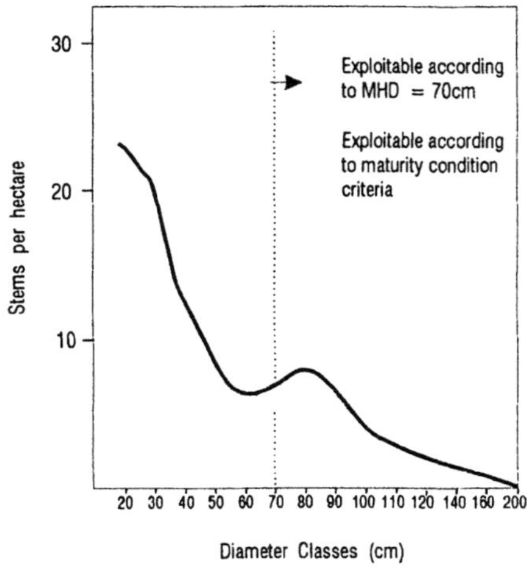

Figure 9. Typical diameter frequency distribution of a gap opportunist species; indicating exploitable portions of the population according to minimum harvestable diameters (e.g. MDH = 70 cm) or maturity condition criteria (Hatched).

Regeneration

The failure to ensure adequate regeneration of commercially valuable species within the context of practised silvicultural systems is a widespread problem (Fox 1976; Synnott and Kemp, 1976; Schmidt, 1987; Wadsworth, 1987; Wyatt-Smith, 1987a, 1987b; Okali and Ola-Adams, 1987; Burgess, 1989; Rietbergen, 1989; Weidelt, 1989; Whitmore, 1991). Following from the disturbance paradigm, successful regeneration is simplistically taken to be a matter of canopy manipulation. The value of canopy opening to induce regeneration has however not lived up to expectation. This practice appears only to accelerate to growth of existing regeneration but does not necessarily induce regeneration of commercial species.

Most of the commercially valuable species belong to the group of relatively fast-growing, but long-lived canopy occupants or emergents. These include many species of the Dipterocarpaceae, Meliaceae (e.g. *Entandrophragma cylindricum*), Podocarpaceae (e.g. *Dacrydium cupressinum*) and Araucariaceae (e.g. *Agathis australis*). This group of species shares a number of characteristics relevant to regeneration and recruitment.

- Early shade tolerance or requirement, resulting in persistent seedlings and saplings which retain the capacity to react to light becoming available at some later stage: 'meristematic dormancy' (Baur, 1964a,b; Fox, 1976; Ecroyd, 1982; Nwoboshi, 1987; Ogden et al., 1987; Ashton, 1988; Weidelt, 1989; Fickinger, 1992).
- Paucity of the intermediate size classes (Cameron, 1954; Baur, 1964a,b; Wardle, 1978; Norton et al., 1988; Appanah and Weinland, 1992; Fickinger, 1992; Lusk & Ogden 1992).
- Overrepresentation of the larger size classes (Fig. 9) as a result of high longevity, coupled with extended and relatively fast stem growth (Baur, 1964a; Norton et al., 1988; Fickinger, 1992).
- Tendency for recruitment to take place away from parent trees under some form of nurse canopy of non-conspecific trees (Cameron, 1954, 1960; Lloyd, 1960; Burgess, 1970; Beveridge, 1973; Fox, 1976; Ecroyd, 1982; Ogden et al., 1987; Ashton, 1988; Norton et al., 1988; Fickinger, 1992; Lusk and Ogden, 1992).
- Dispersal over relatively short distances only (Whitmore, 1991).

In summary, the emerging mode of regeneration does not appear to be pioneer-like gap colonisation, but short distance nurse canopy 'invasion' from the well-represented adult populations. Within this regeneration pattern three features appear to be particularly important in terms of silvicultural intervention:

(I) The phase of juvenile 'suppression' under the nurse tree canopy may be important for cueing the individual to an efficient physiological mode, thereby laying the basis for its longevity and extended fast stem growth (*cf.* Leibundgut 1978; Mayer, 1979). In accordance with the postulates of Backman (1943), too much light too early may result in low stress tolerance and defence capacity, early growth peak, weaker timber and short life. This may then result in susceptibility to fungal attacks (cf. Fickinger, 1992). In consequence, gap opportunist species seem to require sufficient light from above with protection from the sides, but not particularly large gaps, which would favour pioneer species (Brown, 1992; Vooren, 1992; Ådjers et al., 1995; Tuomela et al., 1996).

(II) Sustaining an adequate seedling, sapling or pole bank requires both a spatially well-distributed and temporally well-represented adult population. This spatiotemporal representation is achieved through relative longevity of trees. Adequate regeneration is uncertain with a substantially decimated adult population through exploitation based on relatively low minimum harvestable diameters or short rotations. In this regard, and especially in view of short dispersal distances, *inter alia* Burgess (1989), Appanah and Salleh Mohd. Nor (1991) and Whitmore (1991) stress the importance of leaving sufficient seed trees.

(III) The role of inter- versus intraspecific interactions resulting in an intermixture of nearest neighbour species and age classes in a variably sized mosaic patchwork suggests that recruitment beneath conspecific adults is somehow constrained (e.g. Wills *et al.*, 1997). Bruenig (1996) also emphasised the importance of the filtering effect of the understory matrix in reducing local conspecific overcrowding. Such patterns and processes should caution against producing even-aged stands dominated by fewer species which is attempted in uniform silvicultural systems. The sustainability or success over a number of rotations is uncertain. When we see gap opportunist species as nomadic forest matrix invaders (see below), their long-term successful regeneration depends on the presence of forest matrix patches not overtopped by conspecific adults.

For most species regeneration requires the extended presence of seed trees and suitable establishment space in close proximity. Establishment space is any patch where establishment can take place, i.e. any space below a nurse tree canopy to variably sized canopy gaps. A nurse canopy may be formed simply by one or more non-conspecific trees, or it may additionally require senility of the nurse trees or consist of any overmature pioneer stand. For the attainment of full canopy status it is furthermore of importance that the establishment space is enlarged to allow recruitment into the canopy. The stage and degree of enlargement of this establishment space necessary to allow recruitment into the canopy determines survival and growth performance differentially between and within species. Trees are inter- and intraspecifically differentially adapted or cued in terms of the time span they can tolerate the absence of such an extension. This determines their survival schedule and is partly reflected in their diameter class distributions.

Partly in an attempt to vindicate economically profitable high impact logging and based on an overextended interpretation of the disturbance paradigm, gap opportunist species were treated as pioneers in numerous silvicultural systems applied in tropical moist forests. Often the results were rather unsatisfactory. This points to the reality that the regeneration dynamics of gap opportunist species, notably those in West African forests (e.g. Poorter et al., 1996), is as yet poorly understood (Whitmore, 1991). Under the disturbance paradigm trees species are seen along a unidimensional gradient of declining light requirements; i.e. from pioneer to climax species. Within the confines of this paradigm Swaine and Whitmore (1988) give useful definitions of the related ecological groups. Having served well to improve our understanding, further progress requires us to move beyond the disturbance paradigm. An attempt is made here to present the tentative outline of a multi-dimensional metabolic trade-off paradigm (Figure 10). Unlike the one-dimensional disturbance pradigm, which only involves light as predominant resource factor, the metabolic trade-off paradigm attempts to accommodate many relevant resources (water, nutrients, temperature and light) and environmental as well as endogenous factors (e.g. genetics, heterozygosity, maternal effects) within a series of postulated metabolic trade-off relationships.

I) The main axis of the model relates to the basic trade-off between the capacity to survive stress (S-mode) and growth (G-mode). Along the S-mode specialist to G-mode generalist axis three canopy layer occupation guilds are recognised: forest matrix occupants, canopy dominants and matrix invading supra-canopy occupants. The last mentioned are also known as gap opportunist species and commonly occupy upper canopy and emergent positions. In alignment with this axis, Southeast Asian and West African forests are characterised by the prominent presence of the gap opportunist guild of species, whereas Neotropical forests are dominated by forest matrix occupants (Whitmore, 1991).

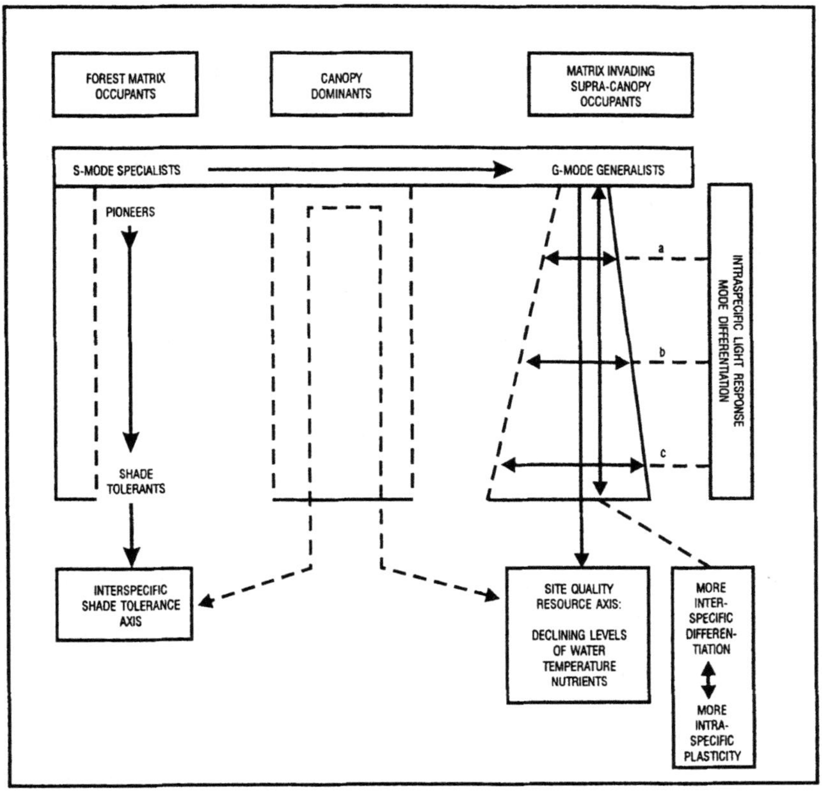

Figure 10. Multiple resource metabolic trade-off paradigm field underlying forest structure and regeneration constraints

- The S-mode specialists predominate in the subcanopy layer or are smaller-statured canopy trees. This axis thus also represents a gradient of increasing tree stature.
  - S-mode specialists are postulated to specialise in stress avoidance at the cost of growth capacity through interspecific specialisation and G-mode generalists sustain growth through stress tolerance achieved by physiological plasticity (a widened metabolic amplitude). Hence the term „generalists“.
  - The axis from forest matrix occupants to matrix invading supra-canopy occupants (gap opportunists) furthermore represents a declining gradient of contribution to species

richness by the successive guilds. This is in line with Gentry (1982) who demonstrated that most of tropical tree species richness emanates from the group of smaller-sized tree species.

II) S-mode specialists are typically differentiated interspecifically along the shade tolerance axis from pioneers to increasingly shade tolerant species. G-mode generalists are postulated to have broader intraspecific amplitudes in respect of light as resource, whereas canopy occupants as a group are considered to be intermediate between these two guilds in this respect. The emphasis of stress avoidance at the cost of growth, which is postulated to be more prevalent towards the forest matrix specialist extreme of the paradigm field (Figure 10), results in an accentuated dependence of site quality on stature. Canopy height accordingly declines strictly congruent with declining site quality; unlike the case for more G-mode generalist species.

III) Gap opportunist species are postulated to invade the forest matrix with a tendency to form gregarious cohorts in the upper canopy or emergent strata. To be successful in doing this they must establish within the forest matrix (have relatively shade tolerant seedlings) and then recruit through this matrix to attain supra-canopy positions. This requires a relatively high growth capacity in order to attain the stature to overtop the forest matrix.

- Recruitment through the forest matrix in variably sized recruitment spaces (overhead gap availability) is facilitated by the capacity for intraspecific light response mode differentiation in response to the stage and amount of light available (differentiation into P – SPP – SP growth strategy types re Figure 8).
- Maintenance of the G-mode for overtopping stature in the face of declining site-based resources (water, nutrients, temperature) is postulated to be achieved via an intra-specifically widened metabolic amplitude (physiological plasticity).
- The stress tolerance capacity allowing growth to be sustained partly insensitive to site-based resource shortages is however paid for by regeneration constraints; the exact nature of which still need to be elucidated (Seydack, 2000a, b and c). Examples here are forest edge / ridge orientated species which are capable of overtopping the forest matrix but experience

regeneration constraints (*Shorea curtisii*: hill forests in Malaysia; *Shorea albida*: Malesian Peatswamp forests; *Acacia melanoxylon* / *Olinia ventosa* in southern Cape Afromontane forests, South Africa). The regeneration constraints of West African gap opportunist species (mainly Meliaceae) are postulated to also lie within this realm of trade-off constraints.

- The regeneration of species with increased G-mode stress tolerance capacity with postulated wide metabolic amplitudes appears to be particularly constrained when confronted with better site conditions (warm, moist, nutrient-rich).
- Along the axis of declining site resource levels (notably moisture and temperature) there is further a postulated gradient from greater prevalence of interspecific differentiation (high species differention) to more intraspecific plasticity (broad intraspecific metabolic amplitudes) but low species differention within the guild of matrix invading gap opportunist species. As an example, the gradient from Lowland Mixed Dipterocarp forests (a), semi-deciduous West African forests (b) to Hill Dipterocarp forests (c) comes to mind.
- Within the gap opportunist species guilds of the forests (a – c), we have interspecific propensities for particular light response modes P – SPP – SP (Figure 8). That is, the light hardwood dipterocarp species of the Lowland Mixed Dipterocarp forests are postulated to have a propensity towards relatively early growth peaks in comparison with the heavy hardwood dipterocarp species. As we move down the site quality resource axis (a to c) the metabolic amplitudes widen and the light response modes (Figure 8) are increasingly differentiated intraspecifically. According to these postulates we therefore expect many gap opportunist species with relatively narrow metabolic amplitudes (interspecific metabolic differentiation) in the Lowland Mixed Dipterocarp forests (a) and fewer species of this guild, but with broader metabolic amplitudes (intraspecific metabolic differentiation) in Hill Dipterocarp to Peatswamp forests (c).
- In parallel with decling average resource levels there is a temporal component, i.e. declining resource levels may often concur with increasing fluctuations in resource availability due to seasonal effects and amplified by site conditions (e.g. ridges). This postulated site resource based gradient appears to be part of a more general gradient in declining species diversity with altitude and latitude.

It is hoped that research and investigation under the broader framework of the multiple-resource metabolic trade-off paradigm may contribute to further our understanding of forest regeneration. In the mean time it is contended that a yield regulation system which is in accordance with the natural disturbance regime, both spatially and over time, should ensure sufficient natural regeneration without special silvicultural inputs. That is, the spatiotemporal proximity of adults and logged areas must be in accordance with the natural grain of disturbance. If the systems of logging do not comply with these requirements, good natural regeneration should not be taken for granted and much silvicultural inputs are required to ensure adequate regeneration (*cf.* Appanah and Weinland, 1992).

Operational economics

Timber harvesting (logging) and silvicultural operations aimed at the improvement of regeneration or productivity are subject to constraints of economics with important consequences in the development of yield regulation systems. Silvicultural operations are often considered economically unaffordable (Whitmore, 1990) or logistically problematic to implement on a large scale. In practice, thus, silvicultural operations are not implemented on a large scale in tropical and subtropical forests (Wyatt-Smith, 1987a, 1987b; Poore, 1989; Synnott, 1989; Uebelhör et al., 1990; Cheah, 1991; Whitmore, 1991). This state of affairs favoured the contention that timber harvesting and the achievement of silvicultural effects should be seen as a combined operation (Bruenig, 1996).

Modern timber removal operations involve high capital investments in mechanisation and high quality road construction. For such operations to be economically viable, large timber volumes have to be logged per hectare (Wyatt-Smith, 1987b; Whitmore, 1991). The minimum profitable volume per unit area is, however, often much lower than believed, as shown by the profitable creaming operations of species with widely dispersed individual trees (cf Palmer, 1975).

The perceived dictates of harvesting economics is one of the main causes for undesirably high harvesting intensities. This results in: I. selection of inappropriate yield regulation approaches and associated excessive canopy opening; II. pressure for unsustainable harvesting levels; III. damage to residual vegetation, including to the regeneration and recruits of desirable timber species and IV. damage to the soil. In addition to the consequences of the high harvesting intensities pursued, logging damage appears to be due to inadequate compliance with rules and regulations by the loggers, to the use of unnecessarily powerful and heavy machinery and the insufficient supervision by the authorities responsible (Wyatt-Smith, 1987b).

Highly mechanised logging with heavy machinery creates conditions not usually encountered in nature (Whitmore, 1991). Particularly troublesome is soil compaction (e.g. Malmer and Grip, 1990) and also disruption of the soil surface, which destroys seedlings, the humus layer and superficial feeding roots and affects the soil seed bank (Whitmore, 1991). In the case of polycyclic management it is especially important that harvesting damage to adolescent trees is kept to a minimum. Failing to do so can result in the loss of volume increment which accumulates between felling cycles (Weidelt, 1989).

Following from the above, two issues of operational economics require careful and innovative conflict reconciliation:

- Logging cost effectiveness and the requirements of optimum yield regulation systems.
- Cost effectiveness of achieving silvicultural effects through integration with timber harvesting.

Economically viable high-tech logging causes considerable damage to the residual stand which is of particular concern with polycyclic systems (Wyatt-Smith, 1987a, 1987 b; Whitmore, 1991). At common cutting rates experienced in Mixed Dipterocarp forests around 10 – 20% of biovolume is removed and the associated extraction damages 20 – 30% of the basal area of the residual stand (Bruenig, 1996). The damage increases rapidly with increasing harvesting intensity. Furthermore, high intensity logging results in substantial canopy destruction which is incompatible with most yield regulation systems attempting to mimic natural forest dynamics and disturbance regimes (Whitmore, 1991). Such heavy canopy opening causes excessive damage and soil exposure and, if regeneration is successful, would eventually amount to a uniform system (Bruenig 1996). Successful forest recovery after logging requires that damage to seedlings, adolescent trees, the soil surface and drainage patterns be minimised (Whitmore, 1991). In this regard, the following considerations are relevant.

- Logging techniques should be adapted to suitably selected yield regulation systems, and not the other way round.

- Pre-harvest climber cutting and even crown lopping should be considered as damage reducing measures under particular circumstances.
- Low-tech logging, the efficiency of which is less dependent on timber volume concentration, should be considered and the removal of timber by animal transport along narrow improvised trails (*cf.* Wyatt-Smith, 1987a).
- Unless effective on-site conversion of large logs can be devised, these will still have to be harvested with heavy machinery (Wyatt-Smith, 1987b).
- Available funds should be directed towards the development and implementation of appropriate logging techniques and measures to reduce felling damage (Hendrison, 1990; Dykstra and Heinrich, 1992).
- Labour intensive, rather than highly mechanised logging, provides scope for the involvement of a larger number of local people, leading to a better base for socio-economic development (Neil, 1981; Ocaña-Vidal, 1992).

Gap opportunist species tend to occur in gregarious patches overtopping the forest matrix. If harvesting is largely confined to the commercial species from this guild, it tends to result in a mosaic of practically clearfelled and untouched patches. The clearfelled patches may be too open for optimal regeneration and recruitment of gap opportunist species requiring some lateral protection (3.2) and in the untouched patches residual recruits may remain without the desirable overhead release. Thus, if commercially harvestable species only cover a small proportion of the forest, (e.g. West African forests) harvesting will only affect a correspondingly small portion and if too concentrated may result in clearfelled parts where regeneration may either fail or when successful may result in unnatural conspecific / intraguild competing cohorts requiring stand improvement liberation thinnings. Under the scenarios described, timber harvesting on its own does not seem to successfully fulfil the role of silvicultural operations which may be deemed necessary.

The challenge of trying to accommodate the conflicting requirements of yield optimisation and operational economics under the described circumstances may be approached in different ways. Three options are given below:

- Restricted high value tree harvesting (target diameters of 60 –100 cm and 80 – 120 cm in low- and high-yielding Dipterocarp forests respectively) with additional release operations of selected potential crop trees (Bruenig, 1996).
- Broader-scale harvesting at relatively low minimum harvestable diameters (down to 60 and 45 cm for dipterocarp and non-dipterocarp species respectively), but then not requiring additional silvicultural operations (Korsgaard, 1992).
- High-value tree harvesting at high harvest maturity thresholds selected according to maturity condition criteria. Due to harvesting at high maturity thresholds and based on the contention that gap opportunist species only have moderate and overhead rather than high light requirements (3.2), silvicultural release operations in addition to harvesting are deemed unnecessary.

In reality post-felling silvicultural treatments have been largely abandoned due to economic constraints. A given sum of money may give better returns in terms of growth increment if invested in reducing logging damage rather than in post-felling silviculture (Whitmore, 1990). In view of the constraints of operational economics usually encountered where tropical / subtropical forests are to be managed, carefully developed yield regulation systems requiring minimal silvicultural intervention, but placing the focus on innovative low impact harvesting and professional harvest tree selection should contribute significantly towards achieving truly sustainable forest management.

Comparative assessment of yield regulation systems

As was noted in Section 2, monocyclic systems involve the harvesting of all harvestable trees during a single operation and relying on seedling regeneration to develop into the next, largely even-aged crop over the rotation period. Under polycyclic systems only a portion of the growing stock above a suitably selected minimum harvest diameter is harvested at shorter intervals, the felling cycle. Advanced residuals which continue to grow and exceed the target diameter are intended to form successive harvests. The resultant forest stands are largely uneven-aged. The earlier selective fellings and logging without particularly heavy machinery resulted in relatively low impacts on the forests. From the late 1940's to late 1960's there was a swing away from polycyclic to monocyclic systems. According to Poore et al. (1989) this was due largely to the realisation that repeated cutting as practised under the polycyclic systems (logging carried out with heavy machinery) would be detrimental to the forest in the absence of intensive control of exploitation damage from logging with heavy machinery. Monocyclic systems also permitted highly profitable economic cuts (Goodland et al., 1990).

Further impetus towards monocyclic systems emanated from the notion that natural forests would have to be domesticated in order to create transformed forests of increased commercial productivity in order to ward off claims on forest areas for non-forest forms of land-use (Lamprecht, 1984, 1989). The dogma of maximising the functions of production led to monocyclic silvicltural systems such as the various forms of shelterwood systems and the early Malayan Uniform System (Bruenig, 1996). Subsequently, however, the growing understanding of forest dynamics led to a shift towards more flexible and adaptable approaches and the prevailing trend thus changed towards polycyclic systems (Poore et al., 1989).

Monocyclic systems, with their emphasis on producing even-aged stands, were favoured initially due to the following considerations (Baur, 1964b):

- The development of mechanical logging methods which for economic working require a high yield from each hectare logged.
- Exploitation damage with mechanical logging to residual trees was bound to be intolerably high for successful polycyclic management.
- Selective logging was perceived to have failed in producing worthwhile regeneration, whereas most commercially desirable species were considered light demanders well suited to regeneration in even-aged stands.

The subsequent shift to polycyclic systems was due to one or more of the following disadvantages or limitations of monocyclic systems:

- Harvesting a very large proportion of the growing stock in a uniform system fails in habitat, soil and water protection (Baur, 1964b; Bruenig, 1996).
- Considering the sensitive nutrient balance which often prevails, cautiously exploited polycyclic systems are more likely to sustain the steady state condition in respect of

accumulation and depletion of nutrients (ex Weidelt, 1989).

- Although high immediate cash flows and rates of volume production are associated with monocyclic systems, their rates of value and quality production is considered comparatively low (Bruenig, 1996).
- Monocyclic systems strongly favour a relatively narrow range of species which may be detrimental in terms of biodiversity values and represents reduced scope for retaining species currently unmerchantable, but for which a market may develop later (Baur, 1964b).
- Substantial opening of the canopy encourages the rapid growth of pioneer species as well as climbers, which, if unattended, may smother the regeneration of the desired species. Tending operations have to be carried out. In the face of the rising cost of labour, systems requiring numerous silvicultural operations cease to be viable and practical (*vide* Parren and De Graaf, 1995).
- Harvesting under uniform systems should only take place if and when adequate regeneration is present. Since it is impractical to delay felling due to inadequate seedling regeneration, such systems (e.g. Malayan Uniform System) were not practised widely (Cheah, 1991).

The Lowland Dipterocarp forests of Malaysia were characterised by having adequate regeneration on the ground in most cases, allowing the implementation of the Malayan Uniform System. In the Hill Dipterocarp forests the adequacy of regeneration was no longer the rule and this resulted in adopting selection cuttings under the Selection Management System (Thang, 1987; Cheah, 1991). The matching of silvicultural systems to prevailing circumstances in this regard tended to be different in West African forests. There the Tropical Shelterwood System, a monocyclic system, was adopted when the existing regeneration and stocking of potential crop trees was insufficient (mainly in the moist semidecidious forests of Ghana and Nigeria). With sufficient regeneration in the moist forests of Ghana a selection system was implemented (Parren and De Graaf, 1995). Application of Tropical Shelterwood Systems was presumably based on the contention that the forest with low levels of regeneration of gap opportunists species (mainly Meliaceae) were „successional" and that the removal of the canopy would induce regeneration of the „light demanding" species. It is not clear from the literature whether the Tropical Shelterwood Systems were actually successful in inducing regeneration significantly more than would have occurred otherwise. According to Wyatt-Smith (1987a) they largely failed in this respect.

As was discussed in Section 3.2 our understanding of the regeneration dynamics of gap opportunist species is rather incomplete. Perhaps this guild of species should rather be seen as forest matrix invaders and not as „long-lived pioneers“ under the disturbance paradigm (3.2; Seydack, 2000b). This would emphasise the importance of the forest matrix in the regeneration dynamics of these species and highlight the potential future problems of any yield regulation system which results in the partial or complete destruction of the forest matrix.

The absence of an effective forest matrix may have a number of detrimental effects regarding the performance of gap opportunist species which typically recruit through a forest matrix:

- Bruenig (1996) cautions against the removal of the subcanopy matrix which is deemed essential to diversify the light climate and the growing space so that the dipterocarp regeneration may differentiate itself while growing through this matrix. The absence of this matrix effect may lead to congested, overcrowded tree stands requiring costly tending to release future crop trees.
- If logging has removed almost all canopy and co-dominant trees, immature residuals will grow through a lower forest matrix than would be the case in a natural forest. Clear bole height and thus total volume of wood production may thus be reduced (Assman, 1970; Cheah, 1991; Poker, 1993; Seydack, 2000c).
- The absence of a species diverse matrix will furthermore result in recruits being exposed to conspecific adult proximity or intra-guild competition within even-aged cohorts. This may have as yet unappreciated negative effects on the recruitment of species which are naturally geared to recruit through a heterospecific, uneven-aged forest matrix (3.1; 3.2).
- Under extreme conditions of high radiation levels early in the development of generalist forest matrix invaders, growth response shifts according to the growth law can possibly occur. This may result in P strategy types (Figure 8) with early growth peaks and early senility (3.2).

Applying polycyclic yield regulation systems results in the avoidance of many if not most of the problems associated with monocyclic systems. However, selection cuttings under the Selection Management System (SMS) as practised in Malaysia represents an example of a high-intervention selection system (relatively low MHD and short felling cycles) and as such has shortcomings of its own. These were summarised following Thang (1987), Appanah and Weinland (1992) and Bruenig (1996):

- The aim of an economic cut of 30 – 45 m^3/ha appears to be incompatible with retaining sufficient seed trees and advanced residuals under most circumstances. Retention of seed trees in practice is generally noted as unsatisfactory. All adult trees which have accumulated to show up as the typical „hump“ in the diameter class distribution of gap opportunist species (Figure 9) are removed.
- Selective logging with the rigid application of a minimum diameter limit as low as 50 or 60 cm removes immature trees before their most productive period of basal area, volume and value increment comes to be realised.
- Under the relevant cutting regime the forests may become poorer in dipterocarps, especially the third and subsequent cut may not be as desired. A modelling exercise indicated that for the SMS at a 35 year cutting cycle, harvesting results in continuous reduction in the harvestable quantity of emergents after every successive cutting (Appanah et al., 1990).
- The increment of residuals turned out to be lower than anticipated. Many areas harvested 20- 40 years ago are now failing to reach maturity for a second harvest. It is possible that a certain proportion of the residuals were not vigorously recruiting, but represented S strategy types *vide* Figure 8 (3.2).
- For the SMS advanced growth is the nucleus of the new timber stand and is expected to grow to exploitable size within 30 years. This requires that the advanced growth is consistently present in adequate numbers and capable of reasonable fast growth. Unfortunately this is often not the case. Particularly in view of the poor representation of the intermediate size classes often encountered in gap opportunist species (see Figure 9), selection cutting with low MHDs result in unsatisfactory residual growing stocks.
- Relatively high levels of logging damage further diminish the residual growing stock.

- Consequently relatively short felling cycles (25 - 40 years) are not associated with sustainable harvests and felling cycles of 60 – 100 years are more realistic if valuable high grade timber is to be produced.

In attempts to address the short-comings and problems of selection cutting systems as practised, Appanah and Weinland (1992) and Cheah (1991) have made recommendations focussing on the reduction of logging damage and tree marking to ensure the retention of seed trees and vigorous residuals suitably spaced over the logged-over forest and also in relation to matrix trees. Furthermore, recommended modifications involve adaptable and suitably selected minimum harvest diameters and felling cycles. Four scenarios are noted in this regard:

- Extension of felling cycles to between 60 – 100 years and MHD's of between 80 – 120 cm for the production of valuable high-grade timber, in combination with liberation thinnings when necessary (Bruenig, 1996).
- Thang (1987) recommended a Variable Management System where felling cycles and cutting diameter limits are variably adjusted depending on stand characteristics. Based on information of the volume and composition of the timber to be cut and on the predominant size class and composition of the regeneration destined to form the new stand, felling cycles and diameter limits would be determined so as to ensure that a new timber crop would have a reasonable chance of developing within the harvest interval.
- Due to the high proportion and clumping of commercial species, low MHDs applied and exploitation with heavy machinery, the Philippine Selective Logging System resulted in a mosaic of virgin forest remnants and basically clearfelled gaps regenerating from seed. This required the conversion to a monocyclic uniform system (Uebelhör et al., 1989). The resultant secondary stands were then treated according the TSI approach (Timber Stand Improvement) which involved the cleaning, weeding and crown thinning around 100 – 125 selected potential crop trees (Uebelhör et al., 1989; 1990).
- For Mixed Dipterocarp forests of Sarawak (Malaysia), Korsgaard (1992) recommended the extension of the felling cycle from 25 - 30 years to 50 - 60 years; the maintenance of strict cutting limits of 60 cm DBH for dipterocarps and 45 cm for other species and low impact logging. It was concluded that a carefully conducted selective harvest according to these

stipulations would pre-empt any requirements for further silvicultural treatment to promote the growth of immature trees or the natural regeneration of commercial species (Korsgaard 1992).

These scenarios embrace a wide range of approaches, but are generally characterised by relative long felling cycles when considered within a selection management framework. Since the intention generally is to realise the growth accumulated between harvesting events as yield, long felling cycles favour economic cuts but may often be associated with harvesting impacts at discord with the natural disturbance regime and forest dynamics, resulting in more even-aged, uniform forests. Particularly long felling cycles with low MHD's are associated with high harvest impacts and may result in more or less even-aged stands. Long felling cycles with high MHD's are likely to be associated with high levels of irretrievable mortality, especially when commercial trees span a wide range of ages or growth response types. Maturity condition harvest tree selection criteria and felling cycles matched to mimic the prevalent disturbance regime are useful to overcome the felling cycle / MHD dilemma under certain circumstances (2.1; 2,3; Figure 9).

It is clearly evident from the comparative assessment here made that the development or selection of a yield regulation system is an optimisation process in respect of numerous trade-offs and within the tension field of economics, sustainability and biodiversity conservation. Depending on various relevant factors any particular yield regulation system will represent a position on the high to low intervention gradient of options i.e. along a continuum of high silvicultural manipulation to naturalistic mimicking of forest dynamics. At the manipulistic end of this continuum, yield realisation is based on inducing a state of net growth accumulation through maintaining stands at relatively reduced growing stocks and trees harvested at relatively low harvest maturity thresholds. High diameter increments are maintained during sub-mature phases through the early release of future crop trees. Stand regeneration is pursued through substantial canopy opening based on the assumption that commercially desirable emergent canopy species are light-demanding long-lived pioneers with large-gap requirements.

At the naturalistic end of the continuum, yield optimisation is pursued through mortality pre-emption by harvesting trees timeously but generally at relatively high harvest maturity thresholds and from fully stocked stands. This approach capitalises on high volume increments through extended growth of trees at large diameters achieved during relative mature growth phases. Stand regeneration relies on small-scale natural renewal processes following from natural gap dynamics deemed adequate also for the regeneration of matrix invading emergent canopy species (e.g. Vooren, 1992).

The appropriate position of any yield regulation system on the manipulistic/naturalistic continuum depends on factors of managerial constraints, the type of forest involved and our understanding of the associated growth patterns, regeneration dynamics and tree diversity / dominance interactions. A naturalistic yield regulation system practised in subtropical uneven-aged multispecies forests in the southern Cape (South Africa) is described in the next section (Seydack et al., 1995).

A naturalistic yield regulaiton system in practice

Yield optimisation

The Knysna forests in the southern Cape, South Africa, are classified as Afromontane by White (1978). These evergreen forests were described as „temperate- form subtropical moist forests" according to Phillips (1931). Laughton (1937), Von Breitenbach (1974) and Geldenhuys (1989) provide detailed descriptions of the forests. Basal areas over a range of forest stands vary between 30 and 45 m^2/ha (>10 cm DBH). A detailed growing stock characterisation is presented in Table 5 (ex Seydack et al., 1995). The forest areas fall under the warm termperate, humid climate region (*vide* Walter, 1979) and grow on nutrient-poor sandstone- derived soils (Van Daalen, 1984).

The objective of an yield regulation system is to maximise sustainable yields, subject ot situational constraints. The yield regulation system developed for the Knysna forests involves mortality pre-emption through the selective harvesting of overmature trees proportional to species-specific turnover (Seydack, 1995, Seydack et al., 1995). The yield optimisation strategy underlying the harvest tree selection under this system will be discussed below. With the possible exception of *Olinia ventosa*, which appears to be the only gap opportunist species in the Knysna forests (Figure 3; 3.2), no regeneration constraints are apparent for the canopy species. No provision is therefore made for silvicultural stand regeneration measures.

Species [a]	Stems per ha > 10 cm [b]	10-30 cm	>30 cm	Percentage composition (> 30 cm)	Mean DBH (> 30 cm) (cm)
Ironwood	138.2	90.0	48.2	27.6	43.4
Yellowwood	106.8	72.0	34.8	19.9	43.6
Cherrywood	87.2	68.1	19.1	109	38.6
White pear	51.8	38.7	13.1	7.5	37.4
Assegai	54.6	43.3	11.3	6.5	37.4
Quar	25.2	16.1	9.1	5.2	40.1
White alder	28.0	19.6	8.4	4.8	43.3
Stinkwood	19.0	11.8	7.2	4.1	41.4
Forest elder	23.6	19.2	4.4	2.5	38.2
Kalander	9.3	6.0	3.3	1.9	56.6
Hard pear	4.8	1.6	3.2	1.8	52.7
Cape beech	8.6	5.9	2.7	1.5	41.3
Cape holly	9.5	7.2	2.3	1.3	41.5
Cape blackwood	11.0	9.9	1.1	0.6	38.7
Above species	577	409	168	96.2	
All species [c]	793	618	175	100.0	41.6

Data from 7383 0.04-ha sample plots: Bergplaas (197), Goudveld (193), Gouna (1155), Diepwalle (2264), Fisantehoek (1377), Bloukrans (695), Lotteirng (1114), Stormsriver (263), Blueliliesbush (95), Witelsbos (30).

[a] Ironwood, *Olea capensis* ssp. *macrocarpa*; Yellowood, *Podocarpus latifolius*; Cherrywood, *Pterocelastrus tricuspidatus*; White pear, *Apodytes dimidiata*; Assegai, *Curtisia dentata*; Quar, *Psydrax obovata* ssp. *obovata*; White alder, *Platylophus trifoliatus*; Stinkwood, *Ocotea bullata*; Forest elder, *Nuxia floribunda*; Kalander, *Podocarpus falcatus*; Hard pear, *Olinia ventosa*; Cape beech, *Rapanea melanophloeos*; Cape holly, *Ilex mitis*; Cape blackwood, *Maytenus peduncularis*.

[b] DBH class

[c] Canopy and subcanopy species

Table 5. Growing stock data of widespread canopy species in southern Cape plateau forests in South Africa (Seydack et al., 1995).

The long-term net growth of a primary forest is zero. However, even when in equilibrium, i.e. when ingrowth equals mortality, there is considerable turnover. For multi-species, multi-aged moist tropical and subtropical forests from various localities of the world, the turnover has the order of magnitude of between 1 and 2% per annum (Crow and Weaver, 1977; Lieberman and Lieberman, 1987; Manokaran and Kochummen, 1987; Primack et al., 1987a; Swaine et al., 1987a,

1987b; Seydack et al., 1990). If the associated mortality can be effectively pre-empted through harvesting of overmature trees with declining vigour and short remaining life expectancy, the growth potential of the site may be optimally utilised. The inherent tendency in a multi-aged, multi-species forest is to concentrate growth on fewer and larger trees (3.1). Thus, in conjunction with the relatively high increments of trees in the larger diameter classes, such forests offer much scope for maximum value increment / volume production with an appropriate harvesting strategy. In this case the yield regulation strategy is characterised by I.) high residual growing stocks; II) the absence of silvicultural operations to enhance the productivity of selected trees and III.) high maturity thresholds of harvestable trees which are identified by maturity condition criteria.

Trees which have not yet reached their final position in the canopy, i.e. partially waiting trees, are slower growing than typical canopy occupants (≥ 30 cm DBH)(Figure 3). These trees do not however use a proportional amount of canopy space. Canopy occupants are relatively insensitive to crowding (3.1; Figure 4a – c) and the productivity per hectare is proportional to stocking (Figures 1 and 2). Age and the associated condition of the canopy-occupying trees more than crowding seems to determine their rate of growth (Table 6).

Active silvicultural release operations in order to accelerate diameter growth rates are not provided for in line with the following considerations:

- The degree and sustainability of attainable growth accelerations of canopy trees are uncertain (3.1) and unlikely to be of a magnitude resulting in release operations to be economically profitable under the prevailing conditions.
- The expendability, i.e. the future commercial utility of presently less desirable species, at the cost of which thinning operations would have to take place, is unclear.
- In view of recruitment constraints near conspecific adults (3.1; Figure 6; also Lübbe and Geldenhuys, 1991), the ecological consequences of interfering with the dynamics of species composition cannot be safely assessed as yet.

		Harvestable trees (a)	Calibration stand (b)	Largest trees in stand (c)
Ironwood	$\overline{X}$	0.924	2.053	2.637
	n	52	560	52
	SD	0.8014	1.4820	1.411
Yellowwood	$\overline{X}$	0.763	1.255	1.950
	n	20	312	20
	SD	0.9364	1.1340	1.3137
Stinkwood	$\overline{X}$	0.554	1.496	1.196
	n	23	212	23
	SD	0.8181	1.4162	1.0276
Cherrywood	$\overline{X}$	0.789	1.983	2.278
	n	9	188	9
	SD	0.7944	1.2158	1.3065
White pear	$\overline{X}$	0.415	0.990	1.000
	n	13	149	13
	SD	0.4100	0.8432	0.7141
Assegai	$\overline{X}$	0.425	1.398	1.387
	n	8	111	8
	SD	0.4432	1.2614	1.2789
Quar	$\overline{X}$	0.378	1.516	1.567
	n	9	115	9
	SD	0.4086	1.0215	0.9367
Hard pear	$\overline{X}$	3.960	4.745	7.600
	n	5	214	5
	SD	1.5469	1.9395	1.7847

Data ex Seydack et al. (1995); n: number of trees involved; $\overline{X}$: mean increment of trees harvested (a), of calibration stand (b) and of (c), the largest trees in (b) equivalent in number to the harvestable trees (a)

Table 6. Comparison of diameter increment rates (cm/year) of trees harvestable according to maturity condition criteria, of the reference stand and of the largest trees (per species; ≥ 30 cm DH)

In forests where trees were and are allowed to reach advanced stages of maturity, slow-growing and thus preferentially harvestable trees can be outwardly identified with maturity condition criteria (Table 7; Seydack et al., 1995). By making use of maturity condition criteria for harvest tree selection, the disadvantages of MHD's can be avoided (2.3). Maturity condition-based harvest tree selection criteria permit the differentiated retention of productive mature trees. This allows trees to fulfil their ecological roles whilst realising the productivity advantages associated with continuing growth of large-sized trees (Figure 9). The harvest tree selection strategy accordingly involves the pre-emption of mortality through the selective harvesting of canopy trees of declining productivity. The harvesting of such trees then opens opportunities for their replacement by waiting recruits (Figure 11). Non-senile canopy occupants are left to continue to grow (value production). Allowing maximum value increment is particularly important in forests with trees of relative small final dimensions.

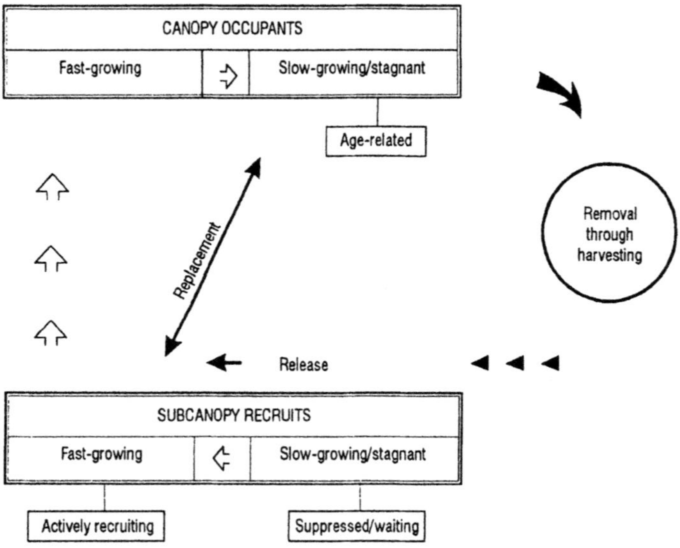

Figure 11. Harvest tree selection strategy (ex Seydack, 1995).

Incidentally, this type of approach to yield regulation was also mentioned by Assmann (1970) as of great potential for natural forests: „When estimating the sustained yield, attention must be paid to the fact that an appreciable increment of natural mixed stands depends on a high content of growing stock". Also: „The increment of such a forest can be maintained by removal of the upper storey trees of declining vigour, i.e. by preventing the upper storey from becoming over-aged". In accordance with the considerations, as discussed, the selected yield optimisation strategy involves the pre-emption of mortality through the selective harvesting of mature trees of declining productivity as identified with maturity condition criteria (5.2: Table 7).

System development

The senility criteria yield regulation system as it was originally termed (Seydack et al., 1995) should preferentially be known as the maturity condition criteria yield regulation system. The term "senility" suggests to many that only trees of very advanced stages of overmaturity and deterioration are harvestable. The harvest maturity threshold can however be optimally selected at relatively earlier stages of maturity. True to its naturalistic nature, maximum sustainable yields are sought; but with minimised artificial impact on the ecology and minimised managerial input for an effective input/output ratio. The development of the maturity condition yield regulation system involved selection of the felling cycle (a), establishment of species-specific maturity condition criteria (b) and the calibration of these criteria to species-specific turnover (c) to form the harvest tree selection criteria.

a) Selection of felling cycle

A 10 year felling cycle was selected. This represents a compromise between cost effectiveness of management, favoured by long felling cycles and the applicable disturbance regime, favouring short felling cycles. Since the disturbance regime of the southern Cape forests is characterised by a predominance of single tree deaths (Geldenhuys and Maliepaard, 1983), a fine grained pattern of age classes results (Midgley et al., 1990). Attempting to partially mimic the natural disturbance regime (*vide* Brown and Press, 1992) therefore requires that the felling cycle is not too long.

b) Establishment of species-specific maturity condition criteria

For this purpose tree condition assessments were undertaken in selected calibration stands which were situated within long-term forest research compartments (Seydack et al., 1995). Increment, ingrowth and mortality data were available for these stands of individually numbered trees. In these calibration stands, trees of the important canopy species (>30 cm DBH) were evaluated and described by way of visual estimation classes in respect of percentage cown dieback, structural damage, degree of stem decay, incidence of agony shoots and any signs of low remaining life expectancy (e.g. partially wind blown or base sporophytes signalling base decay and susceptibility to wind breakage). For each species, correlations were noted between condition parameters and individual tree diameter increments. The increment means of harvestable trees according to maturity condition criteria were clearly lower than the calibration stand means (Table 6). From Table 6 it is furthermore evident that MHD criteria would result in the harvesting of trees with above-average growth rates, which would be decidedly undesirable.

c) Calibration of harvest selection criteria to rates of turnover

Finally, the species-specific set of criteria are established which result in the harvesting of that number of trees which match the inherent turnover of the species, i.e. its rate of mortality. The procedure was to establish those criteria which allowed the identification of the most senile trees equivalent in number to the associated species-specific 10 year mortality rates. Percentage crown dieback was generally the best indicator of low vigour or senility, but the other parameters also tended to be related to senility, depending on species. In this way, the harvest selection criteria were calibrated in relation to species-proportional turnover rates, thereby allowing the harvesting of the most senile trees through periodic mortality pre-emption. With a 10 year felling cycle and a harvest intensity equivalent to the mortality over 10 years, the mortality pre-emption time span equals 10 years. That is, at each harvest the trees with a remaining life expectancy of 10 years or less are harvested. The harvesting of trees on a 10 year felling cycle, but with harvest tree selection criteria calibrated to 1,5 x mortality (= mortality pre-emption span of 15 years) is expected to have a comparably favourable effect on the quality of the timber harvested and the success rate of mortality pre-emption. Harvest selection criteria calibrated to 1,5 x mortality (= turnover) are currently implemented. Equipped with the selected set of harvest selection criteria (Table 7), which represent directives integrating the questions of how many of which trees are to be harvested or retained as residual growing stock, personnel of the indigenous forest planning section set out to identify the trees to be harvested.

Ironwood	(I)	≥ 15 % crown dieback
	(II)	Main shoot lost
	(III)	Fork lost: remaining crown with ≥15% dieback or extremely small
	(IV)	Sporophytes or severe basal rot
Yellowwood	(I)	≥ 25 % dieback
	(II)	Main shoot or fork lost
Stinkwood	(I)	≥ 35 % crown dieback (of total crown)
	(II)	Main shoot or fork broken off/dead: ≥ 10% of remaining crown dying back or dead
Cherrywood	(I)	≥ 15 % crown dieback
	(II)	Crown dieback > 5% and agony shoots and severe stem rot
	(III)	Severe crown damage/loss (main shoot lost; major portion of crown lost) and either (a) ≥ 10% crown dieback, or (b) agony shoots, or (c) severe stem rot
White pear	(I)	≥ 25 crown dieback
	(II)	Lots of agony shoots and >5% crown dieback
	(III)	Agony shoots and ≥ 20% of crown dieback (a) or (b) major crown loss (≥50% of crown: main shoot or fork lost)
	(IV)	Combination a/b of (III)
Assegai	(I)	≥ 35% crown dieback
	(II)	Severe crown loss (≥ 50%: main shoot or fork lost) and ≥ 20% dieback of remaining crown or severe stem rot
Quar	(I)	≥ 25% crown dieback
	(II)	Major crown loss (≥ 50%: main shoot or fork lost)
	(III)	≥ 10% crown dieback and severe stem rot or major branches/fork broken off
	(IV)	≥ 15% crown dieback and bark removed
Hard pear	(I)	≥ 20% crown dieback
Botanical names as per Table 5		

Table 7. 1.5 x mortality rate-calibrated harvest tree selection criteria (Seydack et al. 1995).

In accordance with a 10 year felling cycle, one-tenth of the effective timber harvesting area is covered each year. These areas are subdivided into 10 – 20 m wide strips with rudimentary cutlines. Aided by this subdivision, the respective planning officer inspects all trees with a DBH of 30 cm or larger and marks those which comply with the harvesting criteria. Harvestable trees with large crowns are noted for crown lopping prior to felling. Also, as an integral part of the yield regulation system, provision has been made for the monitoring of stand dynamics in both harvested and unharvested forest stands and responses to harvesting (Seydack, 1991, 1994).

Advantages and limitations

In conclusion, the perceived advantages of the maturity condition criteria yield regulation system are high-lighted.

- Since only trees with visible and defined signs of maturity / senility are harvestable, continued adherence to the system implies that sustainability, as defined, is guaranteed.
- Yield optimisation is pursued and facilitated by:
  I) high volume and value increment of individual trees is attained by allowing extended growth of large trees to proceed;
  II) maximum mortality pre-emption over all merchantable diameter classes;
  III) full stocking results in full site use.
- Since the natural disturbance regime and dynamics of the forests are largely unchanged and ample seed trees remain in the stand all the time, regeneration is ensured, even if regeneration is of an intermittent nature.
- Largely tracking the natural dynamics of the forest facilitates the maintenance of biodiversity.

A number of advantages concerning practical management can be listed.

- Silvicultural operations are not required.
- The system is flexible in that harvesting can be restricted to the currently desirable commercial species without resulting in their depletion (absence of creaming effect').
- Once the criteria have been established, yield regulation is not dependent on pre-marking inventories.

Certain circumstances are perceived to limit the unmodified or widespread application of the maturity condition criteria yield regulation system. When trees are harvested towards the end of their lifespan an increased incidence of timber defects is to be expected in comparison with systems where trees are harvested prematurely. This is particularly problematic in certain types of forest where mature trees of certain species are susceptible to heart rot.

From a managerial perspective, mention must be made that the harvest tree selection process is intensive and demanding and the techniques and economics of spatially dispersed harvesting pose a variety of challenges.

Conclusions

Professionals involved in natural resource management become increasingly aware of the broader content of sustainability and its implications. Sustainability has several components apart from wood production (Goodland et al., 1990) which need to be accommodated in forest management: subsistence base of forest-dependent people, extraction of non-wood products, environmental services provided by forests and biodiversity values. Furthermore, a realism emerges in respect of sustainable timber yield levels from natural forests. Harvesting of only the wood increment since the last cut provides modest yields in comparison to what logging agencies are accustomed to. In other words, ecologically sustainable yields are hardly perceived to meet socio-economic needs (Robinson, 1993).

Any attempts at reconciliation of ecological sustainability of resource use, socio-economic needs and conservation of biodiversity would have to involve the following components.

- Sustainable societies
- Collaborative participatory forest management
- Naturalistic yield regulation systems
- Protected conservation areas

Keeping human population sizes and life styles within the carrying capacity of the environment is an ultimate pre-requisite for sustainable resource use (Goodland et al., 1990; Robinson, 1993). Local communities have short-term subsistence needs, whereas national communities and long-term requirements (future generations) need to be represented by governmental agencies. To be successful, joint forest management ventures require the input from both parties. Without the support by governmental agencies in terms of long-term professional authority, sustainability of resource use and conservation of biodiversity are unlikely to be achievable.

Yield regulation systems are aimed at achieving maximum sustained timber yields within the economics – conservation tension field. There is a shift away from highly interventionist systems, which involve the manipulation of forest structure and function to produce transformed forest systems, to naturalistic systems which attempt to mimic natural forest dynamics. This shift is partly driven by forces of economics (logistics and lack of affordability of manipulative operations) and conservation priorities. There is furthermore an increasing awareness of the risks of manipulative approaches and a realism about what is actually achievable. If we are really serious about mimicking forest dynamics, a further shift away from net-growth yield realisation towards mortality pre-emptive approaches from forests which are structurally closer to the primary state, is required in most circumstances. It is believed that development in this regard can be facilitated by increased understanding of I.) regeneration dynamics (particularly of gap opportunist species); II.) forest productivity (growth patterns in relation to site, inter- and intraspecific competition, climatic cycles and release operations; as well as growth law effects) and III.) the basis of species diversity and dominance. For the operational level more work is required with regard to the development of maturity condition criteria and low-impact exploitation approaches.

Timber harvesting will impact on the mature and over-mature portions of the forest in greater or lesser degrees. In forest biodiversity conservation, there is therefore „no substitute for totally protected areas“ (Struhsaker, 1998). Only the combined existence of production and protected areas represents a balance between utilisation and preservation, which may then form a viable basis for adequate forest conservation.

References

Ådjers, G., Hadengganan, S., Kuusipalo, J., Nuryanto, K. and Vesa, L., 1995: Enrichment planting of dipterocarps in logged-over secondary forests: effect of width, direction and maintenance method of planting line on selected *Shorea* species. *Forest Ecology and Management* 73: 259 – 270.

Alder, D.C., 1992: Simple method for calculating minimum diameter and sustainable yield in mixed tropical forest. In F.R.Miller and K.L. Adam (eds.), Wise Management of Tropical Forests 1992. OFI, Oxford, p. 189 – 200.

Appanah, S. and Salleh Mohd. Nor., 1991: Natural regeneration and its implications for forest management in the dipterocarp forests of Peninsular Malaysia. In A. Gómez-Pompa, T.C. Whitmore and M. Hadley (eds.), Rain Forest Regeneration and Management. Man and Biosphere Series, Vol. 6, UNESCO/Parthenon, Paris, p. 361 – 369.

Appanah, S. and Weinland, G., 1992: Will the management systems for hill dipterocarp forests stand up? *Journal of Tropical Forest Science* 3(2): 140 – 158.

Appanah, S. and Weinland, G., 1993: *Planting Quality Timber Trees in Peninsular Malaysia. A Review.* Malay. For. Rec. No. 38, Forest Research Institute Malaysia, Kepong, Kuala Lumpur, Malaysia, 221 pp.

Appanah, S., Weinland, G., Bossel, H. and Krieger, H., 1990: Are tropical rain forests non-renewable? An enquiry through modelling. *Journal of Tropical Forest Science* 2(4): 331 – 318.

Ashton, P.S., 1988: Dipterocarp biology as a window to the understanding of tropical forest structure. *Annual Review of Ecology and Systematics* 19: 347 – 370.

Assmann, E., 1970: The Principles of Forest Yield Study. Pergamon, Oxford/New York, 504 pp.

Backman, G., 1943: Wachstum und organische Zeit. *Bios*, Leipzig, 15: 1 – 195.

Barbour, M.G., Burk, J.H. and Pitts, W.D., 1987: *Terrestrial Plant Ecology*, Second edition. Benjamin/Cummings, Amsterdam/Sydney, 634 pp.

Baur, G.N., 1964a: *The Ecological Basis of Rainforest Management.* Forestry Commission of New South Wales, Sydney, 499 pp.

Baur, G.N., 1964b: Rainforest treatment. *Unasylva* 18(1): 18 – 28.

Bennis, W.J., 1989: Why leaders can't lead. *Training and Development Journal*, April 1998: 35 – 39.

Beveridge, A.E., 1973: Regeneration of podocarps in a central North Island forest. *New Zealand Journal of Forestry* 18: 23 – 25.

Bloom, A.J., Chapin, III, F.S. and Mooney, H.A., 1985. Resource limitation in plants – an economic analogy. *Annual Review of Ecology and Systematics* 16: 363 – 392.

Boxman, O., De Graaf, N.R., Hendrison, J., Jonkers, W.B.J., Poels, R.L.H., Schmidt, P. and Tjon Lim Sang, R., 1985: Towards sustained timber production from tropical rain forests in Suriname. *Netherlands Journal of Agricultural Science* 33: 125 – 132.

Brown, N.D., 1992: Interaction between forest management for forest production and forest regeneration. In F.R. Miller and K.L. Adam (eds.), Wise Management of Tropical Forests. Proceedings of the Oxford Conference on Tropical Forests 1992. University of Oxford, Oxford, p 11 – 17.

Brown, N. and Press, M., 1992: Logging rainforests the natural way? *New Science* 14: 25 – 29.

Bruenig, E.F. and Poker, J., 1991: Is sustainable utilisation of the tropical evergreen moist forest possible? In W. Erdelen, N. Ishwaran and P. Müller (eds.), Tropical Ecosystems, System Characteristics, Utilization Patterns, Destruction, Conservation Concepts. Verlag Josef Margraf, Saarbrücken – Weikersheim, p. 91 –

106.

Bruenig, E.F., 1996: *Conservation and Management of Tropical Rainforests. An Integrated Approach to Sustainability.* CAB International, 339 pp.

Burgess, P.F., 1970: An approach towards a silvicultural system for the hill forests of the Malay Peninsula. *Malayan Forester* 33(2): 126 – 133.

Burgess, P.F., 1989: Asia. In D. Poore, P. Burgess, J. Palmer, S. Rietbergen and T. Synnott, (eds.), No Timber Without Trees. Sustainability in the Tropical Forest. Earthscan, London, 252 pp.

Cameron, R.J., 1954: Mosaic or cyclic regeneration in North Island podocarp forests. *New Zealand Journal of Forestry* 7: 55 – 67.

Cameron, R.J., 1960: Natural regeneration of podocarps in the forests of the Whirnaki river valley. *New Zealand Journal of Forestry* 8: 337 – 354.

Cheah, L.C., 1991: Sustained yield natural forest management in Peninsular Malaysia: opportunities for more effective forest practices. Malaysian Forestry and Forest Products Research. Proceedings. Forest Research Institute of Malaysia, Kuala Lumpur, p. 1 - 15.

Coley, P.D., 1987: Interspecific variation in plant anti- herbivore properties: the role of habitat quality and rate of disturbance. *New Phytologist* 106: 251 – 263.

Crow, T.R. and Weaver, P.L., 1977: Tree growth in a moist tropical forest of Puerto Rico. Forest Service Research Paper ITF-22, Institute of Tropical Forestry, Rio Piedras, Puerto Rico, 17 pp.

Dawkins, H.C., 1958: The Management of Natural Tropical High Forest with Special Reference to Uganda. Paper No. 34, Imperial Forestry Institute, Oxford University, 149 pp.

Dawkins, H.C. 1959: The volume increment of natural tropical high-forest and limitations on its improvements. *Empire Forestry Review* 38(2): 175 – 180.

De Graaf, N.R., 1986: *A Silvicultural System for Natural Regeneration of Tropical Rain Forest in Suriname.* Agricultural University, Wageningen, 250 pp.

Dickinson, M.B., Dickinson, J.C. and Putz, F.E., 1996: Natural forest management as a conservation tool in the tropics: divergent views on possibilities and alternatives. *Commonwealth Forestry Review* 75(4): 309 – 315.

Dykstra, D.P. and Heinrich, R., 1992: Sustaining tropical forests through environmentally sound harvesting practices. *Unasylva* 169(43): 9 – 15.

Ecroyd, C.E., 1982: Biological flora of New Zealand. 8 *Agathis australis* (D. Don) Lindl. (*Araucariaceae*) Kauri. *New Zealand Journal of Botany* 20: 17 – 36.

FAO, 1988: *Management of Tropical Moist Forests in Africa.* Forestry Paper 88, Food and Agriculture Organization, Rome, 165 pp.

FAO, 1998: *Guidelines for the Management of Tropical Forests. 1. The Production of Wood.* Forestry Paper 135, Food and Agricultural Organization, Rome, 109 pp.

Fickinger, H., 1992: Zur Verjünging einiger Wirtschaftsbaumarten in selektiv genutzten Feuchtwäldern der Republik Kongo. Dissertation. Institut für Waldbau, Georg-August-University, Göttingen, 203 pp.

Fox, J.E.D., 1976: Environmental Constraints on the Possibility of Natural Regeneration after Logging in Tropical Moist Forest. Proc. Division 1, XVI IUFRO World Congress, Norway, Norwegian IUFRO Congress Committee, c/o Forest Research Institute, As-NLH, p 512 – 538.

Geldenhuys, C.J. and Maliepaard, W., 1983: The causes and sizes of canopy gaps in the southern Cape forests. *South African Forestry Journal* 124: 50 –55.

Geldenhuys C.J., 1989: Environmental and Biogeographic Influences on the Distribution and Composition of the Southern Cape Forests (Veld Type 4). Ph.D. thesis, University of Cape Town, 324 pp.

Geldenhuys, C.J., 1997. Growth, ingrowth and mortality patterns over stands and species in the Groenkop forest study site, George. Division of Water, Environment and Forestry Technology, CSIR, Pretoria, 58 pp.

Gentry, A.H., 1982. Patterns of neotropical plant species diversity. *Evolutionary Biology* 15: 1 – 84.

Goodland, R.J.A., Asibey, E.O.A., Post, J.C. and Dyson, M.B., 1990: Tropical moist forest management: the urgency of transition to sustainability. *Environmental Conservation* 17(4): 303 – 318.

Hasse, W.D. and Ek, A.R., 1981: A simulated comparison of yields for even- versus uneven-aged management of northern hardwood stands. *Journal of Environmental Management* 12: 235 – 246.

Hendrison, J., 1990: *Damage-controlled Logging in Managed Rain Forest in Suriname*. Agricultural University, Wageningen, 204 pp.

Hubbell, S.P. and Foster, R.B., 1990: The fate of juvenile trees in a neotropical forest: implications for the natural maintenance of tropical tree diversity. In K.S. Bawa and M. Hadley (eds.), Reproductive Ecology of Tropical Forest Plants. Man and Bisphere Series, Vol. 7, UNESCO / Parthenon, Paris, p. 377 – 341.

Hubbell, S.P., Condit, R. and Foster, R.B., 1990: Presence and absence of density dependence in a neotropical tree community. *Philosophical Transactions of the Royal Society of London* 330: 269 – 281.

Hutchinson, I.D., 1988: Points of departure for silviculture in humid tropical forests, *Commonwealth Forestry Review* 67(3): 223 – 230.

Jonkers, W.B.J., 1987: *Vegetation Structure, Logging Damage and Silviculture in a Tropical Rain Forest in Suriname*. Agricultural University, Wageningen, 172 pp.

Korsgaard, S., 1992: An Analysis of Growth Parameters and Timber Yield Prediction, based on Research Plots in the Permanent Forest Estate of Sarawak, Malaysia. Council for Development Research, Copenhagen, 134 pp.

Lamprecht, H., 1984: Problematik und Folgen der forstlichen Nutzung tropischer Regenwälder. *Spixiana* 10: 97 – 101.

Lamprecht, H., 1989: *Silviculture in the Tropics*. Gesellschaft für Technische Zusammenarbeit, Eschborn, Germany, 296 pp.

Laughton, F.S., 1937: The Silvicuture of the Indigenous Forests of the Union of South Africa with Special Reference to the Forests of the Knysna Region. Sci. Bull. 157, Forestry Series 7, Government Printer, Pretoria, 169 pp.

Leibundgut, H., 1960: Die waldbauliche Planning als Mittel zur Erhöhung des forstwirtschaftlichen Erfolges. *Schweizer Zeitschrift für Forstwesen* 111(11): 548 – 566.

Leibundgut, H., 1978: Über die Dynamik europäischer Urwälder. *Allgemeine Forstzeitschrift* 24: 686 – 690.

Lieberman, D. and Lieberman, M., 1987: Forest tree growth and dynamics at La Selva, Costa Rica (1969 – 1982). *Journal of Tropical Ecology* 3: 347 – 358

Lloyd, R.C., 1960: Growth study of regenerated kauri and podocarps in Russel forest. *New Zealand Journal of Forestry* 8(2): 355 – 361.

Loehle, C., 1988: Tree life history strategies: the role of defenses. *Canadian Journal of Forest Research* 18: 209 – 222.

Lübbe, W.A. and Geldenhuys, C.J., 1991: Regeneration patterns in planted and natural forest stands near Knysna, Southern Cape. *South African Forestry Journal* 159: 43 – 50.

Lusk, C. and Ogden, J., 1992: Age structure and dynamics of a podocarp-broadleaf forest in Tongariro National Park, New Zealand. *Journal of Ecology* 80: 379 – 393.

Maître, H.F., 1987: Natural forest management in Côte d'Ivoire. *Unasylva* 39: 53 – 60.

Maître, H.F., 1992: Recherches sur la dynamique des peuplements arborés en vue de définir une sylviculture de conservation et production durable. In K. Cleaver, M. Munasinghe, M. Dyson, N. Egli, A. Peuker, F. Vencélius (eds.), Conservation of West and Central African Rainforests. World Bank Environment Paper No. 1. Washington D.C., U.S.A., World Bank,

Malmer, A. and Grip, H., 1990: Soil disturbance and losss of infiltrability caused by mechanized and manual extraction of tropical rainforest in Sabah, Malaysia. *Forest Ecology and Management* 38: 1 – 12.

Manokaran, N. and Kochummen, K.M., 1987: Recruitment, growth and mortality of tree species in a lowland dipterocarp forest in Peninsular Malaysia. *Journal of Tropical Ecology*, 3: 315 – 330.

Mayer, H., 1979: Die höchsten Bäume Europas -- waldvegetationsgeschichtliche Vorraussetzungen für ein optimales Wachstum. *Phytocoenologia* 6: 55 – 73.

Midgley, J., Seydack, A., Reynell, D. and McKelly, D., 1990: Fine-grain pattern in southern Cape plateau forests. *Journal of Vegetation Science* 1: 539 – 546.

Neil, P.E., 1981: Problems and Opportunities in Tropical Rain Forest Management. Occasional Paper No. 16, Commonwealth Forestry Institute, Oxford, 126 pp.

Norton, D.A., Herbert, J.W. and Beveridge, A.E., 1988: The ecology of *Dacrydium cupressinum*: a review. *New Zealand Journal of Botany* 26: 37 – 62.

Nwoboshi, L.C., 1987: Regeneration success of natural management, enrichment planting and plantations of native species in West Africa. In F. Mergen and J.R. Vincent (eds.), Natural Management of Tropical Moist Forests. Silvicultural and Management Prospects of Sustained Utilization. Yale University, School of Forestry and Environmental Studies, New Haven, p 71 – 91.

Ocaña-Vidal, J., 1992: Natural forest management with strip clear-cutting. *Unasylva* 169: 24 – 27.

Ogden J., Wardle, G.M and Ahmed, M., 1987: Population dynamics of the emergent conifer *Agathis australis* (D. Don) Lindl. (kauri) in New Zealand. II Seedling population sizes and gap-phase regeneration. *New Zealand Journal of Botany* 25: 231 – 242.

Okali, D.U.U. and Ola-Adams, B.A., 1987: Tree population changes in treated rain forest and Omo Forest Reserve, south-western Nigeria. *Journal of Tropical Ecology* 3: 291 – 313.

Palmer, J.R., 1975: Towards more reasonable objectives in tropical high forest management for timber production. *Commonwealth Forestry Review* 54(3/4) 273 – 289.

Palmer, J.R., 1989: Management of natural forest for sustainable timber production: a commentary. In D. Poore, P.Burgess, J. Palmer, S. Rietbergen and T. Synnott, (eds.), No timber without Trees. Sustainability in the Tropical Forest. Earthscan, London, 252 pp.

Parren, M.P.E., and De Graaf, N.R., 1995: *The Quest for Natural Forest Management in Ghana, Côte d' Ivoire and Liberia,* Tropenbos Series 13, The Tropenbos Foundation, Wageningen, 199 pp.

Phillips, J.F.V., 1931: Forest Succession and Ecology in the Knysna Region. Memoirs of the Botanical Survey of South Africa 14, The Government Printer, Pretoria, 327 pp.

Poker, J., 1993: Struktur und Dynamik des Bestandesmosaiks tropischer Regenwälder – Entwicklungen eines Modellansatzes zur Simulation natürlicher Mischbestände.Mitteilungen der Bundesforschungsanstalt für Forst- und Holzwirtschaft Nr. 174. Hamburg, 258 pp.

Poore, D., 1989: The sustainable management of tropical forest: the issues. In D. Poore, , P.Burgess, J. Palmer, S. Rietbergen and T. Synnott (eds.), No Timber Without Trees. Sustainability in the Tropical Forest. Earthscan, London, 252 pp.

Poore, D., Burgess, P., Palmer, J., Rietbergen, S. and Synnott, T. 1989: *No Timber Without Trees: Sustainability in the Tropical Forest.* Earthscan Publications, London. 252 pp.

Poorter, L., Bongers, F., Van Rompaey, R.S.A.R. and De Klerk, M., 1996: Regeneration of canopy tree species at five sites in West African moist forest. *Forest Ecololgy and Management* 84: 61 – 69.

Primack, R.B., Chai, E.O.K., Tan, S.S. and Lee, H.S., 1987a: The silviculture of dipterocarp trees in Sarawak, Malaysia. I. Introduction to the series and performance in primary forest. *Malaysian Forester* 50(1): 29 – 42.

Primack, R.B., Chai, E.O.K., Tan, S.S. and Lee, H.S., 1987b: The silviculture of dipterocarp trees in Sarawak, Malaysia. II Improvement felling in primary forest stands. *Malaysian Forester* 50(1): 43 – 61.

Rietbergen, S., 1989: Africa. In D. Poore, P. Burgess, J. Palmer, S. Rietbergen and T. Synnott (eds.), No Timber Without Trees. Sustainability in the Tropical Forest. Earthscan, London, 252 pp.

Robinson, J.G., 1993: The limits to caring: sustainable living and the loss of biodiversity. *Conservation Biology* 7 (1): 20 – 28.

Schmidt, R., 1987: Tropical rainforest management. A status report. *Unasylva* 156(39): 2 – 17.

Schmidt, R.C., 1991: Tropical rain forest management: a status report. In A. Gómez-Pompa, T.C. Whitmore and M. Hadley (eds.), Rain Forest Regeneration and Management. Man and Biosphere Series, Vol. 6, UNESCO / Parthenon, Paris, p 181 – 207.

Seydack, A.H.W., van Daalen, J.C., van Dijk, D., Reynell, D., Heyns, H., Jooste, A., Ferguson, M., Pitchford, W., Durrheim, G. and Willems, D., 1990: Yield regulation in the Knysna forests. *South African Forestry Journal* 152: 50 – 61.

Seydack, A.H.W., 1991: Inventory of South African natural forests for management purposes. *South African Forestry Journal* 158: 105 – 108.

Seydack, A.H.W., 1994: Data requirements of sustained yield management for three South African evergreen moist forests. In B. Bredenkamp (ed.), Proceedings of IUFRO S4.02-03 Conference on Minimum Data Requirements for Sustainable Forest Management. Faculty of Forestry, University of Stellenbosch, South Africa.

Seydack, A.H.W., 1995: An unconventional approach to timber yield regulation for multi-aged mutispecies forests. I. Fundamental considerations. *Forest Ecology and Management* 77: 139 – 153.

Seydack, A.H.W., Vermeulen, W.J., Heyns, H., Durrheim, G., Vermeulen, C., Willems, D., Ferguson, M., Huisamen, J. and Roth, J., 1995: An unconventional approach to timber yield regulation for multi-aged mutispecies forests. II Application to a South African forest. *Forest Ecology and Management* 77: 155 – 168.

Seydack, A.H.W., 1997: The challenge of indigenous forest management: from information to implementation. *Southern African Forestry Journal* 178: 47 – 52.

Seydack, A.H.W., 2000a: Invasion ecology of Australian blackwood (*Acacia melanoxylon*) in a South African subtropical forest environment. I. Population structure and growth strategies (in prep.)

Seydack, A.H.W., 2000b: Invasion ecology of Australian blackwood (*Acacia melanoxylon*) in a South African subtropical forest environment. II. Constraints to recruitment (in prep.)

Seydack, A.H.W., 2000c: Invasion ecology of Australian blackwood (*Acacia melanoxylon*) in a South African subtropical forest environment. III. Metabolic trade-off constraints (in prep.)

Silva, J.N.M., De Carvalho, J.O.P., J. do. C.A. Lopes, De Almeida, B.F., Costa, D.H.M., De Oliveira, L.C., Vanclay, J.K and Skovsgaard, J.P., 1995: Growth and yield of a tropical rain forest in the Brazilian Amazon 13 years after logging. *Forest Ecology and Management* 71: 267 – 274.

Sterner, R.W., Ribic, C.A. and Schatz, G.E., 1986: Testing for life historical changes in spatial patterns of four tropical tree species. *Journal of Ecology* 74: 621 – 633:

Struksaker, T.T., 1998: A biologist's perspective on the role of sustainable harvest in conservation. *Conservation Biology* 12(4): 930 – 932.

Swaine, M.D., Hall, J.B. and Alexander, I.J., 1987a: Tree population dynamics at Kade, Ghana (1968 – 1982). *Journal of Tropical Ecology* 3: 331 – 345.

Swaine, M.D., Lieberman, D. and Putz, F.E., 1987b: The dynamics of tree populations in tropical forest: a review. *Journal of Tropical Ecology* 3: 359 – 366.

Swaine, M.D. and Whitmore, T.C., 1988. On the definition of ecological species groups in tropical rain forests. *Vegetatio* 75: 81 - 86.

Synnott, T.J., 1980: Tropical Rain Forest Silviculture, a Research Project. Commonwealth Forestry Institute Occasional Paper. No. 10, Department of Forestry, Oxford University, Oxford, 45 pp.

Synnott, T., 1989: South America and the Caribbean. . In D. Poore, P. Burgess, J. Palmer, S. Rietbergen and T. Synnott (eds.), No Timber Timber Without Trees. Sustainability in the Tropical Forest. Earthscan London, 252 pp.

Synnott, T.J. and Kemp, R.H., 1976: Choosing the best silvicultural system. *Unasylva* 28: 74 – 79.

Thang, H.C., 1987: Forest management systems for tropical high forest, with special reference to Peninsular Malaysia. *Forest Ecology and Management* 21: 3 – 20.

Tuomela, K., Kuusipalo, J., Vesa, L., Nuryanto, K., Sagala, A.P.S. and Ådjers, G., 1996: Growth of dipterocarp seedlings in artificial gaps: an experiment in a logged-over rainforest in South Kalimantan, Indonesia. *Forest Ecology and Management* 81: 95 – 100.

Uebelhör, K., Hernandez, A. and Von der Heyde, B., 1989: Timber Stand Improvement – A chance for the Philippine Dipterocarp forest. In E.F. Bruenig and J. Poker (eds.), *Management of Tropical Rainforests – Utopia or Chance of Survival?* Nomos, Baden-Baden.

Uebelhör, K., Lagundino, B. and Abalus, R., 1990: Appraisal of the Philippine Selective Logging System. Philippine-German Dipterocarp Forest Management Project. Technical Report No. 7. DENR, GTZ, DFS, Manila-Eschborn, 47pp.

Vanclay, J.K., 1989: A growth model of North Queensland rainforests. *Forest Ecology and Management* 27: 245 – 271.

Van Daalen, J.C., 1984: Distinguishing features of forest species on nutrient-poor soils in the Southern Cape. *Bothalia* 15 (1+2): 229 – 239.

Von Breitenbach, F., 1974: *Southern Cape Forests and Trees.* The Government Printer, Pretoria 328 pp.

Vooren, A.P., 1992: Harvest criteria for tropical forest trees. In K. Cleaver, M. Munasinghe, M. Dyson, N. Egli, A. Peuker, F. Vencélius (eds.), Conservation of West and Central African Rainforests. World Bank Environment Paper No. 1., Washington D.C., U.S.A., World Bank, p. 134 – 140.

Wadsworth, F.H., 1987: Applicability of Asian and African silviculture systems to naturally regenerated forests of the Neotropics. In: F. Mergen and J.R. Vincent (Editors), Natural Management of Tropical Moist Forests. Silvicultural and Management Prospects of Sustained Utilisation. Yale Universtiy, School of Forestry and Environmental Studies, New Haven, p.93 – 111.

Walter, H., 1979: *Vegetation of the Earth and Ecological Systems of the Geobiosphere.* Second edition, Springer Verlag, New York, 274 pp.

Wardle, P., 1978: Regeneration status of some New Zealand conifers, with particular reference to *Libocedrus bidwillii* in Westland National Park. *New Zealand Journal of Botany* 16: 471 – 477.

Weidelt, H.J., 1986: Die Auswirkung waldbaulicher Pflegemassnahmen auf die Entwicklung exploitierter Dipterocarpaceen-Wälder. Göttinger Beiträge zur Land- und Forstwritschaft in den Tropen und Subtropen, Vol. 19, George-August-Universität, Göttingen, 166 pp.

Weidelt, H.J., 1989: Die nachhaltige Bewirtschatung des tropischen Feuchtwaldes. Möglichkeiten und Grenzen. *Forstarchiv* 60: 100 – 108.

Westoby, M., 1984: The self-thinning rule. *Advances in Ecological Research* 14: 167 – 225.

White, F., 1978. The Afromontane region. In M.J.A. Werger (ed.), Biogeography and Ecology of Southern Africa.

Dr. W. Junk, The Hague, p. 464 – 513.

Whitmore, T.C., 1990: *An Introduction to Tropical Rain Forests.* Clarendon Press, Oxford, 226 pp.

Whitmore, T.C., 1991: Tropical rain forest dynamics and its implications for management. In: A. Gómez-Pompa, T.C. Whitmore and M. Hadley (eds.), Rain Forest Regeneration and Management. Man and Biosphere Series, Vol. 6, UNESCO / Parthenon, Paris, p. 67 - 89.

Willan, R.L., 1965: Natural regeneration of high forest in Tanganyika. *East African Agricultural and Forestry Journal* 31: 31 – 43.

Wills, C., Condit, R., Foster, R.B. and Hubbel, S.P., 1997: Strong density- and diversity-related effects help to maintain tree species diversity in a neotropical forest. *Proceedings of the National Academy of Science*, 94: 1252 - 1257.

Woell, H.J., 1989. Struktur und Wachstum von kommerziell genutzten Dipterocarpaceenmischwäldern und die Auswirkungen von waldbaulicher Behandlung auf deren Entwicklung, dargestellt am Beispiel von Dauerversuchsflächen auf den Philippinen. (Structure and growth of commercially used Mixed Dipterocarp forests and the results of silvicultural treatment; examples from permanent research plots in the Philippines). Mitteilungen der Bundesforschungsanstalt für Forst-und Holzwirtschaft No. 161, Hamburg, 265 pp.

Wyatt-Smith, J., 1987a: The management of tropical moist forest for the sustained production of timber: some issues. Tropical Forestry Policy Paper No. 4, IUCN/IIED, London, 20 pp.

Wyatt-Smith, J., 1987b: Problems and prospects for natural management of tropical moist forests. In F. Mergen and J.R. Vincent (eds.), Natural Management of Tropical Moist Forests. Yale University, New Haven, p. 6 – 22.

Scenario Planning for Sustainable Forest Management

Klaus von Gadow[1] *and Janna Puumalainen*[2]

Abstract

By anticipating the future in a systematic way forest scenario planning can reduce uncertainty and improve the chance that future developments will agree with specified objectives. Scenario planning models may be developed for large scale applications on a national or subcontinental scale or for use in local planning. This contribution specifically deals with scenario techniques on the forest management unit level. A series of techniques for generating timber harvest scenarios, including age-class simulation, area change models and multi-period harvest scheduling for even-aged forests as well as continuous cover forests are presented with numerical examples, including various optimization techniques. Specific end products require a particular quality of the raw material, and as the end products do become more variable and the processing opportunities more sophisticated, the importance of raw material quality increases. Therefore, not only are total yields and volumes of interest in forest scenario planning, but increasingly also the distribution of tree species, tree dimensions and timber quality. For this reason, a specific application involving the structure of timber product yields is discussed. Finally, scenario techniques for evaluating forestry activities are presented with two examples: a beech forest in Europe and a timber farm in South Africa.

Keywords: *simulation, area change models, product modelling, optimization, harvest scheduling, activity scheduling, continuous cover forest.*

---

[1] Institute of Forest Management, University of Göttingen, Büsgenweg 5, D-37077 Göttingen, Germany
[2] Stora Enso Forest Consulting, Kuparintie 47, 50100 Imatra, Finland

K. von Gadow et al. (eds.), Sustainable Forest Management, 319–356.

Introduction

The purpose of forest scenario planning is to evaluate alternative management options relating to sustainable forest management. Forest scenario planning can reduce uncertainty by anticipating the future in a systematic way, thus reducing the likelihood of unexpected events. It can also improve the chance that future developments will agree with specified objectives. Scenario models may be developed for large scale applications on a national or subcontinental scale (Nabuurs and Päivinen, 1996) or for use in local planning. This contribution specifically deals with scenario techniques on the forest management unit level.

To successfully maintain an industry based on timber products, forest planning must ensure that there are always stands at the right stage of development and in sufficient number to yield the desired product mix coming from the forest. This problem, known as forest-level planning, has been adressed in a great variety of ways. The classical approach involves the calculation of an allowable cutting volume based on the current growing stock and growth rate. For planning purposes one must be able a) to produce accurate descriptions of future management activities and b) to project the development of the forest state. In the past this was achieved by means of yield tables (Schober, 1978). A yield table estimates the development of the remaining stand and the volume of the removed trees for forests which are managed in some standard way. An example of a yield table for beech forests, compiled more than 200 years ago, is shown in Table 1.

A yield table does more than merely estimate timber yields. It produces a description of future management activity, expressed in terms of the volumes removed during thinning operations. It also projects the development of a forest through time, for a given site productivity class.

The yield table format has remained surprisingly constant during the past 200 years (Paulsen, 1795; Hartig, 1847; Weise, 1880; Schwappach, 1890; Gehrhardt, 1930; Wiedemann, 1949; Schober, 1978). This may be the result of perseverance or lack of better tools, but it is likely that the yield table format was a suitable one, meeting the requirements for effective silvicultural planning, especially after the introduction of a system of tree classes and standard thinning grades defined by the Union of Forest Research Stations (Forstliche Versuchsanstalten, 1902).

	remaining stand				*removed trees*	
Age (years)	Stems/ morgen	DBH (inches)	Height (feet)	Volume (*Klafter*)	Stems/ morgen	Volume (*Klafter*)
20	3644	2	12	5,95		
28	911	4	24	11,92	2733	4,46
36	405	6	33	16,39	506	6,62
44	228	8	40	19,86	177	7,17
52	146	10	45	22,35	82	7,15
60	101	12	49	24,34	45	6,83
68	74	14	53	26,32	27	6,41
76	57	16	56	27,81	17	6,16
84	45	18	59	29,30	12	5,83
92	36	20	61	30,30	9	5,56
100	30	22	63	31,18	6	5,34
110	25	24	64	31,78	5	4,89
120	21	26	64	31,78	4	4,71

Table 1. Yield table for a beech forest, on a good site, compiled by Paulsen (1795). Volumes are measured in Klafter per morgen (one Klafter represents approximately 12 m^3, one morgen about 0.4 hectare).

The system of standard thinning grades and yield tables, in combination with periodic inventories of the growing stock, represented a normal scenario of forest development and a simple, effective planning framework.

Timber harvest scenarios

A traditional view of a managed forest is that its development follows a perpetual series of growth cycles which are defined by the rotation. One cycle begins with the establishment of the young trees and ends with the harvest of the mature ones. Intermediate thinnings may improve the value of the final crop. Productivity is measured in terms of the total volume produced, or the mean annual increment at rotation age (MAI_R). Sustained yield control is based on the model of the *Normal Forest*, and the present value of a specified silvicultural program is equal to the sum of the annual net incomes, discounted to the age of establishment[1]. The forest area is not continuously covered by trees and we may thus refer to such a forest as a *discontinuous forest*.

A second type of management which we may call *continuous cover forest* is not uncommon in Central Europe and in tropical regions. Its importance and prevalence are currently increasing in Central Europe. This type is characterized by selective harvesting of

[1] for details concerning the economic criteria refer for example to Clutter et al. (1983) and Chang (1998).

individual trees and the silviculture is appropriately known as *forest gardening*[2]. Forest development has no beginning or end. The forest remains in a state of undefined age, oscillating about a specified level of growing stock (Fig. 1). Harvesting operations may take place at regular intervals and the distinction between thinnings and final harvests is not clearcut. Age-based measures of forest production and valuation, such as mean annual increment or age-based net present value, cannot be used and the appropriate variable for assessing production is the periodic annual increment $\left(PAI_{t_1-t_2}\right)$.

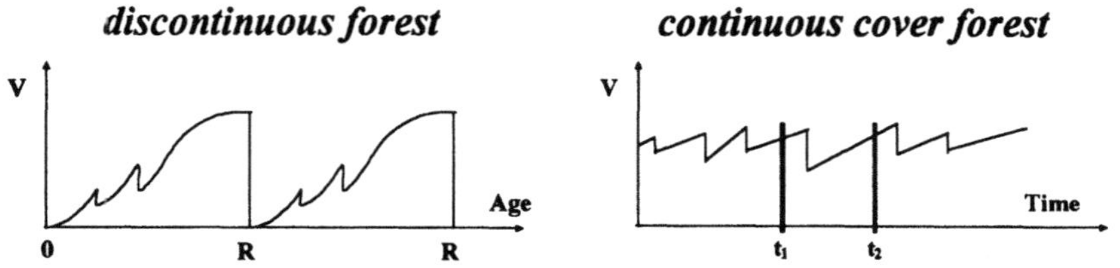

Figure 1. A discontinuous forest develops through a succession of growth cycles defined by a rotation (left). Age and rotation are undefined in a continuous cover forest and the growing stock oscillates about some desired level (right).

A continuous cover forest and its stage of development cannot be defined through age or rotation[3]. The development is determined by intermittent selective harvesting. Abstract forest models like the *normal forests* and other representative forests are difficult to apply considering the great variety of forest states. Consequently, the traditional, simple methods of scenario modelling differ substantially for the two forest management approaches. However, new simulation techniques which generate forest development paths for a limited time period, use quite similar approaches for evaluating future scenarios in both forest types.

Discontinuous forests

The prototype of the discontinuous forest, which develops through a succession of growth cycles defined by a rotation, is very common all over the world. The task of scenario planning is to ensure sustainable production of roundwood and to identify the most desirable development paths for a given site and tree species, including an evaluation of intermediate thinnings. Some of the less sophisticated techniques of scenario planning, like age class simulation, area change modelling and multi-period linear programming, are still quite useful for evaluating sustainable production of roundwood.

---

[2] a translation of the French *jardinage*.

[3] obviously each individual tree has an age, but not the forest as a whole.

The Normal Forest model and the market economy

The model of the *Normal Forest* is an idealized standard which allows comparisons between the actual and the normal age class areas, growing stock volumes, growth rates and harvest volumes. The following example involving the data from two spruce forests in Germany illustrates the principle (Tab. 2).

age class	**normal area (ha)**	**Forest A current area (ha)**	**Forest B current area (ha)**
temporary unplanted	0.0	2.2	0.8
$0 \leq I < 20$	121.0	216.0	18.0
$20 \leq II < 40$	121.0	227.8	60.4
$40 \leq III < 60$	121.0	100.5	12.6
$60 \leq IV < 80$	121.0	139.1	19.0
$80 \leq V < 100$	121.0	81.5	322.3
$100 \leq VI < 120$	121.0	60.0	130.9
$120 \leq VII < 140$	121.0	19.9	283.0
total	847.0	847.0	847.0

Table 2. Current and normal age class distributions of a spruce forest in the Sauerland region/Germany.

Areas with young trees predominate in Forest A, those with older trees in Forest B. Consequently, the sustainable harvest volume in Forest B is currently considerably higher than that in Forest A. The rotation age is 100 years and the MAI_{100} is equal to almost exactly 10m³/year/ha. The data in Table 1 may now be used as a basis for calculating sustainable harvest volumes. The MAI_r is weighted with the average forest age in the formula proposed by Breymann (1868):

$$harvest = MAI_r \cdot \frac{a_w}{(r/2)} \qquad (1)$$

with

harvest = annual harvest (m³/year/ha), r = rotation age,
MAI_r = mean annual increment at rotation age (m³/year/ha),
$aw = \sum_{i=1}^{n} a_i \cdot f_i \Big/ \sum_{i=1}^{n} f_i$ with a_i = age class i midpoint, f_i = age class i area

Thus we obtain for the two forests:

forest	**a_w (years)**	**harvest (m³/ha/year)**
A	45.9	9.2
B	99.1	19.8

The original idea of the normal forest model was to prescribe strict sustainable timber yields for each management unit, which was achieved by means of the periodic block system developed by Hartig (1795) and Cotta (1804). A periodic block is a forest area scheduled for harvest in a given time period. The economic disadvantages of the strict application of the sustainability concept are obvious. To ensure an even flow of roundwood to the market, it may be necessary to harvest young stands prematurely or, when there is an abnormally high proportion of old stands, to leave overmature ones unutilized. Strict sustainability, as defined by the normal forest concept, can only be realized in a supply-oriented, controlled economy. High demand cannot be met because the cutting volume is limited to normal levels. In times of low demand the normal amounts are supplied, irrespective of the financial losses (Fig. 2).

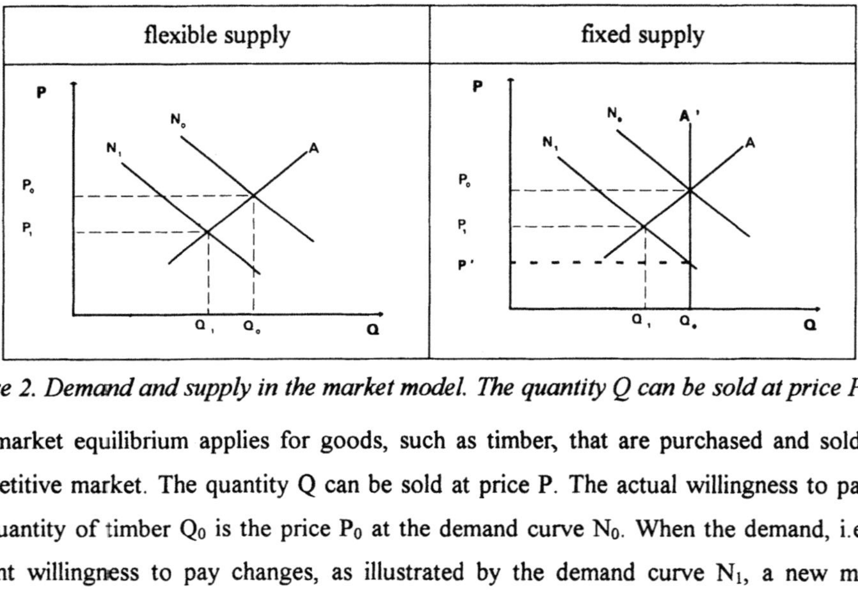

Figure 2. Demand and supply in the market model. The quantity Q can be sold at price P

The market equilibrium applies for goods, such as timber, that are purchased and sold in a competitive market. The quantity Q can be sold at price P. The actual willingness to pay for the quantity of timber Q_0 is the price P_0 at the demand curve N_0. When the demand, i.e. the current willingness to pay changes, as illustrated by the demand curve N_1, a new market equilibrium is established with the supply quantity Q_1 and the price P_1 (Bergen, 1997). Even if the supply is fixed, as implied by the strictly sustained timber yield concept, demand may rise or fall depending on the market. When the demand falls, the price will drop more drastically than in the flexible supply scenario (cf Waggener, 1969).

The model of the normal forest is defined by two conditions: a) a yield model estimating the development of the growing stock over age and b) a rotation. As such, it can be used to generate scenarios which provide useful information to timber companies. The effect of the rotation on the normal growing stock volume and the annual sustainable harvest may be illustrated using an example from the Southern United States (Clutter et al., 1983).

A pulpmill needs 400 000 cunits (about 1.5 million m³) roundwood per year from pine plantations in Southern Georgia, USA. How many acres of forest land are required to secure a continuous harvest of the required timber volume? The most essential tool for addressing this problem is a yield model, such as the one shown in the second column of Tab. 3 (one *cunit per acre* is approximately equivalent to 7.2 m³ per ha).

rotation (years)	growing stock (cunits/ acre)	harvest area (acres/year)	total forest area (acres)	total growing stock (cunits)	annual harvest (proportion of growing stock)	MAI_r (cunits/ac. /year)
(1)	(2)	(3) = 400 000/(2)	(4) = (1)*(3)	(5) = (2)/2*(4)	(6) =(2)*(3)/(5)	(7) = (2)/(1)
6	4.80	83 333.3	500 000.0	1200000	0.33	0.80
8	16.64	24 038.1	192 304.5	1600000	0.25	2.08
10	29.10	13 745.7	137 457.0	2000000	0.20	2.91
12	40.80	9 803.9	117 647.1	2400000	0.17	3.40
14	52.22	7 660.0	107 239.8	2800000	0.14	3.73
16	61.60	6 493.5	103 896.1	3200000	0.12	3.85
18	70.20	5 698.0	102 564.1	3600000	0.11	3.90
20	77.60	5 154.6	103 092.8	4000000	0.10	3.88
22	84.26	4 747.2	104 439.0	4400000	0.09	3.83
24	90.24	4 432.6	106 382.7	4800000	0.08	3.76
26	94.90	4 215.0	109 589.0	5200000	0.08	3.65
28	98.56	4 060.2	113 636.4	5600000	0.07	3.52
30	100.80	3 968.3	119 047.6	6000000	0.07	3.36

Table 3. Hypothetical yield table for pine plantations in Southern Georgia (column 2). The annual cutting area, the required total area, the growing stock volume, the relative harvest rate and the mean annual increment are shown for different rotations.

The growing stock which is available for the final harvest increases with increasing rotation age, which affects the annual sustainable cutting area, the required total area, the total growing stock volume, the relative sustainable harvest rate and the mean annual increment. The MAI reaches a maximum at the age of 18 years. This age represents the minimum-landbase-rotation, i.e. the rotation where the area required for supplying 400 000 cunits per year is at a minimum (which is not necessarily the most economical one).

Age class simulation

The model of the normal forest is a standard which can be used for comparisons of current and normal growing stock volumes. Its value as a scenario model is rather limited. Quite useful, though equally simple, are methods based on age class simulations. In the simplest simulation approach, the forest area is subdivided into *m* age classes each covering an area of

a_{ij} ha in the jth felling period (i=l..m; j=0..n). The available timber volume in the i'th age class is equal to v_{ij} , and the planned total harvest volume for the jth felling period is h_j. The algorithm presented in Fig. 3 may be used to simulate the effect of a given harvest level on the development of the age class distribution of a regional forest resource.

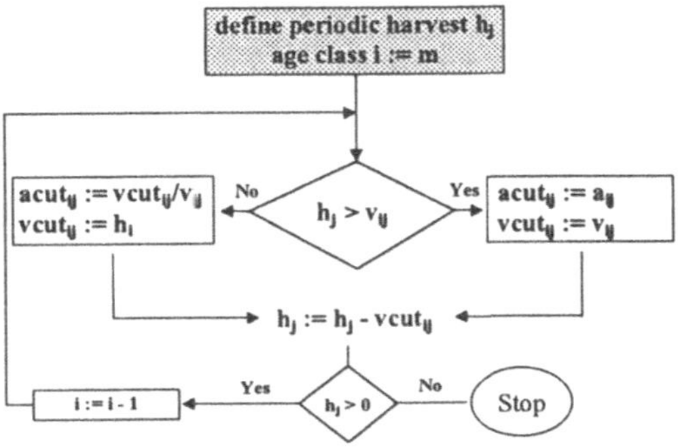

For each harvest period *j*, do:

For each age class i, starting with the oldest one, and while $h_j>0$, do:

a) Calculate the available growing stock volume v_{ij} ;

b) if $h_j \geq v_{ij}$, then harvest the entire growing stock available in age class *i*; else harvest only h_j m³;

c) subtract the volume harvested from h_j

Figure 3. Algorithm for age-class simulation with flowchart (left) and abbreviated pseudocode (right). a_{ij}=forest area available in age class i (i=1..m) and period j (j=1..n); v_{ij}=timber volume available in age class i and period j); $vcut_{ij}$=timber volume harvested in age class i and period j (m³); $acut_{ij}$=harvested area in age class i and period j (ha); h_j =specified harvest volume for period j (m³).

As an example consider a forest in Russia stocked with *Pinus silvestris*. The following empirical yield model for a fully stocked *Pinus silvestris* forest growing on a uniform site of quality class III in northern Russia is available:

$$V(t) = 205.3 \cdot \left(1 - e^{-0.0231 \cdot t}\right)^{2.93} \quad (1)$$

where V(age)=growing stock (m³/ha) at the specified forest age (years).

The model, based on the work by Shvidenko et al. (1995)[4], is used to compile the yield table presented in Tab. 4.

	Age Class Midpoint									
	10	30	50	70	90	110	130	150	170	190
m³/ha	2.0	26.9	67.7	107.4	138.8	161.4	176.8	187.1	193.7	197.9

Table 4. Yield table for a fully stocked Pinus silvestris forest growing on a uniform site of quality class III in northern Russia.

4 Empirical models such as the one described by eq. (1) are not always available and it is often necessary to construct a plausible yield model based on local experience or other scant evidence. Murray and Gadow (1993) developed a yield model which is suitable for highly aggregated applications. The model is derived on the basis of information that is often available, namely estimates of the maximum mean annual increment (MAImax) and the age corresponding to MAImax (the MAI-culmination age).

The forest covers an area of 1000 ha. Assuming a given initial age-class distribution (a_{i0} in Table 2), a constant periodic harvest of 30,000 m³ per 20-year cutting period and immediate regeneration of the felling areas, the age class distributions may be calculated for a series of future cutting periods (Table 5).

	Age Class Midpoint										
	10	30	50	70	90	110	130	150	170	190	Total
a_{i0} (ha)	20.0	300.0	100.0	100.0	200.0	50.0	50.0	100.0	10.0	70.0	1000
a_{i1} (ha)	156.0	20.0	300.0	100.0	100.0	200.0	50.0	50.0	24.0	0.0	1000
a_{i2} (ha)	168.3	156.0	20.0	300.0	100.0	100.0	155.7	0.0	0.0	0.0	1000
a_{i3}(ha)	171.0	168.3	156.0	20.0	300.0	100.0	84.7	0.0	0.0	0.0	1000
a_{i4} (ha)	177.7	171.0	168.3	156.0	20.0	300.0	7.0	0.0	0.0	0.0	1000
a_{i5} (ha)	185.2	177.7	171.0	168.3	156.0	20.0	121.8	0.0	0.0	0.0	1000

Table 5. Simulated development of the age-class distribution of a Pinus silvestris forest growing on a uniform site of quality class III in northern Russia, and assuming a harvest rate of 30,000 m³ per 20-year cutting period.

The development of the resource presented in Table 5 shows the effect of a given harvest level on the age-class distribution. In the specific example, a periodic harvest of 30,000 m³ would result in a wipe-out of the three oldest age-classes already in the second cutting period. The normal age-class area would amount to 200 ha if the rotation of maximum MAI of 100 years is selected. With a mean annual increment of 1.54 m³/ha/year at 90 years, a harvest level of 30 000m³ per year could be sustained indefinitely, assuming that the risks of fire and other disturbances are insignificant.

Obviously, the method involves considerable aggregation over growing sites, forest types and management regimes and the predictions have to be interpreted with the necessary caution. However, an age-class simulation is often the only feasible way to predict the dynamic development of a forest resource for large timber growing regions.

Area change models

Area change models have been used by a number of authors for generating a scenario of forest development. The most famous application is Suzuki's *Gentan* model (Suzuki, 1971). Area change models predict transitions of forest states through time. They project the historical harvesting practice and have been used for timber supply projections in Japan.

We may define $p_i(n)$ as the probability that a randomly selected forest area is in state i at time n and t_{ji} as the conditional probability that state i moves to state j within a given time step, i.e. as the conditional probability that j will be reached provided the system is currently in i. Let $t_{j,1}$ be the probability that a forest stand which belongs to the j'th age class is harvested and thus moves to age class 1. Then $t_{j,j+1}$ is the conditional probability that a stand which belongs to the j'th age class survives and grows into the (j+1)'th age class. Normally, $t_{j,1} + t_{j,j+1} = 1$ and the general matrix of transition probabilities may thus be written as follows:

$$T = \begin{bmatrix} t_{1,1} & t_{1,2} & 0 & 0 & \dots & 0 \\ t_{2,1} & 0 & t_{2,3} & 0 & \dots & 0 \\ \dots & \dots & \dots & \dots & \dots & \dots \\ t_{n-1,1} & 0 & 0 & 0 & \dots & t_{n-1,n} \\ 1 & 0 & 0 & 0 & \dots & 0 \end{bmatrix}$$

Area change models have been used especially in Japan (Konohira and Amano, 1986) and in Europe (Kurth and Dittrich, 1987; Kouba, 1989). Tab. 6 presents an example of a marix of transition probabilities for spruce/beech/fir forests in the Republic of Slovakia.

AGCl	1	2	3	4	5	6	7	8	9	10
1	0.000	1								
2	0.000		1							
3	0.000			1						
4	0.000				1					
5	0.005					0.995				
6	0.115						0.885			
7	0.020							0.980		
8	0.055								0.945	
9	0.040									0.960
10	1.000									

Table 6. Transition probabilities for spruce/beech/fir forests in the Republic of Slovakia for 10 age classes (Kolenka et al., 1996).

The survival probability is defined as the probability that a forest currently in age class j will reach age class j+1 during a discrete time step. Interestingly, the survival probabilities in Tab. 6 do not show a steady decline with increasing age, as in most applications of area change models. The lowest survival rate of 0.885 is expected in age class 6.

The area change approach may have some potential as a scenario modelling tool, e.g. for forecasting the development of age structures on a regional and national scale. However, some work needs to be done to predict the transition probabilities more accurately (Blandon, 1985; Randall and Gadow, 1990). The transition probabilities are not independent of the current age class distribution and this is one of the main problems associated with their use.

Area change models may be suitable for generating scenarios projecting the development of spatial patterns which are affected by natural processes (growth, wind damage) and forestry activities (plantings, thinnings). However, feasible applications may be limited due to the fact that credible transition probabilities are simply not available.

Multi-period harvest scheduling with Linear Programming

Numerous applications of mathematical programming in timber harvest planning have been reported. The models include an objective function that must be maximized or minimized and a set of constraints. The objective may be to maximize the harvest or the discounted cash flows or to minimize the cost of supplying the mill. The constraints are typically formulated so that a smoothing-out of production levels over time and at the same time a specified level of sustainability is achieved. A complete survey of mathematical programming applications to timber harvest scheduling is beyond the scope of this paper[5].

In multi-period harvest scheduling models, the state of the forest at the start of period t+1 is a linear function of the forest state at the start of period t and the areas cut in period t. The application of linear programming to multi-period harvest planning may be illustrated using an example adapted from Buongiorno and Gilless (1987)[6]. Consider a *Eucalyptus grandis* plantation covering an area of 150 ha. 100 ha are 7 years old and the site index is 22.0[7]. The remaining 50 ha, of site index 24.5, are 3 years old. The owner wishes to maximize the volume yield during the next 5 years. The following model can be used to predict total volume yield (V), measured in m^3 per ha for a given site index (SI) and age (t):

$$V = e^{2.83+0.151 \cdot SI} \cdot \left[1 - e^{-0.055 \cdot (t-1)}\right]^{1.19} \qquad (2)$$

Using equation (2), two yield tables can be developed (Tab. 7). The silvicultural regime is clearfelling followed by coppicing. To ensure sustained yields, a strict area control is to be applied by using the rotation of maximum volume production. An additional objective is to increase the total cutting volume by 2% each year.

The first step is to define the model variables. Let X_{ij} be the area cut from the initial age class i (i=1..n) in the cutting period j (j=1..m). Given two initial age classes and five cutting periods, there are ten variables, namely X_{31}, X_{32}, ..., X_{75}. Let v_{ij} be the volume per ha in initial age class i at planning period j. The total volume cut during a specific cutting period is a linear function of the areas cut:

Volume cut = $72.9X_{31} + 99.4X_{32} + \ldots + 183.5X_{75}$.

---

[5] the interested reader may refer to García (1990); Buongiorno and Gilless (1987); Leuschner (1990); Clutter et al. (1983).

[6] see similar applications by Johnson and Scheurman (1977); Hobbelstad and Hofstad (1988); García (1990); Peyron (1993).

age	Site index = 22.0		Site index = 24.5	
	V (m^3/ha)	MAI (m^3/ha)	V (m^3/ha)	MAI (m^3/ha)
1	0.0	0.000	0.0	0.000
2	14.4	7.205	21.0	10.509
3	31.8	10.611	46.4	15.477
4	49.9	12.487	72.9	18.215
5	68.1	13.630	99.4	19.881
6	86.1	14.355	125.6	20.939
7	103.7	14.818	151.3	21.614
8	120.8	15.104	176.2	22.031
9	137.4	15.265	200.4	22.267
10	153.4	15.336	223.7	22.370
11	168.7	15.339	246.1	22.374
12	183.5	15.290	267.6	22.303
13	197.6	15.202	288.3	22.174
14	211.2	15.083	308.0	22.000
15	224.1	14.940	326.9	21.792

Table 7. Two yield tables generated by Eucalyptus volume yield equation (2).

This volume is to be maximized. There are two area constraints, one for each initial age class:

$$X_{31} + X_{32} + X_{33} + X_{34} + X_{35} = 50$$

$$X_{71} + X_{72} + X_{73} + X_{74} + X_{75} = 100.$$

The cutting volume in period j+1 must be 2% more than the volume harvested in period j. For period 2, for example, we could state [volume cut in 2] = [1.02(volume cut in 1)], more specifically:

$$99.4X_{32} + 137.4X_{72} = 1.02(72.9X_{31} + 120.8X_{71}).$$

By a slight margin, the rotation of maximum volume production is 11 years for both age classes. Without jeopardizing the objective and considering the potential economic benefits of lowering the rotation age, it is decided to set the rotation at 10 years. The normal annual felling area is thus 150/10 = 15.0 ha/year and consequently the area harvested may not exceed 75 ha during the 5-year period. The appropriate constraint equation reads as follows:

$$X_{31} + X_{32} + \ldots + X_{75} = 75.$$

The main reason for regulating production in a multi-period optimization model is to maintain sustained levels of timber supply to processing plants. A useful approach is to model relevant aspects of the processing plant subsystem and its interactions with the forestry

[7] Site index is defined here as the stand top height at a reference age of 20 years.

system (see applications by Kallio et al., 1981; Barros and Weintraub, 1982). LOGPLAN is an example of a linear-programming-based model that determines the timber resources that are available to meet mill demands and schedules the necessary harvesting, processing, transportation and regeneration operations in such a way that the cost of meeting mill demands is minimized (Newnham, 1976; 1991).

Continuous cover forests

The large-scale application of new silvicultural systems has become a political reality in Central Europe. This involves a gradual transformation of traditional silvicultural practice towards *Continuous Cover Forestry,* favouring mixed uneven-aged stands, site-adapted tree species and selective harvesting (Griesel and Gadow, 1995). Selective harvesting systems have a long tradition, but the methods of calculating sustainable levels have been practiced only in a few tropical forests (Laughton, 1937; Breitenbach, 1974; Seydack et al., 1995), in some Northern hardwood forests in the USA (Hansen, 1987) and in the European *Plenter* forests (Biolley, 1980; Schütz, 1994).

In a forest managed under the selection system, the stand age is undefined. Forest development does not follow a cyclic harvest-and-regeneration pattern. Instead, it *oscillates* around some ideal level of growing stock, which may be determined in a variety of ways (Otto, 1994; Mlinsek, 1994; Helliwell, 1993). The mean annual increment is not appropriate for measuring productivity and the *Normal Forest* concept and the traditional sustainability criteria are not applicable.

The periodic increment is difficult to estimate in advance. Therefore, to ensure sustained production levels, various techniques have been devised for ensuring sustainable harvests in forest gardening systems. The most common type involves recurring visits to a given compartment and to simply skim off the accumulated increment at each visit. The principle resembles the normal forest and the aim is to ensure that timber can be harvested each year. First, we have to define the number of years between harvests in a particular compartment, r. To ensure sustainable yields, the forest is then subdivided into land parcels each covering one r'th of the total area. Each land parcel represents a state which is characterized by the number of years that have passed since the last harvest. Assuming, for example, a recurrence cycle of 5 years, we can easily map the progression of harvest locations. The following diagram is used to indicate that in harvest period I, a harvest is due in the darkly shaded land parcel where the last visit occurred 5 years ago:

The state-space distribution changes after each time step as a result of the harvest of the oldest parcel (5) and the accumulation of harvestable timber in the other land parcels. In harvest period II the state-space distribution has changed and parcel 4 is up for harvest, and so on. The main problem in a continuous cover forest is to determine which trees are to be removed and which are to remain. The decision may concern tree size or a combination of tree size, species and timber quality. Among the most common methods for estimating harvest levels for continuous cover forests are those that assume some normal residual forest structure. The residual structure may be defined by a diameter distribution norm or a basal area distribution norm.

Residual diameter distribution

The aim of the traditional *Plenterwald* harvesting systems practiced in Switzerland, France, Italy, Slowenia and Germany, is to define an optimum forest structure by using an inverse J-shaped diameter distribution model. The target distribution is re-established by periodic removal of trees in the different diameter classes. An example of the residual diameter distribution approach are the stem number guide curves used in Switzerland (Fig. 4). In theory, the approach is very simple and logical. By comparing the real and the ideal diameter distributions, it is possible to determine the number of stems that should be harvested in each diameter class.

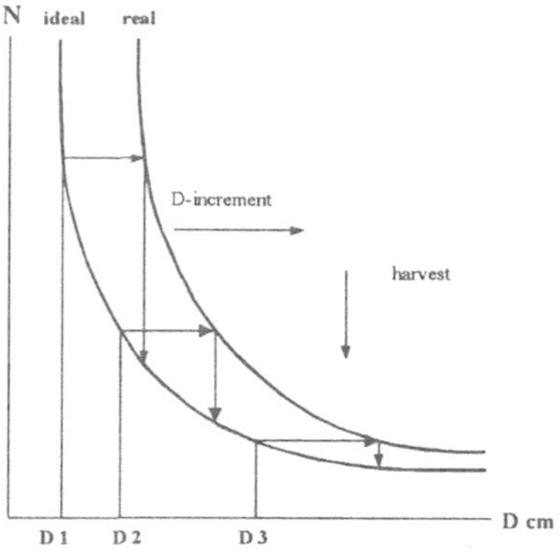

Figure 4. Schematic representation of a real and an ideal diameter distribution in a Plenterwald forest. The ideal guide curve determines the number of trees per diameter class D that need to be removed (after Kurth, 1976).

Meyer (1933) proposed the function $ln(N) = \alpha - \beta \cdot D$ for specifying the ideal diameter distribution. In a *Femel* forest which is characterized by group harvesting and gap regeneration Meyer's β parameter assumes values ranging from 0.08 to 0.15, depending on the maximum diameter of harvestable trees In a *Plenter* forest the β-values are usually much lower, ranging between 0.05 and 0.07. In mixed fir, spruce and beech forests of the Trentino region of the Italian Alps trees are harvested with reference to the model defined by Susmel (1980; see also Virgilietti and Buorgiorno, 1997):

$$N = 600\,;\; G = 0.97 \cdot H\,;\; V = 0.33 \cdot H^2\,;\; D_{max} = 2.64 \cdot H\,;\; k = \frac{4.3}{H^{1/3}}$$

with N=stems per ha with a breast height diameter greater or equal to 17.5 cm; H=the potential tree height, i.e. the maximum dominant height compatible with the ecological characteristics of the site (m); G=basal area (m^2/ha); V= volume (m^3/ha); Dmax=maximum diameter (cm); k = constant ratio between the trees in a given diameter class i and the next class i+1.

Unfortunately, the definition of such diameter distribution norms is not straightforward. There is no generally accepted ideal distribution of tree sizes, but a rather wide range of distributions found in stable *Plenter* and *Femel* forests (Fig. 5).

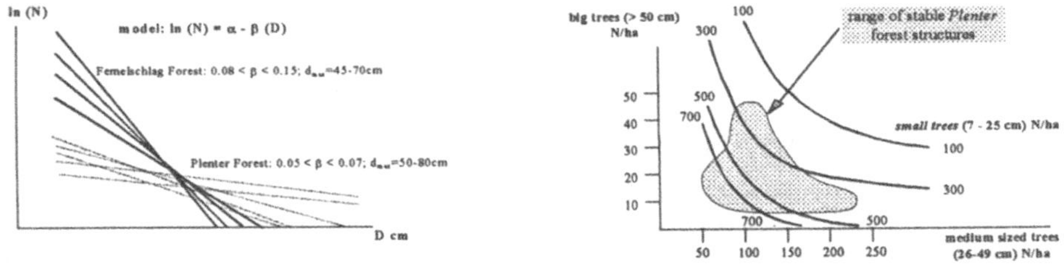

Figure 5. left: wide range of Plenter- and Femel forest structures defined by Meyer's diameter distribution model (after Kurth, 1976). right: range of stable Plenter forest structures found by Mitscherlich (1952).

Obviously, there is no such thing as a unique sustainable forest structure. Stable structures may be found within a wide range of combinations of species mixtures and size classes (cf. Mitscherlich, 1952).

Residual basal area distribution

An alternative criterion for defining the residual forest structure is a basal area distribution norm. An example of this approach is found in the *Northern Hardwoods*, a multi-species forest extending from the central Appalachians northwards towards Canada and from the Atlantic ocean westwards to Minnesota (Eyre, 1980). Hansen (1987) proposed a residual basal area of 19.8 m²/ha for a cutting cycle of 15 years and 19.8 m²/ha for a 30-year cutting cycle. The distribution of the residual basal area over diameter classes is shown in Fig. 6.

	Diameter class cm				
cutting cycle	5-12	13-28	29-43	>43	total
15 years: basal area (m²/ha) =	2.3	5.7	8.0	3.8	19.8
30 years: basal area (m²/ha) =	2.3	5.7	4.6	0.0	12.6

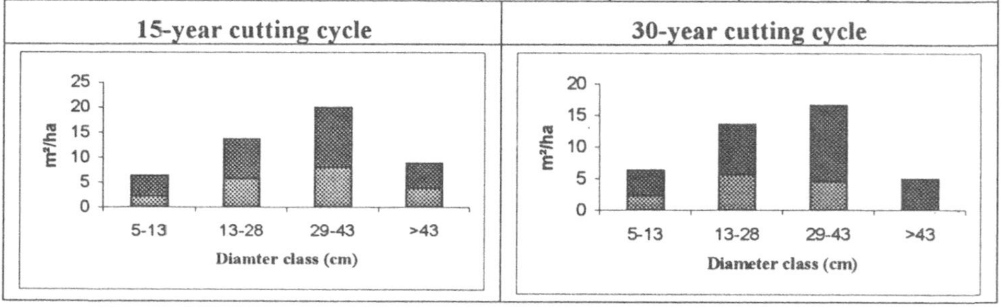

Figure 6. Distribution of the residual basal area over diameter classes for different cutting cycles in the Northern Hardwoods (Hansen,1987). The darker shaded parts of the bars indicates the basal area to be removed.

The basal area norms which are applied in the Northern Hardwoods relate only to size classes. Tree species are not specifically taken into consideration and this may be seen as a shortcoming of the method. Tree species and size are considered simultaneously in the *Knysna* harvest control method.

The indigenous evergreen forests which occur in the Southern Cape coastal belt of South Africa between Mosselbay and Humansdorp, known as the *Knysna forests*, represent a natural resource of great cultural, scientific and economic importance (Laughton, 1937; Breitenbach, 1974; refer to the contribution by Seydack in this volume). In the Knysna forest management model the first step involves a classification of forest types according to the moisture regime which represents a natural classification criterion affecting forest structure, productivity and species distribution (Fig. 7). Then the trees are classified according to species and size using four tree classes.

Figure 7. Schematic representation of two forest types in the Knysna region of South Africa after Breitenbach (1974). Above left: a species-rich medium-moist forest type with dense Trichocladus crinitus undergrowth. Above right: a highly productive moist forest type with relatively few species, including Podocarpus falcatus, P. latifolius, Olea capensis, Ocotea bullata, Vepris undulata, Ilex mitis, Curtisia dentata, Apodytes dimidiata etc. and tree ferns.

The second step involves a classification of different tree species and size classes on the basis of their current and future potential, e.g. I = mature trees of commercially valuable species; II = immature trees of commercially valuable species which will later reach class I; III = other non-commercial tree species; IV = weed trees and undesirable exotic intruders. Finally, a normal basal area distribution will be defined for each forest type and tree class (Fig. 8).

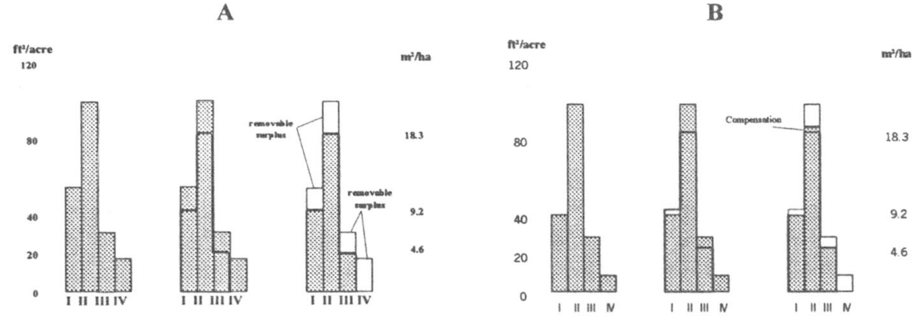

*Figure 8. Basal area normalities for a medium-moist forest. **A**: basal area distribution over tree classes before the harvest, with normal basal area distribution superimposed (centre) and removable surplus (right). **B**: The basal area of tree class I is less than the norm. This is compensated by a temporary increase of the norm in class II. The ingrowth into class I will be greater than normal so that the class I basal area will reach normality sooner. The surplus in tree classes III and IV will be harvested.*

Until recently, it has been rather difficult to carry out effective forest scenario modelling and planning in a continous cover forest. Especially the lack of appropriate simulation models has hampered realistic forecasting of the natural growth and silvicultural activities in uneven-aged mixed forests (Kramer, 1988). Meanwhile the first generation of such simulation models is emerging enabling foresters to generate various forest development scenarios including a variety of forest management options. In many forests, however, forest structures are very heterogenous and variable. It is thus not always possible to classify the forests as clearly as in the Knysna forest example. A detailed description of spatial forest structures further enhances the utility of such simulation models.

An unusual normal forest model

In an even-aged discontinuous forest, the normal forest model is defined by equal areas in the different age classes. In a *Plenter* forest, age is undefined so that areas covered by certain age classes cannot be determined. As an alternative approach, Kurth (1976) proposed a normal forest model which is defined on the basis of diameter classes. Associated with each diameter class is a so-called *temps de passage*, - the time required for a tree to pass through the class (Fig. 9).

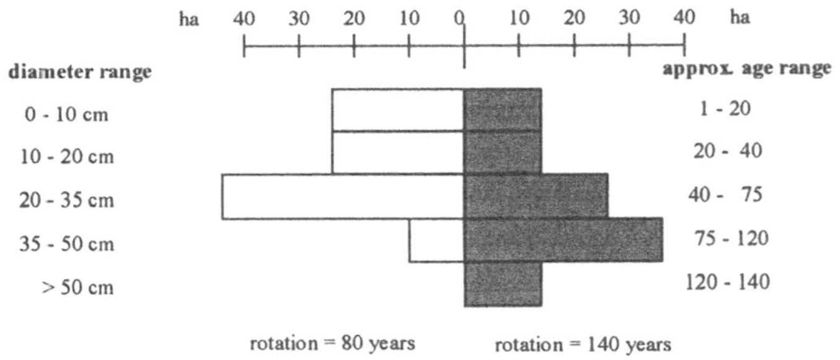

Figure 9. Normal areas and temps de passage for different stages of development in a spruce forest with a dominant height at age 50 of 16.0m, assuming a rotation of 80 years and 140 years (Kurth, 1976).

Each development stage is represented with an area, which corresponds the average time required for passing through a specified diameter class. The calculation of the normal area proportions in the different development stages is based on the proportions of the temps de passage. An example is presented in Tab. 8.

DBH cm	age	temps de passage	normal area at R=80
0-10	1-20	20	$\frac{20}{80} = 0.250$
10-20	20-40	20	$\frac{20}{80} = 0.250$
20-35	40-75	35	$\frac{35}{80} = 0.4375$
35-50	75-120	5	$\frac{5}{80} = 0.0625$

Table 8. Temps de passage and normal area proportions for different development stages and a rotation of 80 years (Kurth, 1976).

Optimizing harvest levels in a continuous cover forest

Sustainable management of a continuous cover forest may imply maximizing volume production while maintaining the forest in a steady state. The theoretical basis for this process is given by the general linear transition matrix approach which assumes that each tree in a finite number of size classes has a known probability of moving to another class, which depends only upon its current size. During a given time step a tree may remain in its class, grow into a higher class or die (Buongiorno and Michie, 1980; Rautiainen, 1999). The basic transition model is shown below:

$$\begin{bmatrix} f_1' \\ f_2' \\ f_3' \\ . \\ . \\ f_m' \end{bmatrix} = \begin{bmatrix} a_1 & 0 & \dots & 0 \\ b_1 & a_2 & \dots & 0 \\ 0 & b_2 & \dots & 0 \\ . & & & \\ . & & & \\ 0 & & \dots & a_m \end{bmatrix} \cdot \begin{bmatrix} f_1 + C \\ f_2 \\ f_3 \\ . \\ . \\ f_m \end{bmatrix}$$

where a_i (i=1..m) defines the probability that a tree in diameter class i will remain in class i and b_i (i=1..m-1) the probability that a tree in diameter class i will advance to class i+1 during the transition period. If the sum of a_i and b_i is less than 1, some of the trees in class i will die during the period. C defines the number of trees that will enter the first diameter class during the period, which is known as the *ingrowth*. The forest is in a steady state if the number of stems in each diameter class remains unchanged, thus if $f_{i,t+1} = f_{i,t} = f_i$. The objective of an optimization model may be to maximize the total harvested volume during each cutting period

$$\sum h_i \cdot v_i$$

where h_i (i=1..m) defines the number of trees cut in diameter class i in a steady state and v_i the per tree volume in class i. The objective is to determine the harvest that maximizes this function while satisfying the equalities and inequalities derived from the transition matrix and the steady state conditions. In a steady state the following constraint applies to each diameter class:

$$f_i = b_{i-1} \cdot (f_{i-1} - h_{i-1}) + a_i \cdot (f_i - h_i)$$

The number of trees that are havested cannot be greater than the number of trees available. Therefore, in addition to the non-negativity conditions ($f_i \geq 0$; $h_i \geq 0$) the following constraints are required: $f_i - h_i \geq 0$. An application of this model is presented by Rautiainen (1999) using data from an uneven-aged *Shorea robusta* forest in Nepal.

A model for any type of forest management

Projecting the development of a continuous cover forest is considerably more difficult than modeling a discontinuous forest with a given rotation and thinning regime. There are two reasons for this. Continuous cover forests are usually complex multi-aged, multi-species structures and the initial forest state is often difficult to describe mathematically. The variety of possible management options is usually great as a result of the great variety of initial forest states. There is also a greater uncertainty about how the structure will be modified by intermediate harvesting operations. Thus, a useful scenario model for any type of forest management would be one that is capable of simulating the development of a continuous cover forest. An example is given by the Model I formulation:

Model Elements

Variables	X_{ij}	= the area (ha) of stand i (i = 1...I) managed with regime j (j = 1...J_i)
Constants	M_{pt}	= the total output of product p (p = 1...P) in period t (t = 1...T)
	A_i	= total area (ha) of stand i
Coefficients	u_{ij}	= the utility value per ha of assigning stand i to regime j
	v_{ijpt}	= the per ha yield of product p in period t, if stand i is managed with regime j

Model Formulation

objective function $\sum_{i=1}^{I} \sum_{j=1}^{J_i} u_{ij} X_{ij} \rightarrow max.$

constraints

$$\sum_{i=1}^{I} \sum_{j=1}^{J_i} v_{ijpt} X_{ij} = M_{pt} \qquad \forall t, p$$

$$\sum_{j=1}^{J_i} X_{ij} = A_i \qquad \forall i$$

and $X_{ij} \geq 0.$

The beauty of this model is the fact that it can be applied in any forest type and management strategy and that it combines stand level objectives and forest level constraints. Consider a forest development with three stands over time. Two levels of planning may be distinguished. Stand level planning attempts to define an optimum management prescription in each of the stands. Forest level planning defines the forest-wide periodic constraints, such as a sustained yield constraint (Fig. 10).

The development of the whole forest is the result of the management actions in each of the three individual stands. One specific space-time pattern of forest development is the result of a specific sequence of management activities. Obviously, there are space-time patterns which may not be very satisfactory because the distribution of either the resource consumption and/or product output through time cannot meet the management constraints.

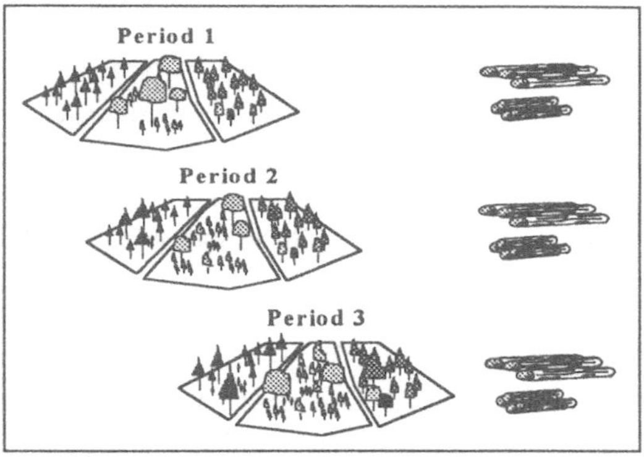

Figure 10. Schematic representation of the development of a forest through time. The logs indicate the periodic total harvests.

A plan may be optimal for individual stands, but this does not necessarily mean that it is feasible for the forest as a whole. How many time-space patterns of forest development can be specified and which one is optimal? Suppose we define one regime for each stand A and B, and two regimes for C. In that case there are two possible time-space patterns of forest development, namely 111 and 112. The number time-space patterns of forest development increases with increasing number of stands and increasing number of possible regimes per stand. Fig. 11 shows six forest maps resulting from the regime specifications: two for A, one for B and three for C.

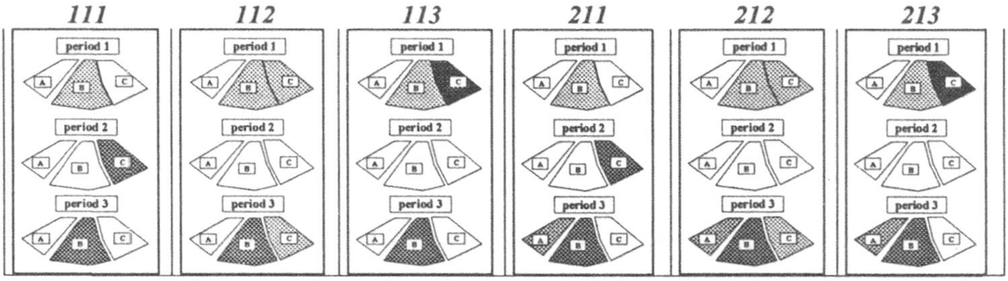

Figure 11. Six possible time-space patterns of forest development resulting from the alternative regime specifications. Dark shadings indicate that timber will be harvested during the specific period.

Options 112, 113, 212 and 213 would be infeasible as no timber is harvested in period 2. One of the maps would be optimal and the task of forest planning is to find it. To be able to implement the method, one must be in a position to generate and evaluate silvicultural development paths. This task is not trivial requiring effective growth modelling tools, realistic specifications of thinning operations and sound evaluation methods.

Linear programming is an elegant method and well suited to compartment-based harvest planning, but it has two serious limitations. First, real life models are so large that it is impossible to run them on a micro computer. Besides this technical limitation, there is an even more serious practical one. The different silvicultural regimes are mutually exclusive so that only one regime per stand is possible. This limitation cannot be imposed in the Model I formulation, unless an integer solution is sought which would be prohibitive in terms of computer time. The general way out of the dilemma is by aggregating stands into larger cutting units. This is a highly simplistic approach if aggregate cutting units include stands located on different growing sites and in different parts of the forest. For a more detailed discussion of the limitations of LP in harvest scheduling, refer to FORPLAN (1986).

Aiming to overcome the two main shortcomings of LP, costly processing and undesirable compartment splitting, Hoganson and Rose (1984) proposed a decomposition method for solving Ware and Clutter's (1971) Model I (see a worked example in Gadow and Bredenkamp, 1992, p. 108 et sqq.). Various other solutions have been offered for similar applications of linear programming to timber harvest scheduling (Siitonen, 1983; García, 1990; Lappi, 1992; Eid, 1993; Peyron, 1993; Pesonen, 1995; Rodriguez, 1996; Hoganson, 1996; Hoen, 1996).

Scenarios of the structure of timber yields

Certain end products require a particular quality of the raw material, and as the end products become more variable and the processing opportunities more sophisticated, the importance of raw material quality increases. Therefore, not only total yields and volumes are of interest in forest scenario planning, but increasingly also the distribution of tree species, tree dimensions and timber quality – irrespective of the selected management regime, i.e. whether management aims at a continuous cover forest or an even-aged discontinuous forest. The tree, timber or wood quality are closely related to the value of a stem.

In even-aged forest management the quality issues are taken care of rather by handling whole stands than by single trees. Thinnings, however, include a component of single tree quality and potential evaluation. In the case of a continuous cover forest, single tree dimensions, current quality and future potential form an important criterion especially for small scale silvicultural planning. Besides the traditional volume in-growth for example from one age class into another, new forest scenario planning approaches take into account so-called quality or value in-growth. Whereas the traditional in-growth tends to be linear within a given time period, the quality or value in-growth may be clearly discontinuous. The reason lies in the quality classification systems. There is normally a certain threshold value, e.g. a minimum log dimension, which a tree needs to satisfy in order to be classified for instance as saw timber. Especially in an even-aged forest a large number of trees are of similar age and size, and thus reach the threshold value at about the same time for instance within one or two succeeding five-year periods. Since the value of pulpwood and saw timber components often differ considerably, the in-growth into the saw timber quality class means also a considerable value increase of a stand. In contrast, as the continuous cover forests consist of trees at different stages of development, the quality and value in-growth are more evenly distributed over time.

The dimension and quality requirements for trees and timber assortments vary between countries, regions, companies and even between the different industrial plants of a given company. The volume of the trees of a forest stand can be more valuable to one producer than to another. This is important especially in medium-scale forest scenario planning when the allocation of different management units or stems into different mills and users are assessed.

Example: timber assortment distribution using dynamic programming

Dynamic programming (DP) is a standard method in optimising various forestry problems. The technique was first introduced by Bellmann (1954), and it is a computational method for multistage decision making. The problem is divided into stages and each stage has various states associated with it. A recursive relationship is then developed to find the optimal policy for each state with n stages remaining, given the optimal policy for each state, (n-1) stages remaining.

The DP approaches have been used both for holistic forest scenario planning (e.g. Brodie et al., 1978 and Valsta, 1986) and for single stage or problem planning. If an optimal division of a single stem into different timber grades is considered, the stages of the problem are usually the stem segments into which the stem has been divided and the state variable is the obtained log length (Pnevmaticos & Mann, 1972). The aim is usually to maximise the value or merchantable volume of a stem. In the current example an application of a forward reaching algorithm of dynamic programming has been used for estimating the proportions of different timber grades (Puumalainen, 1998). The algorithm uses information from tree characteristics, timber grade definitions and current timber prices in order to optimize the cross-cutting pattern. The results are based on the data, which consist of 3,283 Scots pine (*Pinus sylvestris L.*) stems in Finland (Kärkkäinen, 1980). Figure 12 demonstrates how the volumes of different timber grades obtained from the same data vary when different dimension criteria are applied. The prices assigned to the different timber grades have been kept constant and the bulks represent the different cross-cutting scenarios. The dimension and quality criteria used for the alternative "Base" are representative for large forestry companies, which produce fairly homogenous bulk products in Finland.

The effects of varying log length (LI - LIII) and minimum diameter (DI - DIII) constraints of small and specialised mills can be clearly seen in Fig. 12. For instance, by reducing the minimum log length from 37 dm to 22 dm the volume of the branchless log grade, the most valuable grade, has been tripled in some stands (LII vs. LIII). The volume gain obtained by loosening the dimension constraints is generally largest in stands of poor quality. If there is only a short branchless proportion in the trees in a forest stand, the particular timber is more valuable for a producer who can utilise short log lengths instead of the standard lengths.

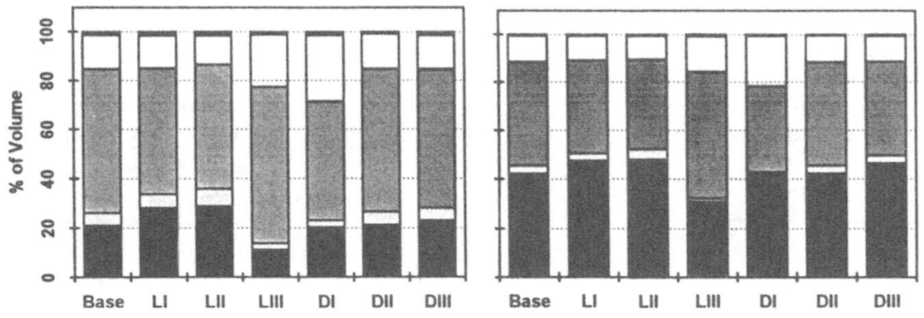

Figure 12. An example of the variation of the distribution of timber grades when different log dimension and quality restrictions are applied. Two different regions in Finland, which include mature pine stands, are presented. On the left hand side each bulk consists of the same 501 and on the right hand side of the same 222 pine stems. Only the log dimension constraints vary between the bulks. The different shades in each bulk represent the volume proportion of the grades. Timber grades (shades) are from the bottom: Grade I (branchless logs), Grade II (logs with living branches), Grade III (logs with dry branches), Pulpwood and Wastewood (Puumalainen, 1998).

The effects of varying log length (LI - LIII) and minimum diameter (DI - DIII) constraints of small and specialised mills can be clearly seen in Fig. 12. For instance, by reducing the minimum log length from 37 dm to 22 dm the volume of the branchless log grade, the most valuable grade, has been tripled in some stands (LII vs. LIII). The volume gain obtained by loosening the dimension constraints is generally largest in stands of poor quality. If there is only a short branchless proportion in the trees in a forest stand, the particular timber is more valuable for a producer who can utilise short log lengths instead of the standard lengths.

Forestry activity scenarios

The scenarios derived from harvest scheduling models are useful when the forestry activities are limited to operations that generate timber output, such as thinnings and clearfellings. They are of limited use in situations where operational scenarios are required that include a variety of other activities, including operations that generate positive environmental effects, such as certain silvicultural activities that can contribute to reducing groundwater nitrate levels (Rothe et al., 1999; Fig. 13). Harvest scheduling has always been a central issue in forest management, but harvest scenarios do not necessarily produce feasible plans. Felling volumes are often specified and assumed to be available at the prescribed time, but land is not the only resource required for producing timber and many additional operations, such as site preparation, pruning and weeding must be carried out besides harvesting to ensure successful forest management.

Fig. 13. Forestry activities consume various kinds of input and produce some desired output.

The central task in medium-term scenario planning is to distribute the work load in such a way that available resources and operational volumes balance, that input and output levels do not fluctuate too drastically and that silvicultural operations are carried out in time and in accordance with specified objectives. In practice, the forest operations required for carrying out certain plans must be scheduled in detail. This task is known as operational forest planning. Forest scenario modeling and strategic planning provide the framework and goals for the medium-term and operational planning.

A feasible scenario of forest development includes the selection of the best possible development path for each stand, subject to constraints relating to the management unit as a whole. Typical constraints are for example the available work capacity in a given time period or the total product mix delivered to the customers. Activity scheduling involves a leveling out of input/output fluctuations and the balancing of available labour capacity and other resources with the desirable operational volume.

Example: A beech forest in Europe

Each compartment within a forest develops over time in response to forestry operations such as plantings, prunings or removals of varying type and intensity. If appropriate tools are available, the characteristics of a given development path (such as the terminal growing stock, the silvicultural costs, the risk of windthrow etc.) can be calculated. Each compartment may develop in a variety of ways and the different development paths will be characterized by different stand attributes, cash flows and risk.

A *scenario* of forest development represents a specific combination of treatment schedules for the different compartments within a specified forest area. A *scenario model* thus embraces all the possible development paths of all the compartments within the forest. The aim of forest scenario modelling is to find the optimum combination of treatment schedules over all compartments. Pukkala and Kangas (1993) present a heuristic optimization method which assumes an additive utility function of the following form:

$$U = \sum_{i=1}^{n} a_i u_i (q_i)$$

where U = total utility for the current combination of treatment schedules
n = number of objectives
a_i = relative weight of objective i ($0 = a_i = 1$; $\sum_{i=1}^{n} a_i = 1$; i=1..n)
q_i = amount of objective variable realized in the current combination of treatment schedules (m^3, DM)
$u_i(q_i)$ = partial utility function of objective i ($0 = u_i(q_i) = 1$)

The relative weights of the different objectives are obtained using n(n-1)/2 pairwise comparisons (Saaty, 1980). The pairs may be lined up in a specific way to avoid biased judgements (Steinmeyer and Gadow, 1994). A graphic interface is useful for indicating the strength of preference for a particular pair item on the computer screen. Pukkala and Kangas (1993) developed a practical technique for estimating the partial utility functions $u_i(q_i)$ for the different objectives. By searching through all the treatment schedules in all the compartments, the minimum and maximum values of each objective variable are found. The range is divided into k intervals of equal width, resulting in k+1 class boundaries which are evaluated, again using the method of paired comparisons. The partial utilities are scaled such that their sum equals 1. Utilities for q_i - values between class boundaries are obtained by linear interpolation. The optimum combination of treatment schedules may be found using a suitable optimization procedure, or simply by sorting.

Consider a forest which is made up of three compartments A, B and C. Compartment A is a 140-year old beech stand approaching maturity. Compartments B is a 60-year old spruce stand requiring regular thinning whereas compartment C, featuring 130-year old spruce, is already beyond its rotation age of 120 years. The compartment details are listed in Table 9.

Table 9. Current compartment details (Ho=stand top height; G=stand basal area).

compt	area (ha)	species	age	Ho (m)	stems/ha	G (m^2/ha)
A	5.0	beech	140	32.0	120	36.0
B	4.0	spruce	60	27.0	650	28.5
C	5.5	spruce	130	36.0	250	43.0

The planning period is 30 years. For this period, different treatment schedules are developed for the three compartments. The different treatment schedules are shown in Table 10.

Table 10. Treatment schedules for the three compartments. t= terminal growing stock at the end of the planning period (m³/ha); p= present value of the cash flows resulting from the treatment schedule (DM/ha).

treatment	description	t	p
A1	selection of elite trees, remove immediate competitors	150	100
A2	three heavy regeneration cuttings	80	120
A3	do nothing, no intermediate harvesting	260	-10
B1	selection of elite trees, remove immediate competitors	260	100
B2	a series of low thinnings	300	90
C1	selection of elite trees, remove immediate competitors	280	150
C2	successive removal of 95% of the current growing stock	20	230

The different time-space patterns of forest development, the terminal growing stock volumes and the area-weighted average present values of the cash flows at the end of the 30-year planning period are listed in Tab. 11.

pattern	T (m³)	P (DM/ha)
A1B1C1	3330.0	119.0
A1B1C2	1900.0	149.3
A1B2C1	3490.0	116.2
A1B2C2	2060.0	146.6
A2B1C1	2980.0	125.9
A2B1C2	1550.0	156.2
A2B2C1	3140.0	123.1
A2B2C2	1710.0	153.4
A3B1C1	3880.0	81.0
A3B1C2	2450.0	111.4
A3B2C1	4040.0	78.3
A3B2C2	2610.0	108.6

Table 11. All the different time-space patterns of forest development, the terminal growing stock volumes (T) and the average area-weighted present values of the cash flows (P) at the end of the 30-year planning period.

The objective is to maintain a high terminal growing stock at the end of the planning period and to maximize the average present value of the cash flows resulting from the treatment schedules in the compartments. The growing stock and the present value objectives were assigned the weights 0.4 and 0.6 respectively. Thus, the following utility function applies:

$$U = 0.4 \cdot u_1(T) + 0.6 \cdot u_2(P)$$

where T = terminal growing stock at the end of the planning period (m³)
P = area-weighted average present value of the cash flows resulting from the treatment schedules in the compartments (DM/ha)
u_1, u_1 = partial utility functions for the two objective variables

The partial utility functions for the two objective variables are defined by the following assessments, obtained by means of paired comparisons:

T	1550	2380	3210	4040
u (T)	0.10	0.64	0.85	1.00
P	78	156		
u (P)	0.20	1.00		

The corresponding graphs are shown in Fig. 14.

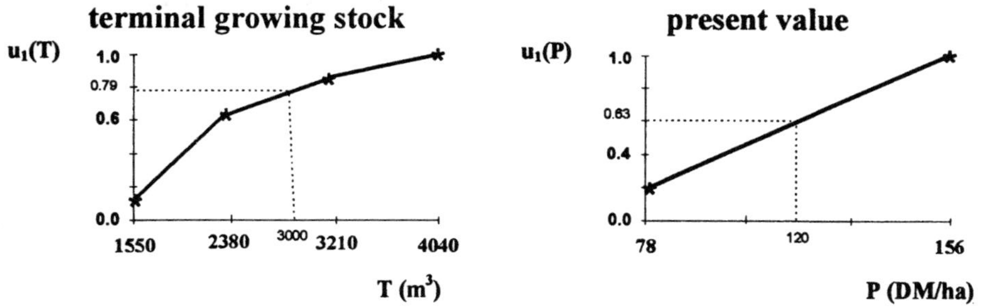

Figure 14. The range of the T- and P-values is divided into 4 intervals of equal width, resulting in five class boundaries which are used as pair items in the method of paired comparisons. Partial utilities for values between class boundaries are obtained by linear interpolation.

The resulting partial utilities for T and P and the total utilities for the different combinations of treatment schedules are listed in Table 12.

pattern	u(T)	u(P)	U
A1B1C1	0.872	0.620	0.721
A1B1C2	0.328	0.928	0.688
A1B2C1	0.901	0.590	0.714
A1B2C2	0.432	0.897	0.711
A2B1C1	0.792	0.691	0.731
A2B1C2	0.100	1.000	0.640
A2B2C1	0.832	0.662	0.730
A2B2C2	0.204	0.973	0.665
A3B1C1	0.971	0.231	0.527
A3B1C2	0.658	0.542	0.588
A3B2C1	1.000	0.200	0.520
A3B2C2	0.698	0.514	0.588

Table 12. Partial and total utilities for the different time-space patterns of forest development.

The optimum by a small margin is the combination A2B1C1. Compartment A should be managed with three heavy regeneration cuttings whereas compartments B and C by using the selection of elite trees and removal of the immediate competitors.

Example: A timber farm in South Africa

Consider an intensively managed timber farm in the Natal Midlands of South Africa covering an area of 127.1 ha. The total area has been subdivided into 12 compartments. There are two working circles. The working circle *pine* includes all stands planted with loblolly pine (*Pinus taeda*) and the working circle *wattle* all stands planted with black wattle (*Acacia mearnsii*). The relevant details of the plantation are shown in Table 13.

Compt	area ha	working circle	stems per ha	age	remarks
A	10.8	Pine	400	25	
B	25.3	Pine	400	22	
C	5.4	Pine	1 100	3	
D	13.2	Wattle	1 500	3	
E	12.7	Wattle	1 461	5	
F	8.5	Pine	1 050	15	never thinned
G	3.9	Wattle	1 356	9	
H	7.6	Wattle	2 000	3	
I	4.2	Pine	400	15	
J	18.3	Pine	400	15	
K	5.0	Wattle	1 380	8	
L	12.2	Pine	650	5	pruned to 3 m

Table 13. Details of the timber farm.

The medium-term scenario planning model may be formulated as follows:

Model Elements

Variables X_{ij} = the area (ha) of stand I (i = 1...I) managed with regime j (j = 1...Ji)

Constants M_{pt} = the total output of product p (p = 1...P) in period t (t = 1...T)

Q_{rt} = the available amount of resource r (r = 1...R) in period t

A_i = total area (ha) of stand i

Coefficients u_{ij} = the utility value per ha of assigning stand i to regime j

v_{ijpt} = the per ha yield of product p in period t, if stand i is managed with regime j

w_{ijrt} = the per ha input of resource r in period t, if stand i is managed with regime j

Model Formulation

objective function $\sum_{i=1}^{I} \sum_{j=1}^{J_i} u_{ij} X_{ij} \rightarrow max.$

constraints

$$\sum_{i=1}^{I} \sum_{j=1}^{J_i} v_{ijpt} X_{ij} = M_{pt} \qquad \forall t, p$$

$$\sum_{i=1}^{I} \sum_{j=1}^{J_i} w_{ijrt} X_{ij} = Q_{rt} \qquad \forall t, r$$

$$\sum_{j=1}^{J_i} X_{ij} = A_i \qquad \forall i$$

and $X_{ij} \geq 0.$

A standard silvicultural regime is prescribed for each working circle. The two sets of operations are presented in Table 14.

Working circle A *pine*

age	activity	symbol	timber yield m^3/ha	labour required (mandays)
1	Regeneration: plant 3x3 m	PP		14/ha
4	1. Prune to 1.5 m	P1		2/ha
4	1.respace 1100 → 650 trees/ha	T1		1/ha
5	2. Prune to 3 m	P2		3/ha
8	3. Prune to 5 m	P3		3.3/ha
12	2. Thin 650 → 400 trees/ha	T2	68.0	0.25/m
13*	2. Thin 650 → 400 trees/ha	T2	77.5	0.25/m
14	2. Thin 650 → 400 trees/ha	T2	86.9	0.25/m
25*	Clearfell	CP	425.9	0.20/m
26	Clearfell	CP	432.6	0.20/m
27	Clearfell	CP	438.8	0.20/m

Working circle B *wattle*

1	Regenerate by line sowing, tending, fertilizer & spacing to 6 000 stems/ha	W1		58.5
2	Slash, thin to 3 000 stems/ha, then to 2 000 stems/ha	W2		44.0
4	Thin to 1 500 stems/ha, prune	W3		4.5
8	Clearfelling & bark stripping	CW	20.9	200
9	Clearfelling & bark stripping	CW	22.7	200
10	Clearfelling & bark stripping	CW	24.2	200
11	Clearfelling & bark stripping	CW	25.3	200

Table 14. Standard silvicultural operations for the two working circles (indicates optimum age to carry out operation).*

The objective is to make sure that input/output fluctuations are smoothed, that the silvicultural time restrictions are observed and that the plan is feasible considering the available mandays. If it is necessary to reduce operational volume, then the reduction must affect the least important operations. Mathematical programming is an appropriate tool for solving the time-space structure and organisation of different activities. In order to directly involve the forestry practice in the planning process, the coefficients of the goal function should be based on the actual silvicultural preferences and possibly determined by the field personnel[9]. The following model provides an appropriate planning framework:

maximize U = 9ACP1 + 9ACP2 + ... + 13LP33, subject to:

(a) compartment areas
CP11 = 5.4; CT11 = 5.4; CP22 = 5.4; CP35 = 5.4; ACP1 + ACP2 = 10.8; BCP3 + BCP4 + BCP5 = 25.3; DCW5 ≤ 13.2; ECW3 + ECW4 = 12.7; HW31 = 7.6; FCP3 + FCP4 = 8.5; GCW1 = 3.9; LP33 = 12.2; HCW5 ≤ 7.6; KCW1 + KCW2 = 5.

[9] Details related to the model concept have been described in Gadow and Bredenkamp (1992, S. 119 ff.)

(b) pine timber
year1: 432.6ACP1 ≤ 2000; year2: 438.8ACP2 ≥ 2000; year3: 425.9BCP3 + 220.5FCP3 ≥ 2000;
year4: 432.6BCP4 + 221FCP4 ≥ 2000; year5: 438.8BCP5 ≥ 2000.

(c) wattle bark
year1: 24.2GCW1 + 20.9KCW1 ≥ 100; year2: 22.7KCW2 ≥ 100; year3: 20.9ECW3 ≥ 100;
year4: 22.7ECW4 ≥ 100; year5: 20.9DCW5 + 20.9HCW5 ≥ 100;

(d) mandays
year1: 86.5ACP1 + 2CP11 + CT11 + 200GCW1 + 4.5HW31+ 200KCW1 ≤ 2200;
year2: 87.8ACP2 + 3CP22 + 200KCW2 ≤ 2200;
year3: 85.2BCP3 + 200ECW3 + 44.1FCP3 + 3.3LP33 ≤ 2200;
year4: 86.5BCP4 + 200ECW4 + 44.2FCP4 ≤ 2200;
year5: 87.8BCP5 + 3.3CP35 + 200DCW5 + 200HCW5 ≤ 2200.

The variables represent the unknown areas of the compartments A to L in which various activities (third column in Tab. 14) are applied during year 1 to 5. Thus ACP1 refers to the hectares in compartment A clearfelled during the first year and similarly LP33 refers to the area in compartment L pruned to 5m during the third year. The solution to the problem was obtained after 17 iterations using the algorithm proposed by Hoganson and Rose (1984). The values of the dual variables associated with the primal harvest level constraints were at first arbitrarily set equal to the corresponding constraint constants. They were re-estimated after each iteration using the smoothing formula proposed by Hoganson and Rose (1984) and a smoothing factor of 0.015. For the purpose of finding a low discrepancy between planned and actual input and output levels, no distinction was made between input constraints and output constraints. The set of labour constraints, for example, was treated like another set of output constraints. The solution details are listed in Table 15.

compt	year 1	2	3	4	working circle
A		fell & plant			Pine
B			fell & plant		Pine
C	thin & prune	prune 3 m			Pine
E			fell & plant		Wattle
F	fell & plant				Pine
G	fell & plant		slash & thin	thin & prune	Wattle
H				fell & plant	Wattle
K		fell & plant		slash & thin	Wattle
L	prune 5 m			thin to 400/ha	Pine
pine timber	1 785.0 m³	1 667.4 m³	1 831.4 m³	829.6 m³	
wattle bark	94.4 t	121.0 t	265.4 t	158.0 t	
mandays	1 328.6	1 403.0	3 138.2	1 757.5	

Table 15. Optimal time-space pattern which was obtained after 17 iterations using the Hoganson-Rose method.

The plan presented in Tab. 15 is practical because the restrictions affecting the whole operational unit have been taken account of. The annual fluctuations in the labour requirements and in the pine timber and wattle bark supplies cannot be reduced further, simply because the problem is so small. With increasing number of compartments and silvicultural alternatives, the periodic input and output fluctuations can be reduced to smaller levels.

Discussion

There seems to be a clear trend towards increasing complexity and higher level of detail in forest scenario models. On the one hand this is the result of the increasing prevalence of uneven-aged mixed forests and on the other hand due to the integration of non-timber values into forest management planning. In the case of mixed uneven-aged forests an appropriate mathematical description of the heterogeneous spatial forest structure represents an important challenge, followed by the estimation of the growth of different forest components as well as the estimation of the effects of different silvicultural options on forest density, spatial structure and residual timber value. These factors must be evaluated at the stand level. They form the essential basis for scenario planning at the level of a management unit. Currently the first generation of operational scenario models is becoming available for Central European conditions (Sterba, 1990; Hasenauer et al., 1994; Kahn, 1995; Pretzsch, 1995; Nagel, 1996; Gadow and Hui, 1999).

The integration of non-timber values into forest scenario planning requires a modification of the traditional objective functions, which used to be strongly associated with economic and productivity objectives. The variety of forest management objectives in general is increasing and thus there may be a variety of objectives within a given optimization problem. The alternative approach is to include non-timber values into planning as a set of restrictions, e.g. as area restrictions for cutting.

When analysing the possible development scenarios of a small forestry enterprise an increasing detail and degree of complexity have to be accounted for. In the bigger companies which operate increasingly within an international context, the timber supply options and related objectives become even more variable. The importance of non-timber values also applies to international companies, which is increasingly considered in scenario planning. As the end products are more sophisticated they also require a certain quality from the raw material. Therefore, it is not only total yields and volumes that are considered, but also the distribution of tree species, log dimensions and quality. Quality aspects are especially important for the mechanical pulp mills and for the sawmilling industry.

Generally, the number of possible time-space options increases as the companies do become more international. They may plan the traditional forestry scenarios for the company own forestry assets in order to supply various mills requiring different types of raw material in different locations. The company's forestry assets may also serve as a buffer for difficult timber supply situations, and part of the forests may be set aside due to specific scenic or conservation purposes. The availability of company-owned forests may be complemented by local timber purchases or imports, and wood of a desired quality may be transferred over national borders into the optimal location within the company. This is usually feasible only in the case of valuable timber assortments due to the transportation costs involved. As the number of logistical options and geographical deviation do become larger, geographical information systems and spatial planning tools gain importance in scenario planning.

Relevant and up-to-date descriptions of current forest states and development scenarios is an interesting challenge for the research in a world characterized by increasing competition and complexity.

References

Barros, O. and Weintraub, A., 1982: Planning for a vertically integrated forest industry. Operations research 30: 1168-1182.

Beese, F., 1996: Indikatoren für eine multifunktionelle Waldnutzung. Forstw. Cbl. 115: 65-79.

Bellman, R.E. 1954. The theory of dynamic programming. Bull. Amer. Math. Soc. 60:503-516.

Bergen, V., 1997: Volkswirtschaftslehre, Teil I. Lecture Notes, University of Göttingen: 62 pp.

Biolley, H., 1980: Oeuvre écrite. Supplement to the Zeitschrift des Schweizerischen Forstvereins, No. 66: 458 p.

Blandon, P, 1985. Forest Economics and the Gentan Probability Distribution. J. Jap. For. Soc. 67 (12): 478-485.

Breitenbach, F. v., 1974: Southern Cape Forests and trees. The Government Printer, Pretoria: 328 p.

Breymann, 1868: Anleitung zur Holzmesskunst, Waldertragsbestimmung und Waldwertberechnung. Wien.

Brodie, J.D., Adams, D.M. & Kao, C. 1978. Analysis of economic impacts on thinning and rotation for Douglas-fir, using dynamic programming. Forest Science 24(4): 513-522.

Brouillet, L., Depierre, A., Demolis, C., Verdot, M., Preney, S., Latreille, C., Giovanini, J.-P., Herbert, I., Nicot, P., u. Allegrini, C., 1991: La Silviculture des Peuplements Réguliers de Hêtre en Franche-Comté - de la Régénération Naturelle à la Première Éclaircie. ONF Bulletin Technique Nr. 22: 9-42.

Buongiorno, J and Michie, B.R, 1980. A matrix model of uneven-aged forest management. For. Sci. 26: 609-625.

Buongiorno, J and Gilless, J.K, 1987. Forest Management and Economics. Macmillan, New York.

Carcea, 1986: Determination of the forest's allowable cut in various countries of the world. IUFRO document, Inst. de Cercetari si amenajari silvice, Romania: 268 p.

Chang, S.J., 1998: A generalized Faustmann model for the determination of optimal harvest age. Can. J. For. Res. 28: 652-659.

Clutter, J.L., Fortson, J.C., Pienaar, L.V., Brister, G.H. u. Bailey, R.L., 1983: Timber Management - A Quantitative Approach. Wiley. S.238-252.

Cotta, H., 1804 : Systematische Anleitung zur Taxation der Waldungen. Sander, Berlin.

Dreyfus, Ph. and Bonnet, F.-R., 1995: CAPSIS – Logiciel de simulation de conduites sylvicoles. Rev. For. Fr. XLVII, No sp: 111-115.

Dunkel, K., Elsasser, P., Oesten, G. u. Roeder, A., 1994: Wertschätzung des Waldes aus der Sicht der Waldbesucher - Ergebnisse einer Zielgebietsbefragung im Pfälzerwald. Mitt. d. Forstl. Vers. Anst. Rheinld-Pfalz 27/94: 1-72.

Dykstra, D.P., 1984: Mathematical programming for natural resource management. McGraw-Hill, New York: 318 p.

Eid, T., 1993: Models for economical forest management planning in Norway. Proc. Symp. Modelling in forest management planning and managerial economics - a critical investigation. Lithuanian Agricultural Academy: 35-43.

Eyre, F.H., 1980: Forest cover types of the US and Canada. Soc. Am. For., Washington D.C.

FORPLAN, 1986: FORPLAN - An evaluation of a forest planning tool. USDA Gen Techn Rep RM-140: 164pp.

Füldner, K., 1995: Strukturbeschreibung von Buchen-Edellaubholz-Mischwäldern. Cuvillier Verlag, Göttingen: 146 S.

Gadow, K. v. and Bredenkamp, B., 1992: Forest Management. Academica, Pretoria.

Gadow, K. v. and Hui, G.Y., 1999: Modelling forest development. *Kluwer* Academic Publisher, Dordrecht: 213 S.

García, O., 1990: Linear programming and related approaches in forest planning. N. Zealand J. of For. Science 20 (3): 307-331.

Gehrhardt, E., 1930: Ertragstafeln für reine und gleichartige Hochwaldbestände von Eiche, Buche, Tanne, Fichte, Kiefer, grüner Douglasie und Lärche. 2. Aufl. Julius Springer Verlag.

Gehrmann, D., 1975: Die Bewertung des Windwurfrisikos der Fichte auf verschiedenen Standortstypen. Mitt. d. Hess. Landesforstverwaltung, No 12.

Griesel, F. and Gadow, K.v., 1995: Naturgemäßer - naturnaher - ökologischer Waldbau - Begriffsbestimmungen, Konzepte und Richtlinien, Anwendungsbeispiele. Internal report, Institute of Forsteinrichtung, Univ. of Göttingen: 20pp.

Hansen, G.D., 1987: Effects of diameter distribution on the growth of northern hardwoods. Can. J. For. Res. 17 (1).

Hartig, G.L., 1795: Anweisung zur Taxation der Forste, oder zur Bestimmung des Holzertrags der Wälder. Heyer, Gießen.

Hartig, Th., 1847: Vergleichende Untersuchungen über den Ertrag der Rotbuche. Verlag Albert Förstner, Berlin.

Hasenauer, H., Moser, M. u. Eckmüllner, O., 1994: Programmbeschreibung. Distanzabhängiger Wachstumssimulator für Mischbestände, MOSES 2.0. Inst. f. Waldwachstumsforschung, Universität f. Bodenkultur, Wien: 36 S.

Helliwell, R., 1993: Newsletter 4. **Continuous Cover Forestry Group, UK.**

Hobbelstad, K. and Hofstad, O., 1988: AVVIRK-3, a model for long-term forest planning. Med. Norsk Institutt for Skogsforskning 41.35: 504-516.

Hoen, H., 1996: Forestry Scenario Modelling for Economic Analysis - experiences using the GAYA-JLP model. In: Päivinen, R., Roihuvuo, L. and Siitonen, M., (eds) 1996: Large-scale forestry scenario models - experiences and requirements. European Forest Institute. EFI Proceedings No. 5: 79-88.

Hoganson, H.M. and Rose, D.W., 1984: A simulation approach for optimal timber management scheduling. For. Sci. 30 (1): 200-238.

Hoganson, H.M., 1996: Using Dtran for the Minnesota GEIS. In: Päivinen, R., Roihuvuo, L. and Siitonen, M., (eds) 1996: Large-scale forestry scenario models - experiences and requirements. European Forest Institute. EFI Proceedings No. 5: 143-152.

Hollenstein, K., 1997: Analyse, Bewertung und Management von Naturrisiken. Hochschulverlag AG der ETH Zürich: 191 p.

Jansen, J.J., 1996: Opbrengsttabel voor de beuk (*Fagus sylvatica*) in Nederland – een OPTAB – modificatie door middel van empirische simulatie, gebaseerd op proefperken uit Noord-Duitsland en Zuid-Zweden. Hinkeloord report No. 16, Landbouwuniversiteit Wageningen: 36 p.

Johnson, K.N. and Scheurman, H.L., 1977: Techniques for prescribing optimal timber harvest investment under different objectives – discussion and synthesis. For. Sci. Monograph 18.

Kahn, M., 1995: Quasikausale Modellierung des Standort-Leistungsbezuges als Voraussetzung zum Aufbau flexibler Mischbestands-Modelle. Forstwissenschaftliches Centralblatt 114: 175-187.

Kallio, M., Propol, A. and Seppälä, R., 1981: A model for the forest sector. In Dantzig, G.B., Dempster, M.A.H. and Kallio, M.J. (ed): Large scale linear programming. Internat. Inst. for Applied Systems Analysis, Laxenburg, Austria: 1055-1100.

Kärkkäinen, M. 1980. Grading of pine sawlog stems. Comm. Inst. For. Fenn. 96(5): 1 - 152. (In Finnish with English summary.)

Kassier, HW, 1976. 'n Oorsig van beplanning en beplanningstegnieke in die bosbou met spesiale verwysing na oesre%oling. Unpublished MSc thesis, University of Stellenbosch.

Konohira, Y and Amano, M, 1986. Methods for the forests allowable cut calculation in Japan. In: Carcea , F (ed): Determination of the forests; allowable cut in various countries of the world. IUFRO, Bucharets: 149 - 157.

Kouba, J., 1989: The theory of an estimate of the development of calamities and of management of the process of forest adjustment to normal forest. Lesnictvi 35 (10): 925-944.

Kramer, H., 1988: Waldwachstumslehre. Verlag Paul Parey, Hamburg und Berlin.

Kroeger, W., 1992: Grundzüge der Sicherheit technischer Systeme. Lecture notes, ETHZ Zurich.

Kurth, A., 1976: Lecture notes, Ertragskunde. Eidgen. Techn. Hochsch., Zürich.

Kurth, H., Gerold, D. and Dittrich, K., 1987: Reale Waldentwicklung und Zielwald – Grundlagen nachhaltiger Systemregelung des Waldes. Wiss. Zeitschr. d. TU Dresden: 121-137.

Lappi, J., 1992: JLP - a linear programming package for management planning. The Finnish Forest Res. Inst., Res. Paper 414: 131 pp.

Laughton, F.S., 1937: The silviculture of the indigenous forests of the Union of South Africa with special reference to the forests of the Knysna region. Sci. Bull. 157, Forestry Series 7, Government Printer, Pretoria: 169 p.

Leuschner, A., 1990: Forest regulation, harvest scheduling and planning techniques. John Wiley: 281 p.

Madrigal Collazo, A., Alvarez Gonzalez, J.G., Rodriguez Soalleiro, R. and Rojo Alboreca, A., 1999: Tablas de producción para los montes españoles. Fundación Conde del Valle de Salazar, Madrid: 253 p.

Meffe, G.K., 1996: Conserving genetic diversity in natural systems. In: Szaro & Johnston, 1996: 41-57.

Meyer, H.A., 1933: Eine mathematisch-statistische Untersuchung über den Aufbau des Plenterwaldes. Schweiz. Zeitschr. Forstwes. 84: 88-103, 124-131.

Mitscherlich, G., 1952: Der Tannen-Fichten-(Buchen)-Plenterwald. Heft 8 d. Schriftenreihe d. Bad. Forstl. Vers. Anst., Freiburg: 42 p.

Mlinsek, D., 1994: Der naturnahe Waldbau - sein kognitiver Weg - eine Herausforderung. Der Dauerwald 10: 35-43.

Murray, D. M. and Gadow, K. v., 1993: A flexible yield model for regional timber forecasting. Southern Journal of Applied Forestry 17 (2): 112-115.

Nabuurs, G.J. & Päivinen, R. (eds.) , 1996. Large scale forestry scenario models – a compilation and review. European Forest Insitute working papers 10. Joensuu, Finland. 174 pp.

Nagel, J., 1996: Anwendungsprogramm zur Bestandesbewertung und zur Prognose der Bestandesentwicklung. Forst u. Holz 51 (3): 76-78.

Newnham, R.M., 1976: Computer may cut costs $1-2/cord. Pulp Pap. Can. 77 (9): 73-79.

Newnham, R.M., 1991: LOGPLAN II – a model for planning logging and regeneration activities. For. Can. Petawawa Nat. For. Inst. Inf. Rep. PI-X-102.

Otto, H.-J., 1994: Die Verwirklichung naturgemäßer Waldwirtschaft in den Niedersächsischen Landesforsten - Chancen und Probleme. Der Dauerwald 10: 3-23.

Paulsen, J.C., 1795: Praktische Anweisung zum Forstwesen. Detmold.

Pearson, S.M., Turner, M.G., Gardner, R.H. and O'Neill, R.V., 1996. An organism-based perspective of habitat fragmentation. In: Szaro & Johnston, 1996: 41-57.

Pesonen, M., 1995: Non-Industrial Private Landowners' Choices of Timber Management Strategies and Potential Allowable Cut - case of Pohjois-Savo. Acta Forestalia Fennica 247: 31 pp.

Peyron, J.-L., 1993: Présentation illustreé d'une méthode de planification de la gestion forestière et de détermination de l'effort de régéneration. Rev. For. Fr. XLV (1): 59-73.

Pielou, E.C., 1961: Segregation and Symmetry in two-species Populations as studied by nearest Neighbour Relations. J.Ecol.49: 255-269.

Pnevmaticos, S.M. & Mann, S.H. 1972. Dynamic programming in tree bucking. Forest Products Journal 22(2): 26-30.

Pretzsch, H., 1993: Analyse und Reproduktion räumlicher Bestandesstrukturen - Versuche mit dem Strukturgenerator STRUGEN. Schriften aus der Forstl. Fak. d. Univ. Göttingen u. d. Nieders. Forstl. Vers. Anst., Band 114, J.D. Sauerländer's Verlag.

Pretzsch, H., 1995: Perspektiven einer modellorientierten Waldwachstumsforschung. Forstwissenschaft-liches Centralblatt 114: 188-209.

Pukkala, T. and Kangas, J, 1993: A Heuristic optimization method for forest planning and decision making. Scand. J. For. Res. 8: 560-570.

Puumalainen, J., 1998: Optimal cross-cutting and sensitivity analysis for various log dimension constraints by using dynamic programming approach. Scand. J. For. Res. 13: 74-82.

Randall, J. H. and Gadow, K. v., 1990: An evaluation of the use of "Gentan" probabilities in forestry planning. South Afr. For. J. No. 154: 27-35.

Rautiainen, O., 1999: Growth dynamics and management of Shorea robusta forests in Southern Nepal. Diss. Faculty of Fortestry, University of Joensuu.

Rodriguez Soalleiro, R. J., Alvarez Gonzalez, J. G. u. Vega Alonso, G., 1994: Pineiro do Pais - Modelo dinamico de crecemento de masas regulares de *Pinus pinaster* Aiton en Galicia. Xunta de Galicia:40 S.

Rodriguez, L., 1996: A Microcomputer Program for Solving Forest Scheduling Problems with Heuristic Approaches. In: Päivinen, R., Roihuvuo, L. and Siitonen, M., (eds) 1996: Large-scale forestry scenario models - experiences and requirements. European Forest Institute. EFI Proceedings No. 5: 153-166.

Rojo, A. and Montero, G., 1996: El Pino Silvestre en la Sierra de Guadarrama. Ministerio de Agriculture, Pesca y Alimentacion, Madrid: 293 S.

Rothe, A., Brandt, S. and Hurler, R., 1999: Waldbewirtschaftung und Nitratbelastung des Grundwassers. AFZ/Der Wald, 10: 531-533.

Rottmann, M., 1986: Wind- und Sturmschäden im Wald. Frankfurt.

Saaty, T.L., 1980: The Analytic Hierarchy Process.McGraw - Hill.

Schober, R, 1978: Ertragstafeln wichtiger Baumarten. Sauerländer's Verlag.

Schütz, J.-Ph., 1994: Geschichtlicher Hergang und aktuelle Bedeutung der Plenterung in Europa. AFJZ 165 (5-6): 106-114.

Schwappach, A., 1890: Wachstum und Ertrag normaler Fichtenbestände. Julius Springer Verlag.

Sedjo, R.A., 1995: Ecosystem Management - an uncharted Path for Public Forests. Resources, (Resources for the Future) No. 121: 10-20.

Seydack, A.H.W., Vermeulen, W.J., Heyns, H.E., Durrheim, G.P., Vermeulen, C., Willems, D., Ferguson, M.A., Huisamen, J. and Roth, J., 1995: An unconventional approach to timber yield regulation for multi-aged, multi-species forests. II. Application to a South African forest. For. Ecol. & Mgmt 77: 155-168.

Shvidenko, A., Venevsky, S., Raille,G. and Nilsson, S., 1995: A system for evaluation of growth and mortality in Russian forests. Water, Air and Soil Pollution 82: 333-348.

Siitonen, M., 1983: A long term forestry planning system based on data from the Finnish national forest inventory. Univ. of Helsinki, Dept of For. Mens. & Mgmt Res. Note 17: 195-207.

Sprugel, D.G., 1991: Disturbance, Equilibrium, and Environmental Variability - what is Natural Vegetation in a Changing Environment. Biological Conserv. 58: 1-18.

Steinmeyer. A. and Gadow, K. v. 1994: Saaty's AHP - dargestellt am Beispiel der Waldbiotopkartierung. Centralblatt f. d. ges. Forstwesen 112 (1): 53-65.

Sterba, H., 1990: Validitätsprüfung des Einzelbaumsimulators für Fichten-Kiefern-Mischbestände. Tagungsbericht. Dt. Verb. Forstl. Forschungsanstalten - Sektion Ertragskunde: 101-117.

Sturm, K., 1995: Möglichkeiten, Probleme und Defizite der Integration naturschutzrelevanter Daten im Rahmen der Forsteinrichtung. Schr. f. Landschaftspfl. u. Natursch. 43: 231-245.

Susmel, L., 1980: Normalizzazione delle foreste alpini – basi ecosistemiche – equilibrio – modelli colturali – produttivita. Liviane editrice, Padova 437 p.

Suzuki, T., 1971: Forest Transition as a Stochastic Process. Mitt. FBVA Wien. Heft 91: 137-150.

Szaro, R.C. and Johnston, D.W., 1996: Biodiversity in Managed Landscapes.Oxford University Press: 778 p.

Thomasius, H., 1988: Stabilität natürlicher und künstlicher Waldökosysteme sowie deren Beeinflußbarkeit durch forstliche Maßnahmen. AFZ No. 43: 1037-1043, 1064-1068.

Ulrich, B., 1987: Stabilität, Elastizität und Resilienz von Waldökosystemen unter dem Einfluß saurer Depositionen Forstarchiv 58: 232-239.

Ulrich, B., 1993:Prozeßhierarchie in Waldökosystemen. Biologie in Unserer Zeit 23 (5): 322-329.

United Nations, 1992: Internationally agreed glossary of basic terms related to desaster management. United Nations Department of Humanitarian Affairs, Geneva.

Valsta, L.T., 1986: Mänty-rauduskoivusekametsikön hakkuuohjelman optimointi. Summary: Optimizing thinnings and rotation for mixed, even-aged pine-birch stands. Folia Forestalia 666. 23 S.

Virgilietti, P. and Buongiorno, J., 1997: Modelling forest growth with management data – a matrix approach for the Italian Alps. Silva Fennica 31 (1): 27-42.

Volz, K.-R., 1995: Zur ordnungspolitischen Diskussion über die nachhaltige Nutzung der Zentralressource Wald. Forst u. Holz 50 (6): 163-170.

Waggener, T.R., 1969: Some economic implications of sustained yield as a forest regulation model. Contribution No. 6, Univ. of Washington: 22 pp.

Waldherr, M., 1997: Risikoverluste und Erntealter. AFZ/Der Wald: 206-207.

Ware, G.O. and Clutter, J.L., 1971: A mathematical programming system for the management of industrial forests. For. Sci. 17: 428-445.

Weise, W., 1880: Ertragstafeln für die Kiefer. Julius Springer Verlag.

Wiedemann, E., 1949: Ertragstafeln der wichtigsten Holzarten. M.& H. Schaper, Hannover.

GPSR Compliance

The European Union's (EU) General Product Safety Regulation (GPSR) is a set of rules that requires consumer products to be safe and our obligations to ensure this.

If you have any concerns about our products, you can contact us on ProductSafety@springernature.com

In case Publisher is established outside the EU, the EU authorized representative is:

Springer Nature Customer Service Center GmbH
Europaplatz 3
69115 Heidelberg, Germany

Batch number: 10815974

Printed by Printforce, the Netherlands